William C. Haneberg

Computational Geosciences with Mathematica

William C. Haneberg

Computational Geosciences with Mathematica

With 297 Figures and a CD-ROM

Dr. William C. Haneberg
Haneberg Geoscience
10208 39th Avenue SW
Seattle WA 98146
USA
E-mail: *bill@haneberg.com*

Library of Congress Control Number: 2004106660

ISBN 3-540-40245-4 **Springer Berlin Heidelberg New York**

Springer is a part of Springer Science+Business Media
springeronline.com

Printed in Germany

Cover design: E. Kirchner
Production: A. Oelschläger
Typesetting: F. Herweg, Germany
Printing: Mercedes Druck, Berlin
Binding: Stein + Lehmann, Berlin

Printed on acid-free paper 32/3141/AO – 5 4 3 2 1 0

Preface

Mathematica® is a comprehensive mathematics package that can be used to perform numerical calculations, manipulate symbolic expressions, develop complicated computer programs, and create sophisticated scientific graphics. My objective in writing *Computational Geosciences with Mathematica* was to show how the program can be applied to solve a wide range of problems of interest to geologists, geomorphologists, hydrologists, geophysicists, and other geoscientists. As such, it is partly a textbook on quantitative geoscience and partly a manual showing how *Mathematica* can be used to solve some problems of interest to geoscientists. It is written at a level appropriate for graduate students embarking on quantitative research projects, professors who are interested in learning about new approaches to quantitative problem solving, and practicing geoscientists with an interest in *Mathematica*. While some of the material is more advanced than that taught in typical undergraduate geology programs in the United States, much of it will be accessible to motivated junior and senior students.

It has long puzzled me that, while advanced computational tools such as *Mathematica* have been available for 15 years or more, many geoscientists (geologists in particular) seem to be stuck in a spreadsheet rut. While spreadsheets manipulate rows and columns of numbers adequately, they are not well suited for much more than simple arithmetic. Still, their popularity persists and, in my opinion, continues to make life more difficult for students or professionals trying to solve quantitative geoscientific problems. I began using *Mathematica* in 1989 and it has become an indispensable computational and graphics tool in my research and professional practice. Gerard Middleton seems to share a similar view of spreadsheets to his book *Data Analysis in the Earth Sciences Using Matlab*, and I am glad to see that I am not alone in that regard. Inexpensive student versions make Mathematica particularly well suited for students in computer methods or quantitative geology classes.

The subject matter and examples in *Computational Geoscience with Mathematica* were drawn largely from my experience as an applied researcher in engineering geology and hydrogeology, university instructor, and consulting geologist. I have tried to include a broad range of topics, but there are many geoscientific and mathematical topics that are not covered. Fractals, wavelets, and geostatistics, for example, are all topics that can be fruitfully addressed *Mathematica*, but these are topics that either fall well outside my range of experience or the space available in this book. I hope that, rather than sparking complaints, their omission will motivate specialists in those fields to fill the void.

Mathematica is published by Wolfram Research, Inc., and is currently (summer 2003) in version 5.0. For more information, contact the company at:

Wolfram Research, Inc.
100 Trade Center Drive
Champaign, IL 61820-7237 USA

(217) 398-0700
info@wolfram.com
www.wolfram.com

This book would not have been written without the support and encouragement of my wife Lisa. Others who deserve a measure of credit (but none of the blame for any mistakes) include Arvid Johnson and Paul Potter who, respectively, taught me how to formulate geological problems in terms of mechanics and statistics. John Hawley and the late Frank Kottlowski hired me and provided a fertile environment for professional growth at the New Mexico Bureau of Mines and Mineral Resources, a division of New Mexico Tech. The late Allan Gutjahr, also at New Mexico Tech, was a fountain of statistical wisdom during our carpooling years. Mike Whitworth, Marshall Reiter, Dave Love, Laurel Goodwin, Peter Mozley, and many other colleagues introduced me to a fascinating array of technical topics during my time in New Mexico. Finally, Wolfram Research generously provided copies of *Mathematica* and allowed me to use a pre-release version of *Mathematica* 5.0 during the writing of this book.

William C. Haneberg
Port Orchard, Washington
August 2003

About the Author

William C. Haneberg is an engineering geologist, hydrogeologist, and accredited *Mathematica* consultant living and working in the Seattle area. Before moving to the Pacific Northwest to establish a consulting practice in 1999, he served as Senior Engineering Geologist and Assistant Director of the New Mexico Bureau of Mines and Mineral Resources in Socorro and Albuquerque, New Mexico. He has also taught undergraduate and graduate classes in geology, hydrology, geophysics, and geological engineering at New Mexico Tech and Portland State University. Dr. Haneberg is an author or co-author of more than 25 papers and co-editor of two multi-author monographs (*Clay and Shale Slope Instability*, published by the Geological Society of America, and *Faults and Fluid Flow in the Shallow Subsurface*, published by the American Geophysical Union). He earned a Ph.D. in geology from the University of Cincinnati, where his dissertation research concerned precipitation induced pore pressure increases in potentially unstable slopes. For additional information, please visit www.haneberg.com.

Contents

1 Introduction to *Mathematica*

1.1 What is *Mathematica*?

Mathematica is a computer program (or, more correctly, a system of computer programs) for performing mathematical operations such as symbolic manipulation, numerical calculations, graphics, and programming. *Mathematica* is sometimes described as a computer algebra system, probably in deference to the strong symbolic but somewhat limited numerical capabilities of its earliest versions, but it has evolved into system that is now a very practical tool for large scale numerical calculations. Its applications can range from simple calculations to complicated programs, and *Mathematica* can be a powerful tool for quantitative geoscientists such as geologists, geographers, geophysicists, hydrologists, and oceanographers. Because it is a general mathematics system rather than a task-specific application, however, its utility may not be apparent at first glance. It does not, after all, solve specific geoscientific problems any more than do programming languages such as C or FORTRAN or spreadsheet programs. Over the years I have encountered many geoscientists who know that Mathematica exists, but are not sure what it does with regard to the specific problems they face in their research or professional practice. The short answer is that it does nothing geoscientific in particular, but has the potential to do just about anything that a user can imagine. *Computational Geosciences with Mathematica* is intended to help fill that gap a by illustrating how *Mathematica* can be used to formulate, solve, and visualize a variety of problems of interest to geoscientists.

Mathematica was first published by Wolfram Research in 1988 and the latest version, 5.0, was released during Summer 2003. This book was written using versions of *Mathematica* ranging from 4.1 through 5.0, and care has been taken to ensure that the examples will work with both version 5.0 and its immediate predecessor, version 4.2. Version 5.0 is available for Windows 98/Me/NT 4.0/2000/XP and several versions of the Unix (including Macintosh OS X) and Linux operating systems. Specialized versions are available for web applications, grid computing using computer clusters or multiprocessor computers, and student use.

The functionality of *Mathematica*, which includes many standard packages with specialized functions in areas such as statistics and graphics, can be expanded with packages available through Wolfram Research, other commercial sources, or the public domain. Packages that may be of interest to geoscientists, but which are not covered in this book, include a database access kit, digital image processing, experimental data analysis, real time 3-D graphics, fuzzy logic, neural networks, signal processing, time series analysis, and wavelets. Readers interested in those packages should contact Wolfram Research for more information about their capabilities and availability.

1.2 Getting Help

Mathematica offers several kinds of documentation and online help. One source is *The Mathematica Book* (Wolfram, 1999), the current edition of which is written for version 4.0. A paper copy is included with professional (but not student) versions of *Mathematica*. *Mathematica* also offers a Help menu that includes a Help Browser (which can access a digital copy of *The Mathematica Book*), a ten-minute tutorial, a hyperlink to a web information center maintained by Wolfram Research (http://library.wolfram.com/infocenter), and a hyperlink to the general Wolfram Research web site (http://www.wolfram.com). If you know the name of a function, typing **?***function* then pressing Enter will return a brief description of the function with a hyperlink to more information. If you only know part of the function name, typing a question mark followed at least the first letter of the function name and then pressing Command-k will return a list of *Mathematica* functions that match the letters typed.

There are a number of good introductory books about *Mathematica*, in particular *The Beginner's Guide to Mathematica Version 4* (Glynn and Gray, 2000). Another useful source of information is the comp.soft-sys.math.mathematica newsgroup, which can be accessed through http://groups.google.com. For those wishing to retain expert help, Wolfram Research maintains a list of accredited *Mathematica* consultants.

1.3 Installing and Running Mathematica

Virtually any modern personal computer or workstation should have enough memory and computational speed to run *Mathematica* although, as with all software, the more memory and the faster the processor the better. The examples in this book were all developed using an iMac computer with a 700 MHz G3 processor, 512 Mb RAM, and Macintosh OS X. Free 30 day trial versions of *Mathematica* are available from Wolfram Research (http://www.wolfram.com).

If you are installing *Mathematica* on your own personal computer, insert the CD and follow the directions that appear on your screen. Start up *Mathematica* as you would any other program. If you are using *Mathematica* over a computer network, consult your system administrator or help desk for details.

The Getting Started directory in the *Mathematica* help browser offers several introductory lessons that may be useful for first-time users. To access them, select the Help Browser from the Help menu. The Help Browser item Tour includes a 10 minute introduction that covers many features of *Mathematica*, and Getting

Started offers an introduction to entering and executing basic *Mathematica* commands.

Although it may not be apparent, *Mathematica* consists of two programs: a kernel that performs calculations and a front end that handles input and output. Although it is possible to use *Mathematica* with a front end on one computer and the kernel on another, this book assumes that both are running on the same computer. When you click on the *Mathematica* icon, you will start the front end. The first time you execute a statement from the front end, there will be a short pause while the kernel starts.

To install the CompGeosci.m package included with this book, copy it from the CD to one of the directories along *Mathematica*'s default file search path. This will differ among operating systems and versions. To obtain a list of the default paths, type **`$Path`** and press the Enter key. Some of the directories listed may be accessible only to system administrators on multi-user systems, in which case the package may have to be installed in a local user library. On the system being used to write this book, the package has been put into the directory /Users/bill/Library/Mathematica/Applications. Consult your system administrator or help desk for guidance if you are using a multi-user system.

1.4 How the Book is Organized

Each of the chapters in *Computational Geosciences with Mathematica* was prepared as a *Mathematica* document known as a notebook. Therefore, readers with copies of *Mathematica* can open the accompanying digital versions of the chapters and follow the calculations as they read. Notebooks can contain combinations of text, mathematical input and output, graphics, and even sound. Mathematical variables are generally denoted by *italics*, whereas *Mathematica* functions and references to specific variables within the *Mathematica* examples are denoted using the same **`courier`** font that *Mathematica* uses for its default input and output. Appendix A is a list of *Mathematica* functions included in the CompGeosci package accompanying this book and Appendix B (CD only) is an overview of color in *Mathematica* graphics.

The beginning of each chapter notebook includes a series of **`Needs`** statements, which tell *Mathematica* which of its standard packages will be used in that notebook. If you are following the examples on your own computer, make sure to execute the **`Needs`** statement before any others in the notebook. Many of the examples in this book make use of data sets contained on the accompanying CD. If you plan to follow the examples, you can copy the data files to a directory on your hard disk or access them directly from the CD. You will, however, have to change the file path given in the examples to correspond to the location on your computer.

1.5 A Brief Tour of *Mathematica*

1.5.1 Symbolic and Numerical Operations

When *Mathematica* is started with its default settings, two things appear: a blank window named Untitled and a Basic Input palette filled with mathematical operators and Greek letters. *Mathematica* accepts expressions written either in standard text or using operators and symbols pasted from the Basic Input palette. To enter an expression, position the cursor near the top of the blank window, click the mouse to make it active, type in a simple expression, and press either Enter or Shift-Return. Using 2/3 as an example, the result is:

```
In[1]:= 2/3
```

Out[1]= $\frac{2}{3}$

After pressing enter, the intial expression is assigned an input number (in this case 1) and a corresponding output line is shown immediately below. *Mathematica* distinguishes between exact integer expressions and approximate numerical expressions, and therefore returned a value of 2/3 rather than 0.666667. Important irrational numbers such as π are also manipulated as symbols unless *Mathematica* is forced to assign a numerical approximation. Purely symbolic expressions can also be used, for example

```
In[2]:= a/b
```

Out[2]= $\frac{a}{b}$

Input and output numbers are reset each time the *Mathematica* kernel is started. Therefore, if you start *Mathematica,* save and close the window, and then open a new window the input and output numbers will continue in sequence because the kernel was not restarted.

One of *Mathematica*'s strengths is its ability to perform symbolic manipulation, for example algebra and calculus. It can find symbolic solutions to many kinds of equations, for example

```
In[3]:= Solve[a/b == 4, b]
```

Out[3]= $\{\{b \to \frac{a}{4}\}\}$

Likewise, the solution to $3\,x + 7 = 18$ is

```
In[4]:= Solve[3 x + 7  == 18, x]
```

Out[4]= $\{\{x \to \frac{11}{3}\}\}$

Note that multiplication can be specified using an asterisk (**3 * x**), by placing a space between two variables (**3 x**), or by using the multiplication operator from the

Basic Input palette (**3 × x**). As discussed in Chapter 3, matrix and vector multiplication is slightly more specific and the multiplication operators cannot be switched indiscriminantly. The same approach works for sets of equations

```
In[5]:= Solve[{2 x + 6y == 18, 7 x - 8 y == 7}, {x, y}]
```

$$\text{Out[5]= } \left\{\left\{x \to \frac{93}{29}, y \to \frac{56}{29}\right\}\right\}$$

and equations involving real numbers

```
In[6]:= Solve[a/b == 4.0, b]

Out[6]= {{b → 0.25 a}}
```

Solve is one of *Mathematica*'s standard functions, which all begin with uppercase letters and have arguments enclosed in square brackets. There are hundreds of standard functions, and hundreds more in packages accompanying the standard *Mathematica* distribution. They are listed alphabetically in *The Mathematica Book* and can also be viewed using the Help Browser. *Mathematica* uses curly braces, **{}**, to enclose lists of expressions or variables such as the lists of two equations and two variables above. It can also evaluate just about any derivative or integral that is likely to be included in standard mathematical references. A simple example, the derivative of $x^2$with respect to x, is

$$\text{In[7]:= } \partial_x x^2$$

```
Out[7]= 2 x
```

Integrating the result to recover the original expression,

$$\text{In[8]:= } \int 2\,x dx$$

$$\text{Out[8]= } x^2$$

The derivative and integral symbols were pasted into the *Mathematica* notebook by clicking on the Basic Input palette. If the limits of integration are specified, *Mathematica* will also calculate a definite integral.

$$\text{In[9]:= } \int_a^b 2x dx$$

$$\text{Out[9]= } -a^2 + b^2$$

Say we know the values of a and b. They can be substituted into the result above using a replacement rule specified with the **/.** operator. For example, if $a = 3.0$ and $b = 7.2$

```
In[10]:= % /. {a → 3., b → 7.2}

Out[10]= 42.84
```

Using the replacement rule evalutes the expression with $a = 3.0$ and $b = 7.2$ only in this instance, and does not permanently change the value of the expression. The

% sign is shorthand for the previous output, and **%%** is shorthand for the output line before that. Output lines in general can be referenced using either **%***n* or **Out [***n***]**, where *n* is the output line number. Alternatively, the the definite integral could have been evaluated numerically by using real numbers for the limits of integration.

```
In[11]:= ∫_3.^7.2 2 x dx
Out[11]= 42.84
```

The = sign is used to permanently assign values to variables. Variables can be numerical values

```
In[12]:= x = 7.2
Out[12]= 7.2
```

lists or tables of values

```
In[13]:= data = {1.2, 4.6, 9.2, 4.9}
Out[13]= {1.2, 4.6, 9.2, 4.9}
```

or the results of operations

```
In[14]:= solution = Solve[3 z == 4.2, z]
Out[14]= {{z → 1.4}}
```

Once a value is assigned to a variable name, it can be used like any other variable. For example,

```
In[15]:= √x
Out[15]= 2.68328
```

because we previously assigned the value of 7.2 to **x**. To ensure that it does not cause confusion further on, we can also clear the value of **x**.

```
In[16]:= Clear[x]
```

In can sometimes be desirable to suppress output, which can be done with a semicolon.

```
In[17]:= sinx = Sin[10. °];
```

In this case, a result is calculated and assigned to the variable name **sinx** but is not displayed. Entering the variable name will display the result

```
In[18]:= sinx
Out[18]= 0.173648
```

Like other computer languages, *Mathematica* requires angular measurements to be specified in radians. The built-in variable **Degree** is a conversion factor ($\pi/180$)

that converts angular measurements in degrees to radians. To convert radians to degrees, divide by **Degree**. *Mathematica* also recognizes commonly used mathematical symbols such as π, i, ∞, and e.

There are several methods that can be used to force *Mathematica* to return a numerical approximation of an exact integer. First, an integer expression can be followed by the expression **//N**.

```
In[19]:= 2/3 //N

Out[19]= 0.666667
```

Another way to force numerical output is to use *Mathematica*'s **N** function.

```
In[20]:= N[2/3]

Out[20]= 0.666667
```

A third way to force numerical output is to make at least one of the integers into a real number by adding a decimal point.

```
In[21]:= 2/3.

Out[21]= 0.666667
```

Mathematica will approximate a value for irrational numbers such as π

```
In[22]:= N[π]

Out[22]= 3.14159
```

or e

```
In[23]:= N[e]

Out[23]= 2.71828
```

If asked to give a numerical value for the imaginary number i, *Mathematica* returns

```
In[24]:= N[i]

Out[24]= 0. + 1. i
```

Mathematica's early versions used text input and output of expressions, but recent versions have included sophisticated mathematical notation and typsetting capabilities. The result is that many *Mathematica* functions can be specified using fairly traditional mathematical notation or simple text-only input. For example, the derivative and integral above can also be expressed as

```
In[25]:= D[x^2, x]

Out[25]= 2 x
```

and

```
In[26]:= Integrate[2 x, x]
```

Out[26]= x^2

The definite integral of 2 x from $a = 3.0$ to $b = 7.2$ can be specified as

```
In[27]:= Integrate[2 x, {x, 3.0, 7.2}]

Out[27]= 42.84
```

Likewise, the square root of 2.8 can be represented by either

In[28]:= $\sqrt{2.8}$

```
Out[28]= 1.67332
```

or

```
In[29]:= Sqrt[2.8]

Out[29]= 1.67332
```

or

In[30]:= $2.8^{1/2}$

```
Out[30]= 1.67332
```

Special symbols such as π, i, and e can be represented using the text equivalents **Pi**, **I**, and **E**.

1.5.2 Vector and Matrix Operations

Mathematica treats vectors of symbols, integers, and real numbers as lists and matrices as lists of lists. A list of data might be

```
In[31]:= data = {1.2, 4.8, 2.8, 7.2, 9.1, 6.5}

Out[31]= {1.2, 4.8, 2.8, 7.2, 9.1, 6.5}
```

whereas one list is used to represent each row of a matrix using a **Table**.

```
In[32]:= m = {{a, b}, {c, d}}

Out[32]= {{a, b}, {c, d}}
```

Elements of lists or tables can be isolated using either **Part** or double square brackets [[]]. The first element in the second row of **m** is

```
In[33]:= Part[m, 2, 1]

Out[33]= c
```

or, equivalently,

```
In[34]:= m[[2, 1]]

Out[34]= c
```

Matrices can also be filled with values following some functional relationship by using the **Table** function.

```
In[35]:= Table[i * j, {i, 1, 3}, {j, 1, 3}]

Out[35]= {{1, 2, 3}, {2, 4, 6}, {3, 6, 9}}

In[36]:= MatrixForm[%]
```

$$Out[36]= \begin{pmatrix} 1 & 2 & 3 \\ 2 & 4 & 6 \\ 3 & 6 & 9 \end{pmatrix}$$

Matrices can be displayed in more traditional form using **//MatrixForm** or **MatrixForm[]**.

```
In[37]:= m// MatrixForm
```

$$Out[37]= \begin{pmatrix} a & b \\ c & d \end{pmatrix}$$

```
In[38]:= MatrixForm[m]
```

$$Out[38]= \begin{pmatrix} a & b \\ c & d \end{pmatrix}$$

They can also be constructed by clicking on the matrix button in the Basic Input palette. Many of *Mathematica*'s functions are listable, meaning that they can be applied to lists (or lists of lists). To calculate the square root of each element in **data**, for example, apply the square root function to the entire list.

In[39]:= $\sqrt{\text{data}}$

```
Out[39]= {1.09545, 2.19089, 1.67332, 2.68328, 3.01662, 2.54951}
```

Squaring the list returns the original values.

In[40]:= $\%^2$

```
Out[40]= {1.2, 4.8, 2.8, 7.2, 9.1, 6.5}
```

Chapter 3 discusses mathematical operations on matrices, including dot and cross products.

1.5.3 2-D and 3-D Graphing

Mathematica contains functions for 2-D and 3-D graphing of functions, lists, and arrays of data. The following statement plots sin x over the range of $0 \le x \le 2\pi$.

```
In[41]:= Plot[Sin[x], {x, 0, 2 π}]
```

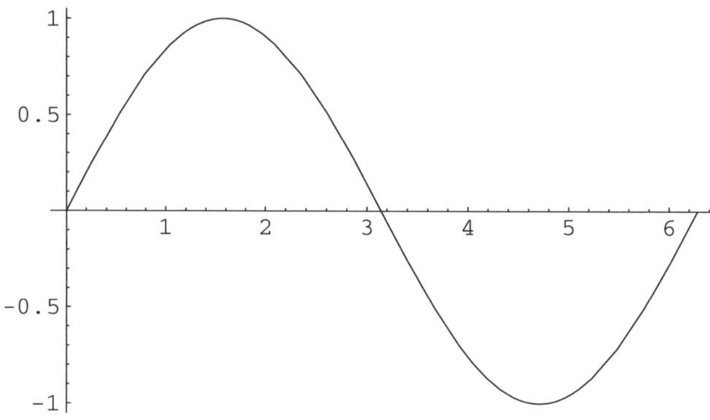


```
Out[41]= -Graphics-
```

This statement adds a title and labels to the two axes.

```
In[42]:= Plot[Sin[x], {x, 0, 2 π}, PlotLabel- > "Example Plot",
           AxesLabel → {"x", "sin x"}]
```

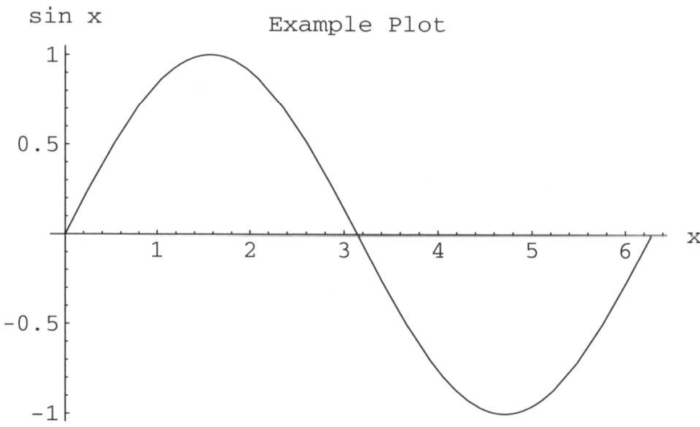


```
Out[42]= -Graphics-
```

A different function, **ListPlot**, is used for lists of data. If a list of single values is given, for example the list **data** defined above, **ListPlot** will assume that they are dependent variables and that the independent variable has the values 1, 2, 3 ...

In[43]:= **ListPlot[data, PlotStyle- > PointSize[0.02]]**

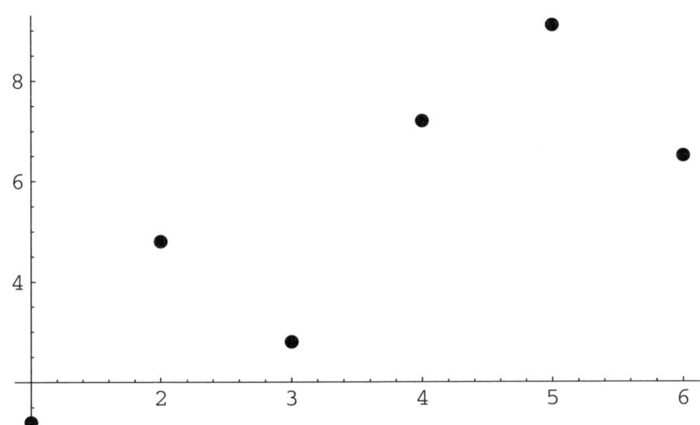

Out[43]= -Graphics-

The dots can be connected using the option **PlotJoined → True**.

In[44]:= **ListPlot[data, PlotJoined → True]**

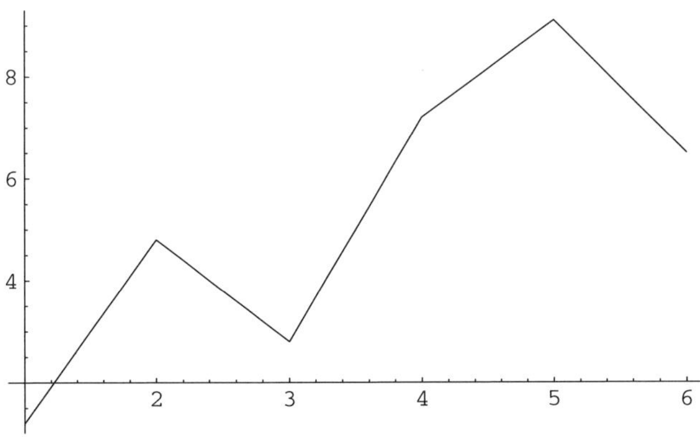

Out[44]= -Graphics-

Here is a data list with x and y values.

In[45]:= **{{1., 0.8}, {2.9, 0.7}, {3.2, 0.7}, {4.2, 0.5}}**

Out[45]= {{1., 0.8}, {2.9, 0.7}, {3.2, 0.7}, {4.2, 0.5}}

In this case, *Mathematica* plots the first element of each pair as the independent variable and the second element as the dependent variable.

```
In[46]:= ListPlot[%, PlotJoined → True]
```

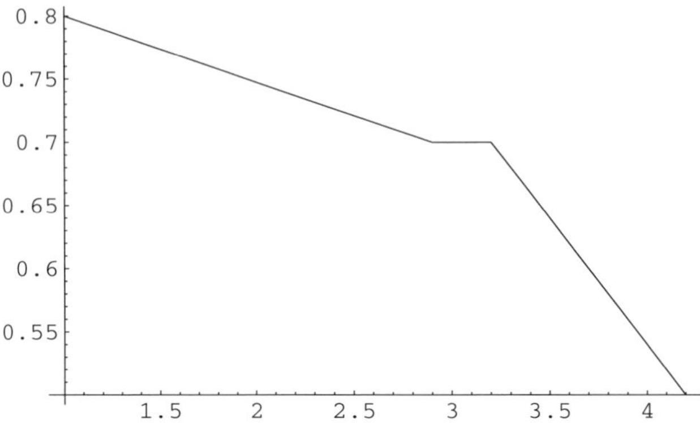


```
Out[46]= -Graphics-
```

Functions of two variables can be visualized as 3-D surface plots, contour plots, or density plots.

```
In[47]:= Plot3D[Sin[x] Sin[y], {x, 0, 2 π}, {y, 0, 2 π},
          ColorOutput → GrayLevel]
```

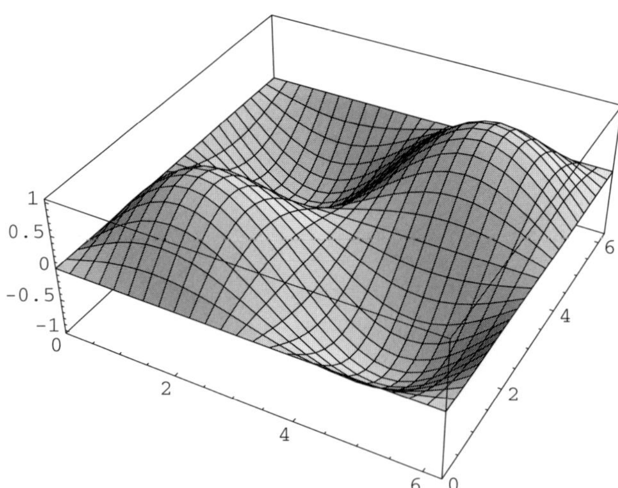


```
Out[47]= -SurfaceGraphics-
```

As with other *Mathematica* functions, options can be used to control the details of the plots. The plot below sets the number of points at which the function is evaluate to 50 instead of the default value of 25.

```
In[48]:= Plot3D[Sin[x] Sin[y], {x, 0, 2 π}, {y, 0, 2 π},
           ColorOutput → GrayLevel, PlotPoints → 50]
```

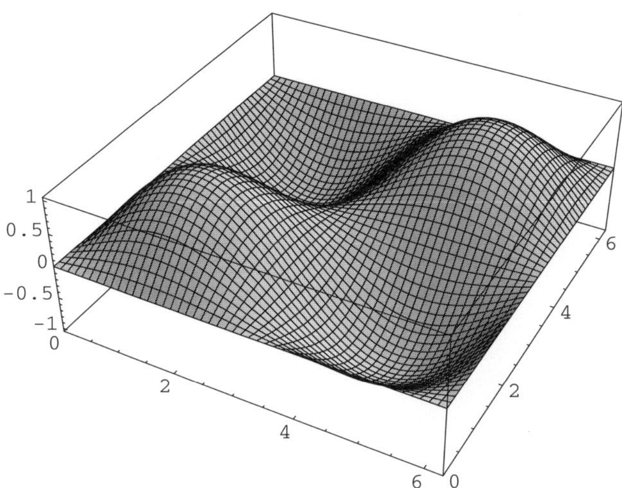


```
Out[48]= -SurfaceGraphics-
```

and this one removes the mesh.

```
In[49]:= Plot3D[Sin[x] Sin[y], {x, 0, 2 π}, {y, 0, 2 π},
           ColorOutput → GrayLevel, Mesh → False]
```

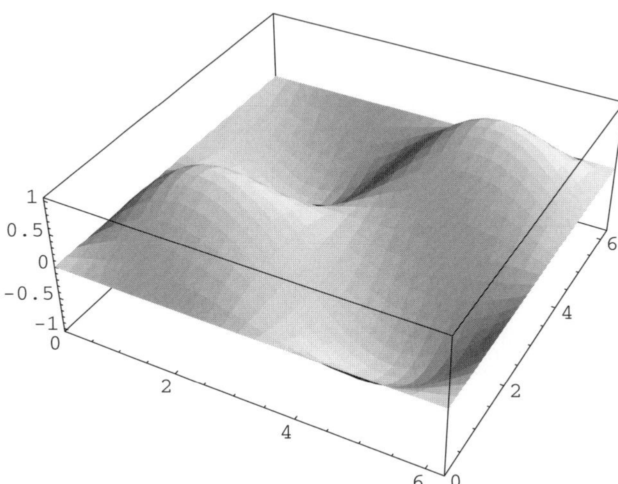


```
Out[49]= -SurfaceGraphics-
```

The **Plot3D** default is to shade surfaces using three simulated colored light sources (rendered here using gray levels; see Appendix C for a detailed discussion of color and lighting). Setting **Lighting → False** removes the lighting and shades the surface according to its height.

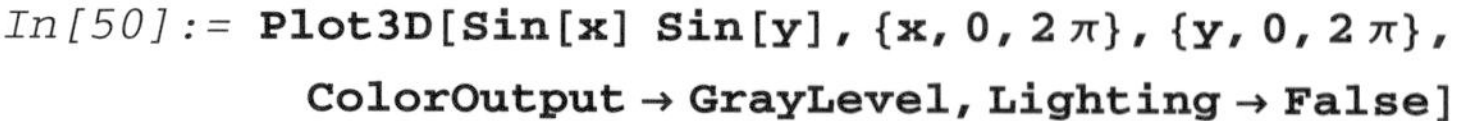

```
In[50]:= Plot3D[Sin[x] Sin[y], {x, 0, 2 π}, {y, 0, 2 π},
           ColorOutput → GrayLevel, Lighting → False]
```

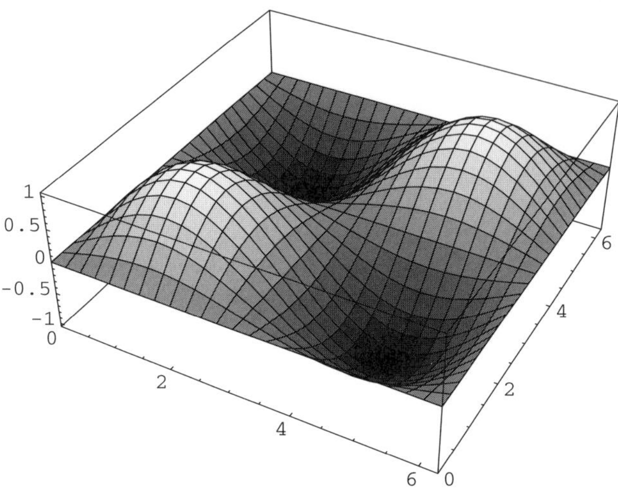


```
Out[50]= -SurfaceGraphics-
```

Setting **Shading → False** produces a wire-mesh plot and using **Hidden-Surface → False** renders the surface transparent.

```
In[51]:= Plot3D[Sin[x] Sin[y], {x, 0, 2 π}, {y, 0, 2 π},
           ColorOutput → GrayLevel, Shading → False,
           HiddenSurface → False]
```

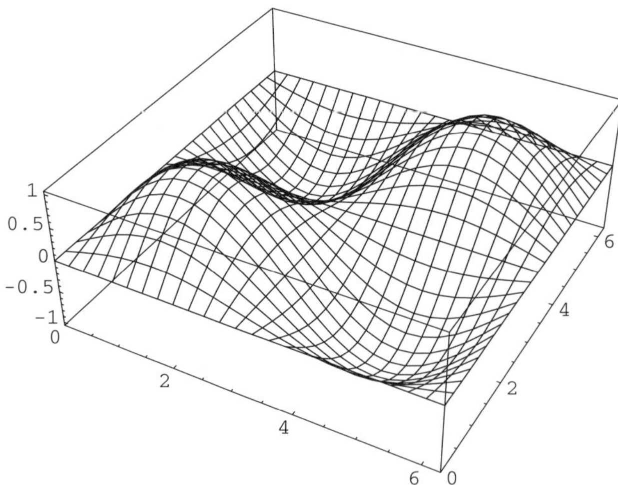


```
Out[51]= -SurfaceGraphics-
```

To see a complete list of the options available for any *Mathematica* function, use **Options**[*function_name*].

```
In[52]:= Options[Plot3D]

Out[52]= {AmbientLight → GrayLevel[0], AspectRatio → Automatic,
    Axes → True, AxesEdge → Automatic, AxesLabel → None,
    AxesStyle → Automatic, Background → Automatic,
    Boxed → True, BoxRatios → {1, 1, 0.4},
    BoxStyle → Automatic, ClipFill → Automatic,
    ColorFunction → Automatic,
    ColorFunctionScaling → True,
    ColorOutput → Automatic, Compiled → True,
    DefaultColor → Automatic, DefaultFont :→ $DefaultFont,
    DisplayFunction :→ $DisplayFunction, Epilog → {},
    FaceGrids → None, FormatType :→ $FormatType,
    HiddenSurface → True, ImageSize → Automatic,
    Lighting → True, LightSources → {{{1., 0., 1.},
        RGBColor[1, 0, 0]}, {{1., 1., 1.},
        RGBColor[0, 1, 0]}, {{0., 1., 1.},
        RGBColor[0, 0, 1]}}, Mesh → True,
    MeshStyle → Automatic, Plot3Matrix → Automatic,
    PlotLabel → None, PlotPoints → 25,
    PlotRange → Automatic, PlotRegion → Automatic,
    Prolog → {}, Shading → True,
    SphericalRegion → False, TextStyle :→ $TextStyle,
    Ticks → Automatic, ViewCenter → Automatic,
    ViewPoint → {1.3, -2.4, 2.},
    ViewVertical → {0., 0., 1.}}
```

The function **ContourPlot** works in a similar manner, but with different options.

```
In[53]:= ContourPlot[Sin[x] Sin[y], {x, 0, 2 π}, {y, 0, 2 π},
           ColorOutput → GrayLevel]
```

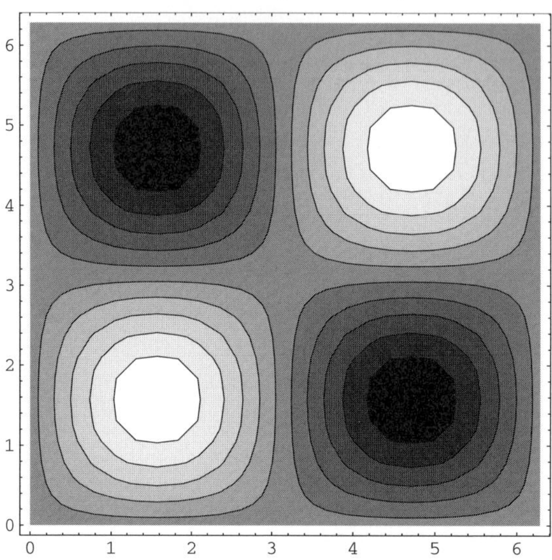

```
Out[53]= -ContourGraphics-
```

Here is the same function plotted with 3, instead of the default 10, contours.

```
In[54]:= ContourPlot[Sin[x] Sin[y], {x, 0, 2 π}, {y, 0, 2 π},
           ColorOutput → GrayLevel, Contours → 3]
```

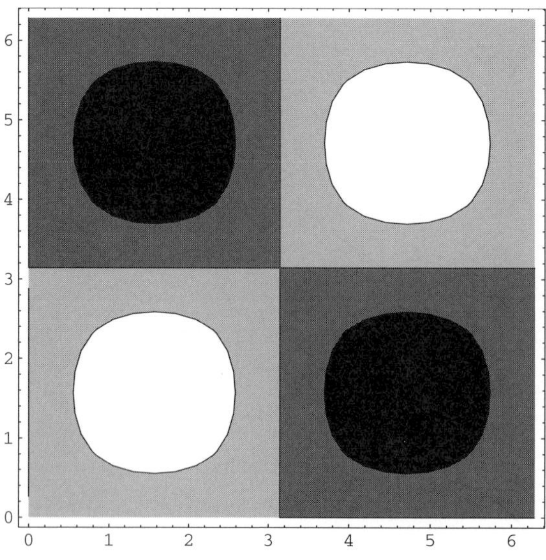

```
Out[54]= -ContourGraphics-
```

This time, without shading the contour intervals.

```
In[55]:= ContourPlot[Sin[x] Sin[y], {x, 0, 2 π}, {y, 0, 2 π},
           ColorOutput → GrayLevel, ContourShading → False]
```

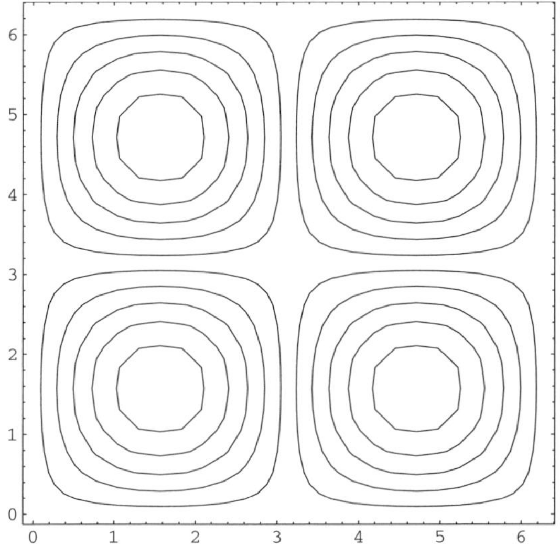

```
Out[55]= -ContourGraphics-
```

Density plots display a function of two variables using continuous shades or colors instead of contour intervals. Here is one with the default mesh

```
In[56]:= DensityPlot[Sin[x] Sin[y], {x, 0, 2 π}, {y, 0, 2 π},
           ColorOutput → GrayLevel]
```

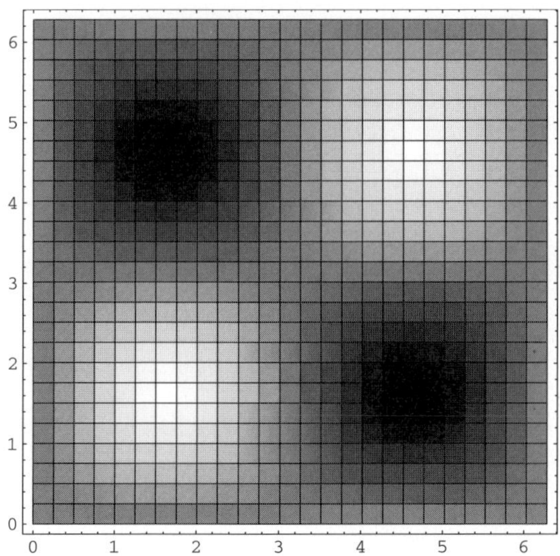

```
Out[56]= -DensityGraphics-
```

and one without the default mesh.

```
In[57]:= DensityPlot[Sin[x] Sin[y], {x, 0, 2 π}, {y, 0, 2 π},
           ColorOutput → GrayLevel, Mesh → False]
```

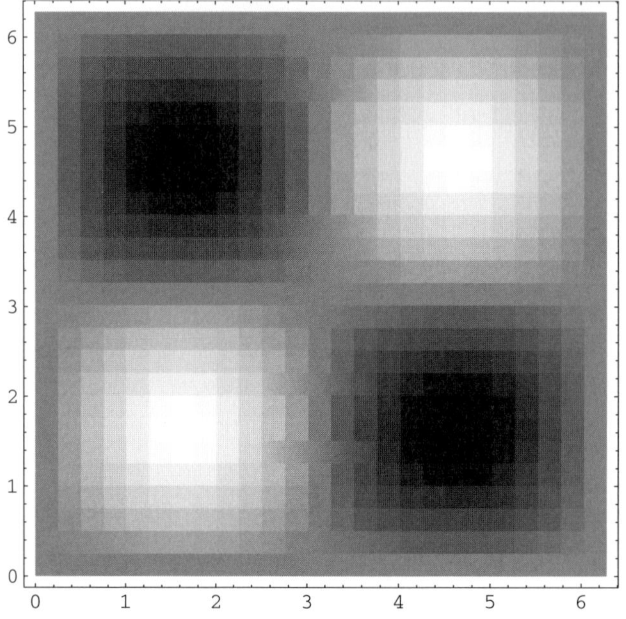


```
Out[57]= -DensityGraphics-
```

The functions **ListPlot3D**, **ListContourPlot**, and **ListDensityPlot** produce similar plots from 2-D matrices or arrays of data. None of these three accept independent values, and the horizontal coordinates are simply row and column numbers. To illustrate, first fill a table with values of sin x sin y over the ranges $0 \leq x \leq 2\pi$ and $0 \leq y \leq 2\pi$, with grid increments of $\pi/10$. The semi-colon is used to suppress the output of the resulting table.

```
In[58]:= Table[Sin[x] Sin[y], {x, 0, 2 π, π/10}, {y, 0, 2 π, π/10}];
```

Then, plot the results using **ListPlot3D**.

```
In[59]:= ListPlot3D[%, ColorOutput → GrayLevel]
```

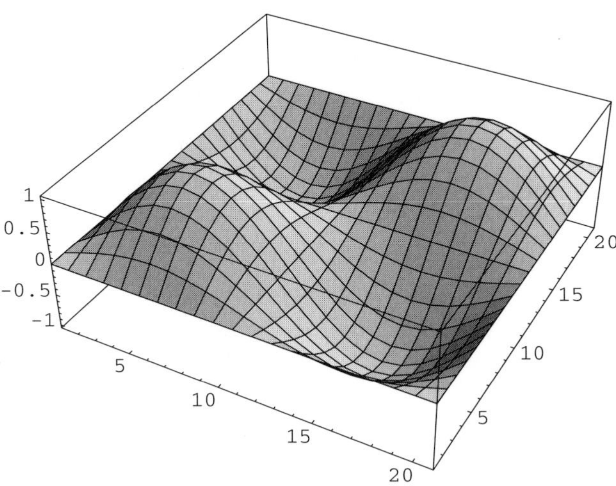


```
Out[59]= -SurfaceGraphics-
```

To change the horizontal coordinates from row and column numbers, use the **MeshRange** option.

```
In[60]:= ListPlot3D[%%, ColorOutput → GrayLevel,
           MeshRange → {{0, 2 π}, {0, 2 π}}]
```

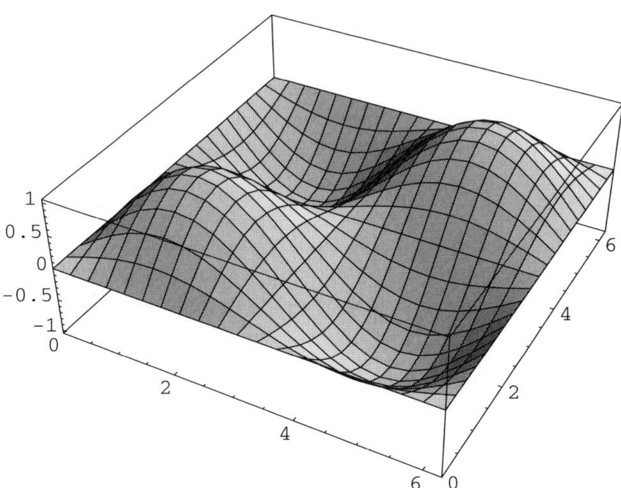


```
Out[60]= -SurfaceGraphics-
```

1.5.4 User-Defined Functions

Mathematica allows the definition of functions that can range from one-line statements to complicated programs involving logical operations, calculations, and graphics. A simple example of a function to calculate the square of a numer is:

```
In[61]:= x2[x_] := x^2
```

The combined colon and equal sign, **:=**, delays the assignment of the value $x^2$to **x2** until the function is executed, and is therefore different than **x2** =x^2. Once a function is defined, it can be used just like any of the built-in *Mathematica* functions.

```
In[62]:= x2[9.5]

Out[62]= 90.25
```

An equivalent way to accomplish the same thing is to use the **Function** function

```
In[63]:= Function[x2, x^2]

Out[63]= Function[x2, x^2]

In[64]:= x2[5]

Out[64]= 25
```

or, using *Mathematica* shorthand,

```
In[65]:= (#1^2)&

Out[65]= #1^2&

In[66]:= %[5]

Out[66]= 25
```

The shorthand version can produce very compact programs and is often used by expert *Mathematica* programmers, but can also be very difficult for others to read and understand.

Mathematica contains a variety of functions useful for flow control in longer programs – for example **If**, **Do**, **While**, and **For** – that can be used for traditional procedural programming. It also contains functions such as **Map** and **Apply** that can be used for functional programming. Here are four different ways to calculate the sines of a table of real numbers:

```
In[67]:= values = Table[x, {x, 10.°, 40.°, 10.°}]

Out[67]= {0.174533, 0.349066, 0.523599, 0.698132}

In[68]:= Map[Sin, values]

Out[68]= {0.173648, 0.34202, 0.5, 0.642788}

In[69]:= Sin[values]

Out[69]= {0.173648, 0.34202, 0.5, 0.642788}
```

```
In[70]:= Table[Sin[values[[i]]], {i, Length[values]}]

Out[70]= {0.173648, 0.34202, 0.5, 0.642788}

In[71]:= Do[values[[i]] = Sin[values[[i]]],
           {i, Length[values]}]; values

Out[71]= {0.173648, 0.34202, 0.5, 0.642788}
```

Complicated multi-line programs can be built up using the **Module** function, which allows for local variables and functions to be defined. Here is a simple application of **Module** that takes a list of data, determines its length, calculates its mean and standard deviation, and then returns all three results in a list.

```
In[72]:= Example[data_] :=
           Module[{len, mean, dev},
             len = Length[data];
```

$$\text{mean} = \frac{1}{\text{len}}\sum_{i=1}^{\text{len}} \text{data}[[i]];$$

$$\text{dev} = \frac{1}{\text{len}-1}\sqrt{\sum_{i=1}^{\text{len}} (\text{data}[[i]]-\text{mean})^2};$$

```
             Return[{len, mean, dev}]
           ]
```

Using the previously defined list **data**, the results are:

```
In[73]:= Example[data]

Out[73]= {6, 5.26667, 1.30833}
```

Outside of the module, however, the variables **len**, **mean**, and **dev** have no values.

```
In[74]:= {len, mean, dev}

Out[74]= {len, mean, dev}
```

1.5.5 Data Import and Export

Mathematica can import and export many common kinds of data and image files. The general **Import** function can in most cases recognize file types and make the appropriate conversions. **Import** assumes that any file name ending in .dat contains rows and columns of data. For example, the file example.dat contains four rows each consisting of four columns of data.

```
In[75]:= Import["/Users/bill/Mathematica_Book/example.dat"]

Out[75]= {{1, 2, 3, 4}, {5, 6, 7, 8}, {9, 10, 11, 12}, {13, 14, 15, 16}}
```

If **List** is specified as the file format, however, *Mathematica* will treat the data as a single list.

```
In[76]:= Import["/Users/bill/Mathematica_Book/example.dat",
           "List"]

Out[76]= {1, 2, 3, 4, 5, 6, 7, 8, 9, 10, 11, 12, 13, 14, 15, 16}
```

The file path name can be pasted into an **Import** statement by selecting Get File Path... from the Input menu. The same syntax works for graphics files.

```
In[77]:= Import["/Users/bill/Mathematica_Book/pako.jpg"]

Out[77]= -Graphics-
```

Using the syntax above, *Mathematica* will use the file suffix to identify the file format. See the *Mathematica* documentation for information on files without suffixes. Graphics files do not appear until they are specifically shown using the **Show** function.

```
In[78]:= Show[%]
```

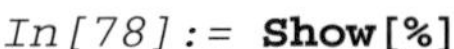

```
Out[78]= -Graphics-
```

The **Export** function works similarly to the **Import** function except that both an expression (the data or image to be exported) and a file name must be specified.

1.5.6 *Mathematica* Packages

Mathematica functions and programs can be stored as text files known as packages and loaded when needed. The standard distribution of *Mathematica* includes dozens of packages with special functions for algebra, calculus, graphics, linear algebra, numerical mathematics, and statistics. To see a complete list of the standard packages accompanying *Mathematica*, bring up the Help Browser window, choose Add-ons

& Links in the far left column, then Standard Packages in the middle column. The right column will contain a list of directories, each of which contains several add-on packages that can be loaded whenever they are needed. Additional packages are available from Wolfram Research, from other commercial developers, and in the public domain (generally downloadable from the internet). This book includes a package named CompGeosci, which contains a number of functions for specialized plots and calculations as well as color functions that are useful for color graphics. Users can also write their own packages, although the details of package writing are beyond the scope of this book.

Mathematica packages can be loaded in two ways. The first is to use **<<** *package_name*, which loads the specified package. This is generally not the recommended method because problems can arise if a package is loaded more than once during a *Mathematica* session. The preferred method is to use **Needs**[*package_name*], which loads parts of the package as needed and will not load part of a package more than once.

Package names can be specified either using their complete file path or, if they are located along one of *Mathematica*'s default file paths, using their directory (context in *Mathematica* terms) and package name. For example, to load the package DescriptiveStatistics from the Statistics directory (context) located along one of the default file paths, enter

```
In[79]:= Needs["Statistics`DescriptiveStatistics`"]
```

Note that the ` character is not a single quotation mark! It is the character located beneath the tilde (~) character in the upper left hand corner of most keyboards.To see a listing of the default file path for the installation of *Mathematica* on your computer,type **$Path** and press Enter.To see a list of the packages that have been loaded during a given *Mathematica* session, type **$Packages** and press Enter.

1.6 References and Recommended Reading

Glynn, J. and Gray, T., 2000, *The Beginner's Guide to Mathematica Version 4*: Cambridge University Press.

Wolfram, S., 1999, *The Mathematica Book (4th ed.)*: Cambridge University Press.

2 Special Plots for Geoscience Data

2.1 *Mathematica* Packages You Will Need

Be sure to execute the following statement to ensure that you will have available all of the add-on and book-specific *Mathematica* functions used in this chapter.

```
In[1]:= Needs["Statistics`DataManipulation`"]
        Needs["CompGeosci`"]
```

Computer Note: The CompGeosci package will load correctly only if it is located in one of the directories in *Mathematica*'s standard file path. Execute the statement **$Path** to see a list of the default paths on your computer and place the file CompGeosci.m in one of those directories. The specific file paths may differ from one operating system to another. See Chapter 1 for more information about installing the CompGeosci package.

2.2 Overview

Although *Mathematica* can easily plot functions and lists of data in Cartesian (x-y) and polar (θ-r) coordinate systems, it does not include functions to make several kinds of plots that are of particular interest to geoscientists. This chapter shows how to produce stem plots of discretely sampled data, rose plots of 2-D orientation data, ternary (triangular) plots, stereographic and equal-area projection plots of 3-D orientation data, box-and-whisker plots of cumulative statistics, and borehole log plots in which the independent variable of depth is plotted vertically rather than horizontally. Several examples include step-by-step instructions to illustrate the use of *Mathematica* graphics primitives to develop custom plots.

2.3 Stem Plots

Stem plots are often used to visualize discretely sampled data, for example peak annual discharges of rivers or data collected at regular or irregular intervals along a traverse. *Mathematica* does not include a stem plot function, but it is easy to construct one.

2.3.1 Importing the Data

Start by importing some data to plot, in this case from an external file containing.precipitation measurements for Eureka, California between 1949 and 2002.

```
In[2]:= data = Import[
          "/Users/bill/Mathematica_Book/EurekaPrecip.dat"];
```

> **Computer Note:** It isn't practical to invite each purchaser of this book to my office for a personal demonstration on my trusty iMac. If you are following this example on your own computer, therefore, you will have to change the file path names of data sets to correspond to the locations of the files on your computer.

The data file contains a total of

```
In[3]:= Length[data]

Out[3]= 54
```

records, each of which in turn consists of 14 items: the year, 12 monthly precipitation values, and the total annual precipitation. Here is the first row of **data**:

```
In[4]:= data[[1]]

Out[4]= {1949, 1.63, 6.09, 6.94, 0.41, 2.56,
          0.06, 0.16, 0.02, 0.5, 2.03, 3.23, 4.49, 28.12}
```

Let's say that our objective is to make a stem plot of the total annual precipitation for each year. To do this, we will need to isolate the year and total annual precipitation data from the 12 monthly values that separate them in each row of the data set.

```
In[5]:= annualdata = Table[{data[[i, 1]],
          data[[i, 14]]}, {i, Length[data]}];
```

2.3.2 Creating the Stem Plot

The first step in creating a stem plot is to create a table of lines representing the stems. The *Mathematica* syntax for a line is **Line[{{**x_1, y_1**}, {**x_2, y_2**}}]**; hence, the statement below creates a table filled with lines that run from the x axis to the y value of each annual precipitation value.

```
In[6]:= Table[
          Graphics[
            Line[{{annualdata[[i, 1]], 0.},
                {annualdata[[i, 1]], annualdata[[i, 2]]}}]],
          {i, Length[annualdata]}
        ];
```

As discussed in Chapter 1, the **Show** function must be used in order to display the lines. This may seem awkward, but there are many times when it is useful to be able to create a series of graphics objects without showing each on invidually and then superimposing all of them with one **Show** command. The following statement shows all of the stems and sets the option **Axes** to true (the default for general graphics objects is false):

```
In[7]:= Show[%, Axes → True,
          AxesLabel → {"Year", "Precipitation"}]
```

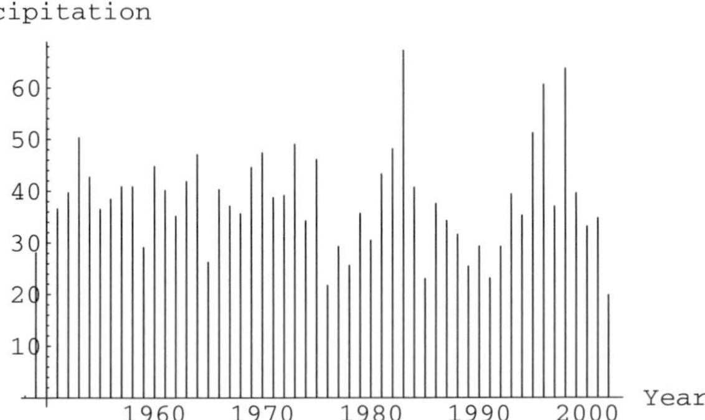


```
Out[7]= -Graphics-
```

The previous two statements could have been combined to produce and display the stem plot in one step. The position of the data points can also be emphasized by adding a ball to the end of each stem, which is most easily accomplished using the **Point** graphics function. The statement below sets the point size to 0.02 (relative to the size of the entire graphic) then, as with the stems, creates a table full of points.

```
In[8]:= Graphics[
          {PointSize[0.02],
            Table[
              Point[{annualdata[[i, 1]], annualdata[[i, 2]]}],
              {i, Length[annualdata]}
            ]
          }
        ];
```

The points can be added to the existing stem plot using **Show**, noting that the **Axes- > True** option is not carried through and must again be specified.

```
In[9]:= Show[%, %%, Axes → True,
          AxesLabel → {"Year", "Precipitation"}]
```

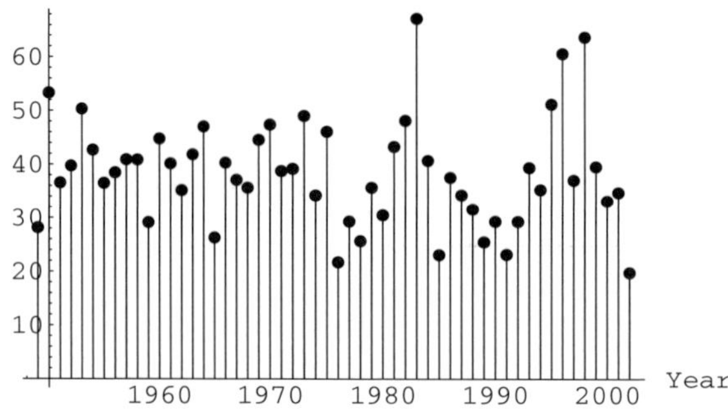

Out[9]= -Graphics-

Mathematica automatically places the y axis at a value of $x = 1950$ and labels the x axis in 10 year intervals, which leaves the 1949 stem to the left of the y axis. This can be easily changed using the **AxesOrigin** option.

```
In[10]:= Show[%, AxesOrigin → {1948, 0}]
```

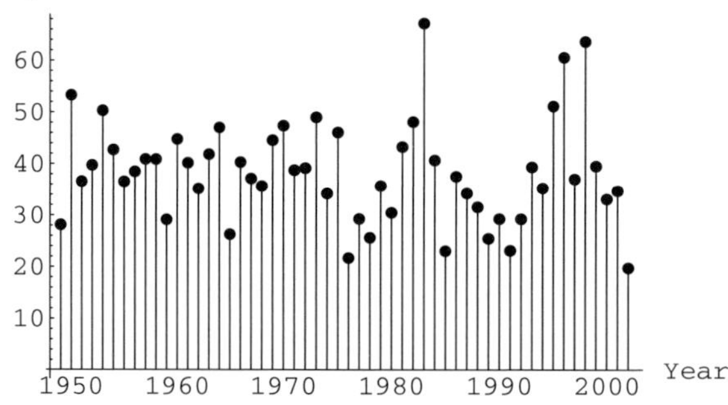

Out[10]= -Graphics-

The function **ListStemPlot** in the CompGeosci.m package accompanying this book draws stem plots in one step. Data can be supplied to **ListStemPlot** as either a series of y values, in which case the x values will be assumed to be 1, 2, 3... (just as in the built-in **ListPlot** function), or as a series of x, y pairs.

```
In[11]:= ListStemPlot[data, 0.02,
           AxesLabel → {"Year", "Precipitation"},
           AxesOrigin → {1948, 0}]
```

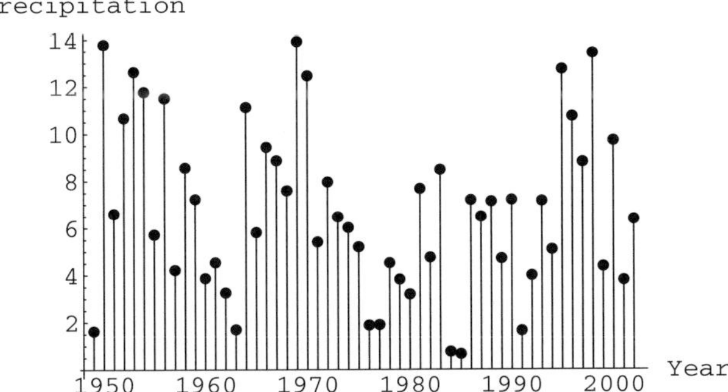

The second argument in **ListStemPlot**, in this case 0.02, is used to set **PointSize** within the function. Set it equal to zero to produce a stem plot without balls.

2.4 Rose Plots

Two-dimensional orientation data, for example paleocurrent directions measured in the field or linear fracture trace orientations measured from aerial photographs, can be shown using **rose plots**. They are, in effect, circular histograms that can be assembled using a combination of simple *Mathematica* graphics objects.

In many cases of interest to geoscientists, the data shown on rose plots is bi-directional. Fracture trace orientations, like the strike of a dipping plane, can be measured in either of two directions that differ by 180°. Some paleocurrent indicators, for example wave ripple marks and elongate clast orientation, provide bi-directional data. Other paleocurrent indicators, for example current ripple marks, imbricated clasts, and flute casts yield unidirectional data. Therefore, it will be convenient to be able to produce rose plots that can display either uni-directional or bi-directional data. We will begin with a rose plot of bi-directional data.

2.4.1 Importing the Data

First, read in a data file containing the orientations (in degrees) of elongated clasts as seen in a thin section of a fault rock. Following geoscientific convention, the bi-directional orientations are recorded so that they fall into the two upper compass quadrants (*i.e.*, azimuths of 270° to 360° and 0° to 90°).

```
In[12]:= data = Import[
         "/Users/bill/Mathematica_Book/cataclasite_data.dat",
         "List"]
```

```
Out[12]= {26., 48., 335., 337., 347., 330., 77., 10., 27., 324.,
          335., 330., 47., 347., 291., 326., 325., 31., 82., 46.,
          75., 300., 11., 4., 342., 357., 316., 326., 37., 26.,
          334., 307., 345., 336., 53., 339., 63., 341., 332., 44.,
          292., 358., 33., 359., 324., 12., 358., 350., 339., 55.,
          9., 290., 16, 1., 314., 281., 343., 76., 15., 51.}
```

The **List** format specification was used because, lacking any information to the contrary, **Import** assumes that any file with a .dat extension contains multiple rows and columns of data. It explicitly assigns each value to its own row (meaning that each value is put inside its own set of curly brackets). Using **List** forces *Mathematica* to import the values as a single list rather than a list of one element lists.

2.4.2 Creating the Rose Plot

The first step in creating a rose plot is to count the number of data points falling into bins of a fixed angular width, in this case 30°

```
In[13]:= bincts = BinCounts[data, {0, 360, 30}]

Out[13]= {11, 10, 5, 0, 0, 0, 0, 0, 0, 5, 10, 19}
```

The zeroes in the middle of the list represent the lower compass quadrants, for which no values were recorded. It will also be helpful to represent the number of bins, the width of each bin, and the maximum radius of the bins with their own variables.

```
In[14]:= binlen = Length[bincts]

Out[14]= 12

In[15]:= binwidth = 360./binlen

Out[15]= 30.

In[16]:= maxbinrad = Max[bincts]

Out[16]= 19
```

The second step is to represent each bin as a segment of a disk with a radius proportional to the numbers of data points in the bin. This is done using **Disk**. The example below plots an angular segment of a disk that is centered at (0,0), has a radius of **bincts[[1]]**, and ranges in angle from 0° to 30°. The axes are added to illustrate that the arc does indeed have a radius of 11, and the **PlotRange** and **AspectRatio** options are used to ensure that the height:width ratio of the plot is not distorted.

```
In[17]:= Show[
           Graphics[
             Disk[{0., 0.}, bincts[[1]], {0. °, 30. °}]
           ], Axes → True, PlotRange → {{-11, 11}, {0, 11}},
           AspectRatio → 1/2.
         ]
```

-10 -5 5 10

2 4 6 8 10

```
Out[17]= -Graphics-
```

At this point, it is important to think about sign conventions for angles. *Mathematica*, like virtually every mathematical textbook and computer program, conventionally measures angles positive-counterclockwise from the positive x axis. In most geoscientific applications, however, angles are conventionally measured positive-clockwise from North (which is, to add another layer of convention, usually shown towards the top of the page). In order to plot the orientation data according to geoscientific convention, then, it will be necessary to a) rotate the data by +90° and b) reverse the sign of each value. In *Mathematica* angular convention, the orientation measurements fall within the range of 180° to 0°. The wedge representing the first bin is now:

```
In[18]:= Show[
           Table[
             Graphics[
               Disk[{0., 0.}, bincts[[1]], {(90. - 30.) °,
                 (90. - 0.) °}]
             ], {i, binlen}
           ], Axes → True, PlotRange → {{-11, 11}, {0, 11}},
           AspectRatio → 1/2.
```

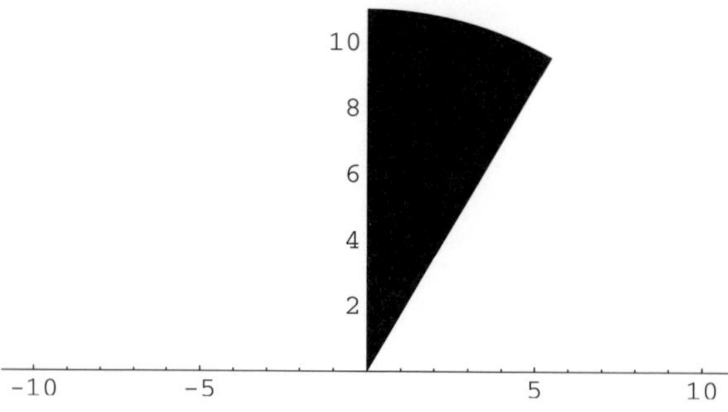


```
Out[18]= -Graphics-
```

Now that the first bin is plotted according to geoscientific convention, the next step is to plot all of the bins that have non-zero values by filling a table with wedges and then showing them.

```
In[19]:= Show[
           Graphics[
             Table[
               Disk[{0., 0.}, bincts[[i]],
                 {(90. - i * binwidth) °,
                 (90. - (i - 1) * binwidth) °}],
               {i, binlen}
             ]
           ],
           Axes → True, PlotRange → {{-maxbinrad, maxbinrad},
           {0, maxbinrad}}, AspectRatio → 1/2.
         ]
```

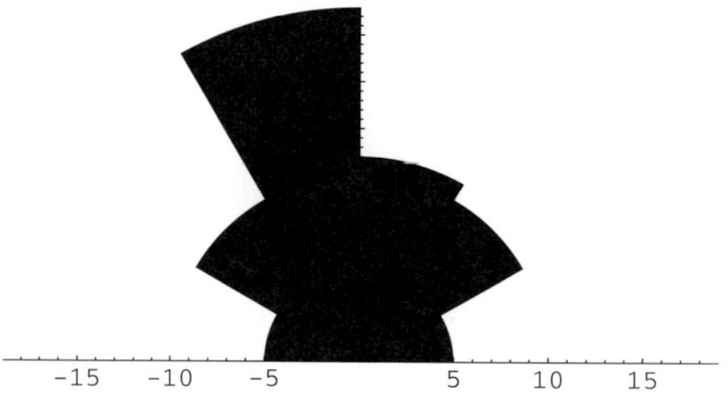


```
Out[19]= -Graphics-
```

We can also dress up the plot by adding some radii, for example in increments of 5, by first creating a table of graphics objects

```
In[20]:= Graphics[
           Table[Circle[{0., 0.}, r, {0., 180. °}], {r, 5, 20, 5}]
         ];
```

and then showing them along with the previous plot. Note that **PlotRange** was changed in order to show the outermost radius (the previous plot range was 19).

```
In[21]:= Show[%%, %, PlotRange → {{-20, 20}, {0, 20}}]
```

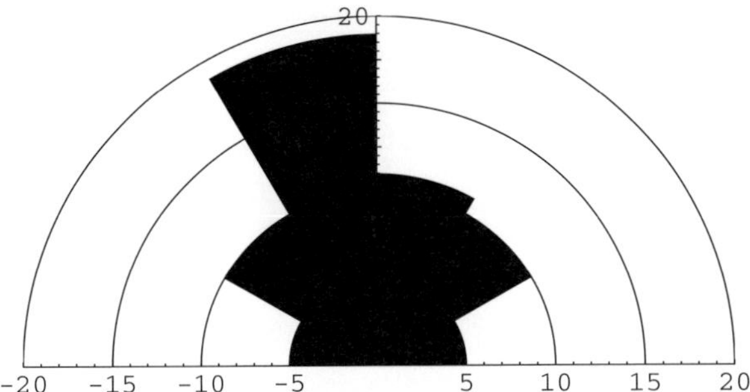


```
Out[21]= -Graphics-
```

Bi-directional rose plots are often drawn with both directions shown. This can be accomplished by adding a second table of wedges in which the reference direction is –90° rather than +90°. The plot range and aspect ratio are changed accordingly, and that the table of radii is incorporated into the list of graphics objects. Also note that, because **Graphics** is now being supplied with a list of objects instead of a single table as in the previous examples, the list must be enclosed in curly brackets { }. Failure to do so will produce an error message but no plot.

```
In[22]:= Show[
           Graphics[{
               Table[
                 Disk[{0., 0.}, bincts[[i]],
                   {(90. - i * binwidth) °,
                   (90. -  (i - 1) * binwidth) °}],
                 {i, binlen}
               ], Table[
                 Disk[{0., 0.}, bincts[[i]],
                   {(-90. - i * binwidth) °,
                   (-90. -  (i - 1) * binwidth) °}],
                 {i, binlen}
               ],
               Table[Circle[{0., 0.}, r], {r, 5, 20, 5}]
             }
           ],
           Axes → True, PlotRange → {{-20, 20}, {-20, 20}},
           AspectRatio → 1., Ticks → None
         ]
```

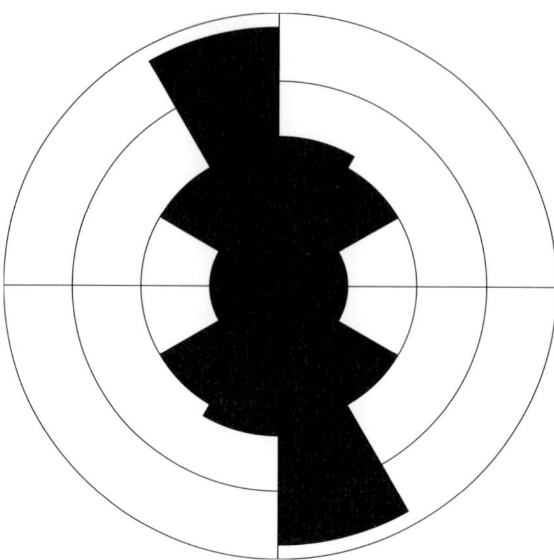

`Out[22]= -Graphics-`

Bi-directional or uni-directional rose plots can be drawn in one step using the function **ListRosePlot[data, ΔΘ, Δr, grayshade]** contained in the CompGeosci.m package. The required input data are a list of orientation values, the angular bin width $\Delta\theta$, the radial increment Δr used when drawing the circular grids, and a valid but optional gray level specification As illustrated below, **ListRosePlot** automatically adds radii in 30° increments and some labels to the four primary direction azimuths.

```
In[23]:= ListRosePlot[data, 30. °, 5., GrayLevel[0.5]]
```

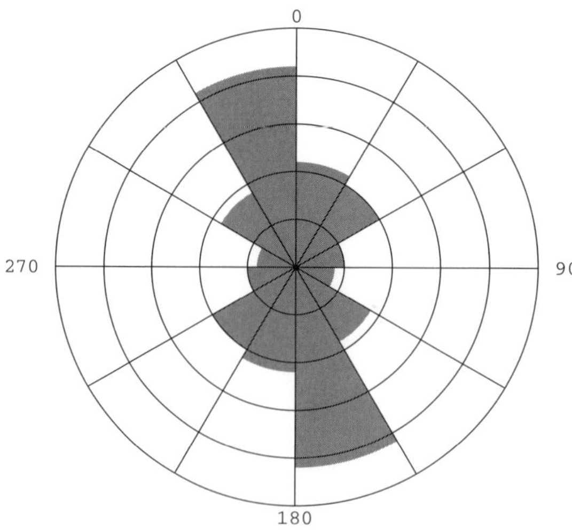

`Out[23]= -Graphics-`

ListRosePlot determines whether to make a bi- or uni-directional rose plot by checking to see if any of the input azimuths have values between 90° and 270°. If the answer is no, a bi-directional plot is drawn. If the answer is yes, a uni-directional plot is drawn.

> **Computer Note:** Obtain a unidirectional data set (preferably by going into the field or lab and measuring something!) and use it to test **ListRosePlot**.

2.5 Ternary Plots

Ternary plots are useful when data can be grouped into three categories, for example the quartz, feldspar, and lithic fragment components of a sandstone. The values used in ternary plots must sum to 1 for each point, so only two of the three values are actually needed to plot a point. The **ListTernaryPlot** function in the CompGeosci.m package is built up following the same strategy that we used for the stem and rose plots. Graphics objects representing the triangular frame, three sets of dashed grid lines, and the data points are all defined and then superimposed using **Show**. Calculation of the coordinates for the three sets of grid lines and the data points is interesting but lengthy, so we will not follow the procedure step-by-step. Instead, interested readers are referred to the file containing the CompGeosi.m package.

The syntax for the ternary plotting function is **ListTernaryPlot[data, labels, pointsize, pointshade]**, where **labels** is a list of three strings that are used to label the three vertices. The third and fourth arguments, **pointsize** and **pointshade**, are optional with default values of 0.02 and **GrayLevel[0]**, respectively. In functions where arguments are optional, *Mathematica* determines the number of arguments that have been specified and begins dropping them from right to left. Therefore, even though pointsize is optional, it must be specified if **pointshade** is to be listed as an option. Otherwise, *Mathematica* will interpret the intended **pointshade** value to be the missing **pointsize** and an error will occur.

In the example below, the data are mean quartz, lithic, and feldspar compositions for three offshore Cenozoic sandstones studied by Marsaglia (2003). The points are black because an optional **GrayLevel** value is not given and the point shade defaults to black.

```
In[24]:= data1 = {{0.22, 0.32, 0.46}, {0.17, 0.32, 0.51},
            {0.19, 0.36, 0.44}}

Out[24]= {{0.22, 0.32, 0.46}, {0.17, 0.32, 0.51},
            {0.19, 0.36, 0.44}}
```

```
In[25]:= plot1 = ListTernaryPlot[data1, {"Q", "F", "L"}]
```

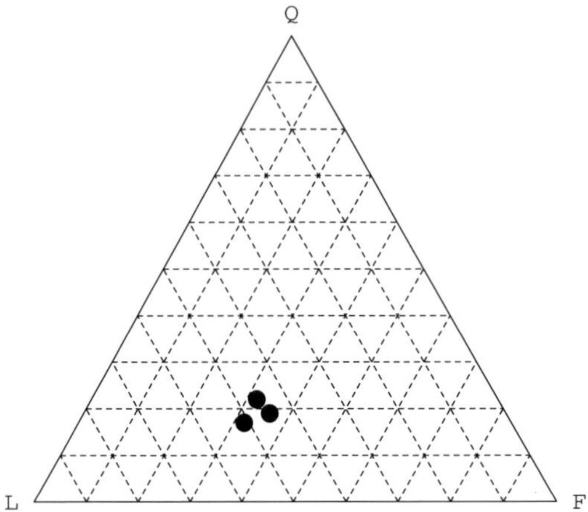


```
Out[25]= -Graphics-
```

The second set of mean quartz, feldspar, and lithic percentages, also from Marsaglia (2003), are from onshore streams and beaches. We will plot them using gray symbols to distinguish them from the offshore sand compositions, which is accomplished by adding the optional **GrayLevel[0.6]** to the list of arguments.

```
In[26]:= data2 = {{0.02, 0.23, 0.75}, {0.47, 0.39, 0.14},
           {0.3, 0.54, 0.06}}

Out[26]= {{0.02, 0.23, 0.75}, {0.47, 0.39, 0.14},
           {0.3, 0.54, 0.06}}

In[27]:= plot2 = ListTernaryPlot[data2, {"Q", "F", "L"},
           0.02, GrayLevel[0.6]]
```

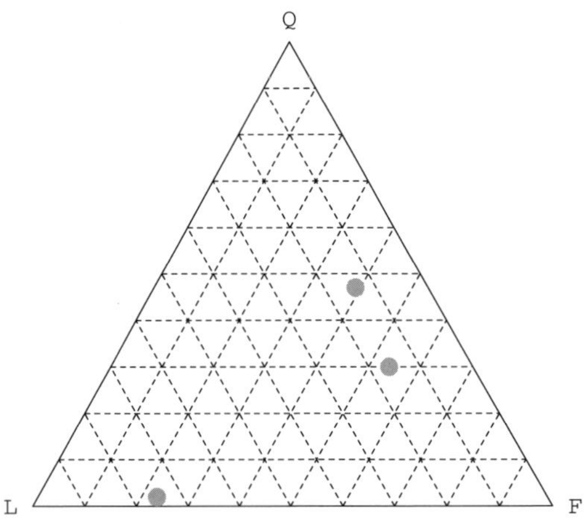


```
Out[27]= -Graphics-
```

Now, the two ternary plots can be superimposed to illustrate the compositional differences.

```
In[28]:= Show[plot1, plot2]
```

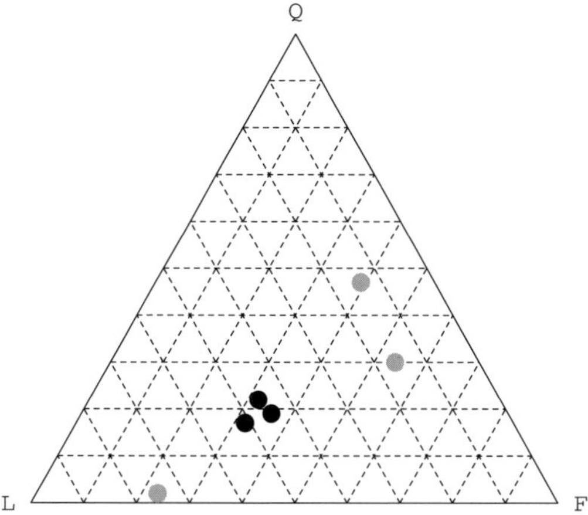


```
Out[28]= -Graphics-
```

2.6 Stereographic Projections

Stereographic projections are widely used to display and analyze angular relationships among 3-D lines and planes. They are particularly well suited for problems in which the angular relationships between planes or lines is important, because the angles between elements in a stereographic projection are the same as those between the elements being represented. For this reason, stereographic projections are sometimes referred to as **equal angle projections** in order to constrast them to the **equal area projections** described in the next section. The cost of preserving angular relationships, however, is that area is not preserved. This can be demonstrated by obtaining a copy of a Wulff stereographic net from a structural geology textbook or lab manual such as Marshak and Mitra (1998, p. 146). A 10° by 10° area at the edge of a Wulff stereographic net will be much larger than a 10° by 10° area near the center of the net. Despite their utility in problems involving angular relationships, the fact that they do not preserve area makes stereographic projections useless for applications involving the statistical analysis of orientation data (*e.g.*, contouring the density of points per unit area of the projection).

The geometric foundations of stereographic projections are described in structural geology textbooks and lab manuals such as Twiss and Moores (1992) and Marshak and Mitra (1998). Hobbs *et al.* (1976) contains a particularly thorough explanation of stereographic projections. The necessary mathematics are not complicated, involving only some basic trigonometry, but the variety of conventions used to denote the attitude of planar and linear elements can complicate the task of writing a general stereographic plotting routine. A structural geologist, for example, could equally correctly denote the strike and dip of a single dipping plane as (S45°W, 45°NW), (225°, 45°NW), or (225°, 45°). The last example is given using the **right-hand rule** that is described in many structural geology textbooks, which is convenient for computer applications because it allows input and output to be completely numerical. Using the right-hand rule, the strike is chosen so that the plane dips to the right when an observer is looking in the direction of the strike. The implication of this is that the angle **from** the chosen strike direction **to** the dip direction will always be 90° measured in a clockwise direction. Another possibility is to describe the attitude of the plane using the plunge and azimuth of its dipline, which is (45°, 315°). Dipline orientations are also convenient for computer applications because, like strikes and dips specified using the right-hand rule, they do not need non-numerical information added to eliminate ambiguities.

As illustrated in Marshak and Mitra (1998), the stereographic projection of a plane with a dip angle of ϕ is a circular arc (sometimes referred to as a cyclographic trace) with a radius of $r_{plane} = \tan\phi + \tan(\pi/4 - \phi/2)$ and a center located $\tan\phi$ from the center of the projection, measured in a direction opposite to that of the dipline azimuth. The stereographic projection of a line plunging at an angle δ is a point located at radius $r_{point} = \tan(\pi/4 - \delta/2)$ from the center of the circle, measured in the direction of the azimuth of the point. Both of these formulae assume that the stereographic plot has a maximum radius of 1.

The CompGeosci.m package that accompanies this book contains two functions to plot stereographic projections of lines and planes. The function **`ListStereoArcPlot[data, arcshade, arcdash, opts]`** constructs a stereographic plot from a list of strikes and dips. **`ListStereoArcPlot`** requires that strikes and dips be specified using the right-hand rule, with the strike listed first and the dip listed second (as in the example below). The arguments **`arcshade`** and **`arcdash`** specify the gray level and dashing of the great circle traces, with default values of black and no dashing. The last argument, **`opts`**, allows the user to specify the plot range and aspect ratio. As with **`ListTernaryPlot`**, *Mathematica* begins elminating optional arguments from right to left if the number of optional argument specified is less than the total number of options.

2.6.1 Stereographic Projections of Planes

The data set below consists of the strikes and dips of 14 joints measured at an outcrop of basalt during an engineering geologic mapping project. All of the measurements are given in degrees; therefore, they must be converted to radians before being plotted. The easiest way to do this is with the **`Degree`** constant built into *Mathematica*.

```
In[29]:= data = {{342., 75.}, {148., 50.}, {290., 80.},
           {15., 62.}, {333., 65.}, {15., 75.}, {31., 65.},
           {319., 66.}, {312., 67.}, {349., 89.9}, {359., 89.9},
           {105., 85.}, {323., 82.}, {350., 89.9}} °;
```

Notice that several of the dip angles are listed as 89.9°. This is because the plotting routines must calculate the tangent of the dip angle, and the tangent of 90° is:

```
In[30]:= Tan[90 °]

Out[30]= ComplexInfinity
```

Reducing the 90° dip angles by an imperceptible amount alleviates the complex infinity result and allows the arcs to be plotted. Used as input for **ListStereoArcPlot**, which is included in the CompGeosci.m package, they produce the following stereographic projection:

```
In[31]:= lineplot = ListStereoArcPlot[data, GrayLevel[0.3],
           Dashing[{0.005}]]
```

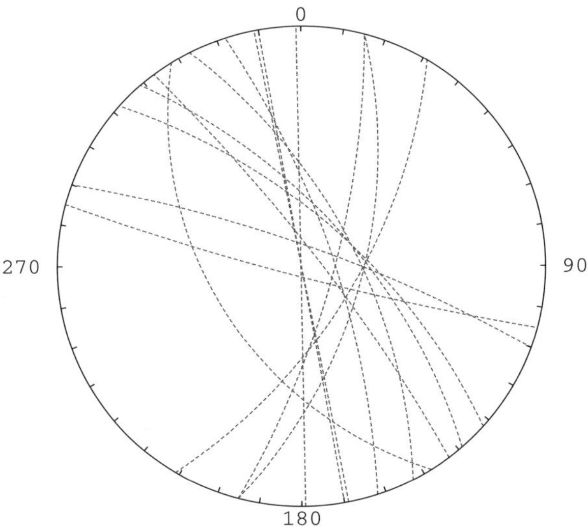


```
Out[31]= -Graphics-
```

2.6.2 Stereographic Projections of Lines

Another way to represent the planes is using the stereographic projections of their diplines, which are especially easy to calculate if the strikes are specified using the right-hand rule. Because **data** has already been converted from degrees to radians, there is no need to do so again.

```
In[32]:= diplinedata = Table[{data[[i, 2]], data[[i, 1]] + π/2},
           {i, Length[data]}];
```

Here is a plot of the diplines made using the function **ListStereoPointPlot** from the CompGeosci.m package, specifying a point size of 0.03 and open (rather than filled) circles for the points.

```
In[33]:= pointplot = ListStereoPointPlot[diplinedata,
          GrayLevel[0.], 0.03, "open"]
```

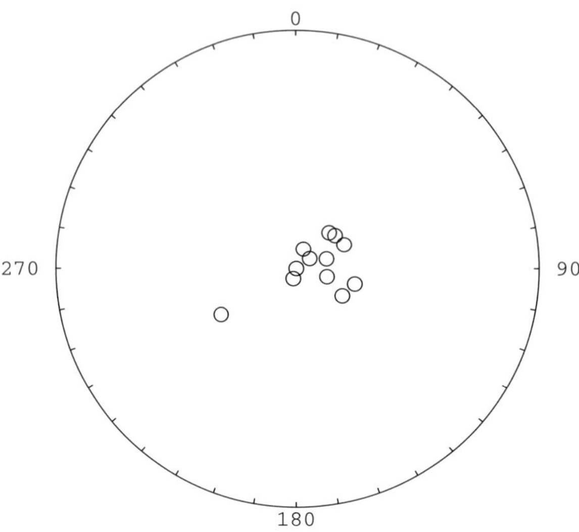


```
Out[33]= -Graphics-
```

And, to demonstrate that the diplines were calculated correctly, a superposition of the two plots showing the dipline projections exactly at the midpoints of the plane projections:

```
In[34]:= Show[lineplot, pointplot]
```

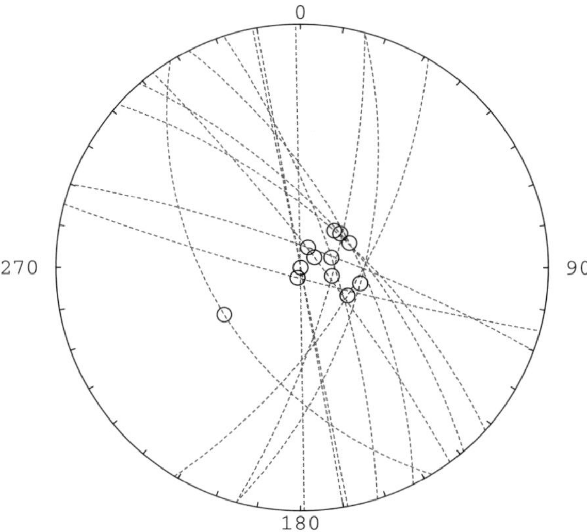


```
Out[34]= -Graphics-
```

The same **ListStereoPointPlot** function can be used to plot other linear elements represented by plunge and azimuth angles, for example 3-D paleoflow indicators, striations, lineations, and poles to planes (see the Computer Note below).

Computer Note: It is often useful to plot poles to planes, which are the lines normal to the planes. Write a short *Mathematica* function to calculate the orientation of the pole from the strike and dip. Using the definition of the right-hand rule and a sketch to illustrate the geometric relationships should help to define the problem.

2.7 Equal Area Projections

Equal area projections, as their name implies, preserve projected areas rather than angular relationships and provide an alternative to stereographic (equal angle) projections in cases where the statistical analysis or contouring of orientation data is the primary concern. The preservation of area can be demonstrated using a Schmidt equal area net from a structural geology lab manual such as Marshak and Mitra (1988, p. 146). Mark off several 10° by 10° areas on different parts of the net. All of them will be, within the limits of experimental accuracy, identical. Likewise, line segments of a given angular dimension will be equal in length regardless of their position on a Schmidt equal area net.

The equal area projection of a line plunging at an angle δ is a point located at radius $r_{point} = \sqrt{2}\sin(\pi/4 - \delta/2)$ from the center of the circle, measured in the direction of the azimuth of the point (Marshak and Mitra, 1998; Hobbs *et al.*, 1976). It is more difficult to calculate the equal area projection of a plane than the stereographic projection of a plane because the arc of the former is a portion of an ellipse rather than a circle. In practical terms, though, the increased difficulty does not matter because equal area plots are generally not used to plot arcs representing planes. Instead, they are almost always used to plot poles to planes or diplines that may be contoured or used as the starting point for statisical analyses using techniques developed specifically for data on a sphere (Fisher *et al.*, 1987).

2.7.1 Equal Area Projections of Lines

Below is an equal area plot of the dip lines contained in the list **diplinedata**. The function used to create the plot, **ListEqualAreaPointPlot**, takes the same arguments as **ListStereoPointPlot**.

```
In[35]:= ListEqualAreaPointPlot[diplinedata,
           GrayLevel[0], 0.03]
```

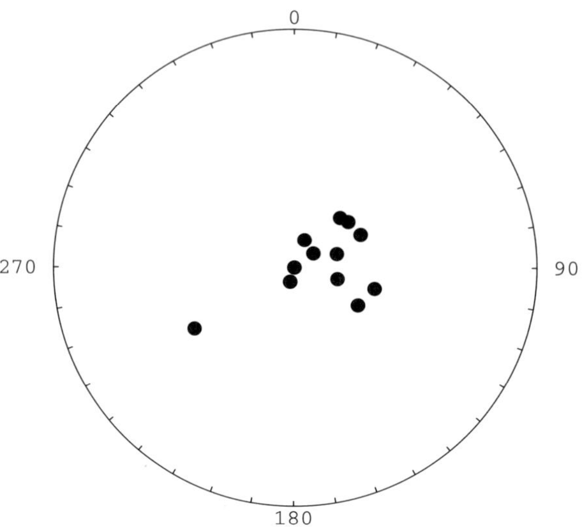


```
Out[35]= -Graphics-
```

The difference between the stereographic (equal angle) and equal area projections of the diplines can be illustrated by using **Show** to superimpose the two plots. The filled circles are the equal area projections and the open circles are the stereographic projections.

```
In[36]:= Show[%, pointplot]
```

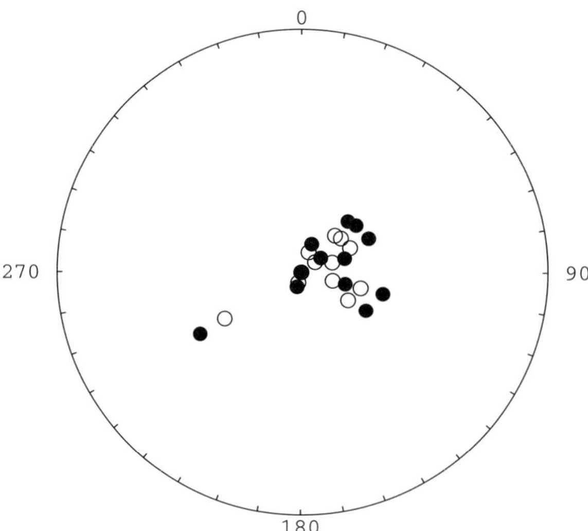


```
Out[36]= -Graphics-
```

2.7.2 Contouring Equal Area Projections

One of the principal uses of equal area projections is to analyze the angular distribution of large numbers of linear elements. These can be elements that are actually linear— for example, elongated mineral grains or clasts in a metamorphic rock, crystals in glacial ice, fault plane striations, fold axes— or elements such as dip lines or poles that are unique linear representations of planes. One way to summarize distributions of large numbers of points on an equal area net is to contour their density as described in Fisher *et al.* (1987), Marshak and Mitra (1998), and many other references.

Several different contouring methods exist, each of them with its own advantages and disadvantages. The most sophisticated contouring method in general use, and one that is is particulary well suited for computer implementation, was proposed by Kamb (1959). Most equal area contouring methods use the percentage of the data falling into overlapping counting circles with an a small area, say 1% of the entire plot (Marshak and Mitra, 1988). The size of the counting circles is not related to the size of the data set being contoured, and in that respect is arbitrary. Kamb's method, in contrast, chooses the size of the counting circles such that it is extremely unlikely that any circle would contain no data points if the data were randomly oriented. Therefore, the complete absence of points or the occurrence of many more points than would be predicted to occur on average in any given counting circle has statistical significance. The calculations used to determine the counting circle area in Kamb's method draw upon the **binomial probability distribution**, which is discussed in Chapter 4. At this point, it will suffice to know that the necessary conditions are met if the mean of the binomial distribution used to calculate the circle area is three times its standard deviation, or $\mu = 3\,\sigma$. Written in terms of the variables used in equal area net contouring, $\mu = N\,A$, where N is the number of data and A is the area of each counting circle, and $\sigma = \sqrt{N\,A\,(1-A)}$. Solving $\mu = 3\,\sigma$ for r, the radius of the counting circles is found to be $r = 3/\sqrt{\pi\,(9+N)}$. To further emphasize the statistical significance of the number of points falling into each counting circle, the contour interval is conventionally chosen to be a multilple of the standard deviation.

To illustrate the use of Kamb's method, first import a data set. In this example, we will use the orientations of clasts measured in a glacial till (Aber, 1988)

```
In[37]:= data = Import[
            "/Users/bill/Mathematica_Book/till.dat"] °;
```

Now, create an equal area plot of the orientation data.

```
In[38]:= pointplot = ListEqualAreaPointPlot[data,
            GrayLevel[0], 0.02, "filled"]
```

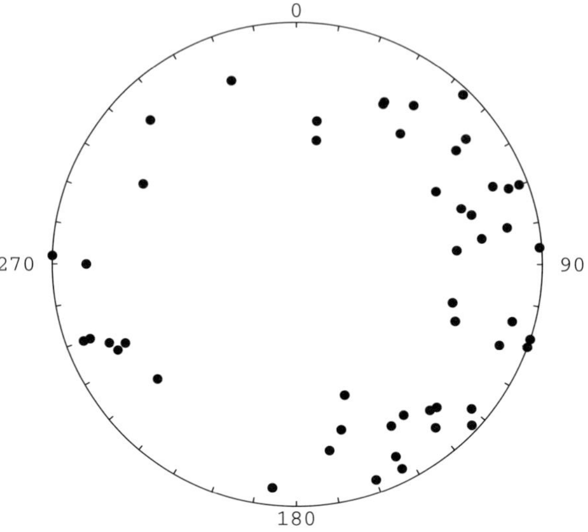


```
Out[38]= -Graphics-
```

The next step is to determine the radius of the counting circles using the formula presented above. In this example, there are

```
In[39]:= Length[data]

Out[39]= 50
```

data points and the radius of the counting circles is therefore

$$\text{In[40]:= } \mathbf{r} = \frac{\mathbf{3.}}{\sqrt{\pi\,(\mathbf{9} + \mathbf{Length[data]})}}$$

```
Out[40]= 0.220354
```

The counting circles are located on a grid with centers spaced r units apart in the $x$and y directions. The following statement illustrates the grid of counting circles by plotting 1) a table of black disks representing the points in the counting circle grid, 2) a table of gray counting circles with radii of 0.220354, and 3) a heavy black circle representing the boundary of an equal area plot with a radius of 1. The variable Δ is used to ensure that the distribution of counting circles is symmetric about the center of the equal area plot. **Floor[x]** returns the largest integer that is less than or equal to **x**.

```
In[41]:= Δ = r * Floor[1/r];
         Show[
           Graphics[
             {Table[Disk[{x, y}, 0.025], {x, -Δ, Δ, r},
                 {y, -Δ, Δ, r}],
               GrayLevel[0.4],
               Table[Circle[{x, y}, 0.220354], {x, -Δ, Δ, r},
                 {y, -Δ, Δ, r}],
               GrayLevel[0], Thickness[0.01],
               Circle[{0., 0.}, 1.]
             }
           ], AspectRatio → 1.
         ]
```

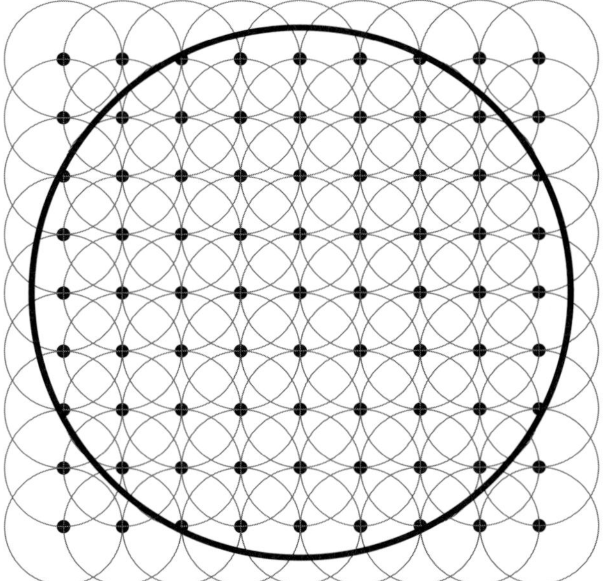

```
Out[41]= -Graphics-
```

Some of the counting circles intersect the edge of the equal area plot and four fall completely outside of the plot. *Mathematica* performs contouring on rectangular areas, so it is not possible to simply discard the circles lying completely outside of the equal area plot. Instead, they will be hidden by placing a circular mask over a square contour plot. The counting circles that straddle the equal area net boundary pose a more difficult problem because the number of points falling within the circle must be adjusted to compensate for the fact that only part of the counting circle is within the equal area plot. The Kamb contouring routine in the CompGeosci.m package accomplishes this by calculating the fraction of the counting circle that falls within the equal area plot boundary and then dividing the number of points in the circle by that fraction. For example, if 1/3 of a particular counting circle falls within the equal area plot then the number of points is multiplied by 3. Once a grid of values is generated, a polynomial surface passing exactly through all of the points is obtained using *Mathematica*'s **ListInterpolation** function and the result is contoured using **ContourPlot** using a 50 by 50 grid of interpolated values. The use of an interpolated surface produces smoother contours than would be obtained by using **ListContourPlot** to contour the results at their original grid spacing of *r*. Finally, a mask is placed over the contour plot to hide the points falling outside the equal area plot boundary. The function **ListKambPlot** is fairly long and includes two supporting functions, so it is not listed here. The functions, can, however be inspected by opening the CompGeosci.m package as a *Mathematica* notebook or with a text editor. As illustrated below, **ListKambPlot** takes as its arguments a data set consisting of (plunge, azimuth) pairs and a contour interval scaling factor. All contours are plotted in multiples of the standard deviation of the binomial distribution used to determine the counting circle area.

```
In[42]:= contourplot = ListKambPlot[data, 1.]
```

```
N  = 50

μ  = 7.62712

σ  = 2.54237

CI = 1. σ
```

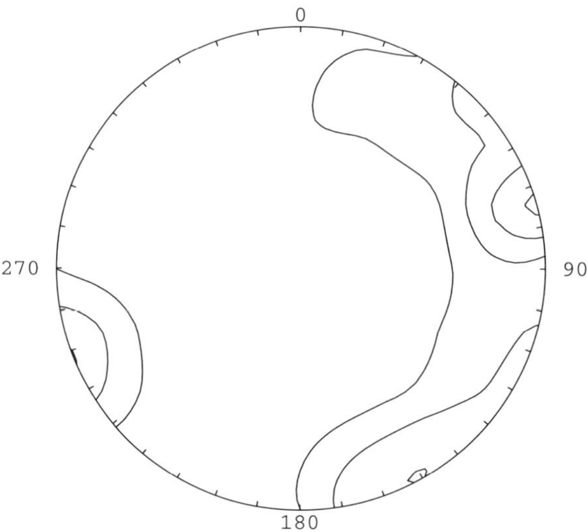


```
Out[42]= -Graphics-
```

As usual, **Show** can be used to combine the point and contour plots to see how well the contours agree with any visible clusters.

```
In[43]:= Show[contourplot, pointplot]
```

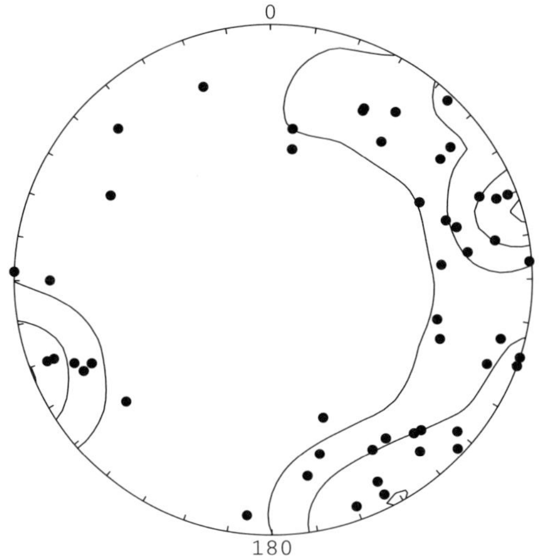

```
Out[43]= -Graphics-
```

Computer Note: Modify the **ListKambContourPlot** function to create colored contour plots. To do this, open the CompGeosci.m package as a *Mathematica* notebook or in a text editor, copy the function and rename it (so as not to destroy the original function!), and change the options in the **ContourPlot** portion of the function. Use **ColorFunction → Hue** or one of the color functions described in Appendix B (located on the CD accompanying this book).

2.8 Box and Whisker Plots

Box and whisker plots are often used to summarize the cumulative distribution of data sets. The CompGeosci.m package accompanying this book includes the function **ListBoxWhiskerPlot**, which takes the cumulative statistics for one or more data sets and draws a box and whisker plot. To illustrate, we will first need two data sets to compare.

```
In[44]:= data1 = {3.42, 4.21, 3.97, 4.7, 4.02, 4.65, 4.5, 5.19, 4.2,
           4.45, 3.64, 4.57, 4.84, 3.92, 3.85, 3.76, 4.1, 4.06,
           3.08, 4.26, 4.31, 4.66, 4.27, 4.35, 3.69}

Out[44]= {3.42, 4.21, 3.97, 4.7, 4.02, 4.65, 4.5,
           5.19, 4.2, 4.45, 3.64, 4.57, 4.84, 3.92, 3.85, 3.76,
           4.1, 4.06, 3.08, 4.26, 4.31, 4.66, 4.27, 4.35, 3.69}
```

```
In[45]:= data2 = {3.79, 2.63, 3.36, 4.9, 3.22, 2.17, 4.91, 2.68,
           3.55, 4.09, 5.09, 4.33, 3.73, 2.7, 3.11, 2.57, 3.7,
           5.03, 3.3, 3.46, 4.61, 3.71, 4.55, 3.79, 3.09}

Out[45]= {3.79, 2.63, 3.36, 4.9, 3.22, 2.17, 4.91,
           2.68, 3.55, 4.09, 5.09, 4.33, 3.73, 2.7, 3.11, 2.57,
           3.7, 5.03, 3.3, 3.46, 4.61, 3.71, 4.55, 3.79, 3.09}
```

Next, we will need a way to calculate the cumulative statistics (sometimes referred to as **percentiles** or **quantiles**). **ListBoxWhiskerPlot** requires five values for each data set: its minimum; its 25th, 50th, and 75th percentiles; and its maximum. The *n*th percentile of a data set is the value to which *n* percent of the data are less or equal. The following routine (which is **not** in the CompGeosci.m package) takes a list of data and returns the five values.

```
In[46]:= Percentiles[indata_] :=
          Module[{len, minval, maxval, pct25, pct50, pct75, data},
            len = Length[indata];
            data = Sort[indata];
            minval = Min[data];
            pct25 = data[[Round[len/4.]]];
            pct50 = data[[Round[len/2.]]];
            pct75 = data[[Round[3. len/4.]]];
            maxval = Max[data];
            Return[{minval, pct25, pct50, pct75, maxval}]
          ]
```

For example, the minimum; 25th, 50th, and 75th percentiles; and maximum of **data1** are:

```
In[47]:= Percentiles[data1]

Out[47]= {3.08, 3.85, 4.2, 4.5, 5.19}
```

The function **ListBoxWhiskerPlot** takes as its arguments a list of data sets and an optional scaling parameter that controls the width of the boxes. The default value of the scaling parameter is 0.1.

```
In[48]:= ListBoxWhiskerPlot[
           {Percentiles[data1], Percentiles[data2]},
           0.2, FrameLabel → {"data set", "percentiles"}]
```

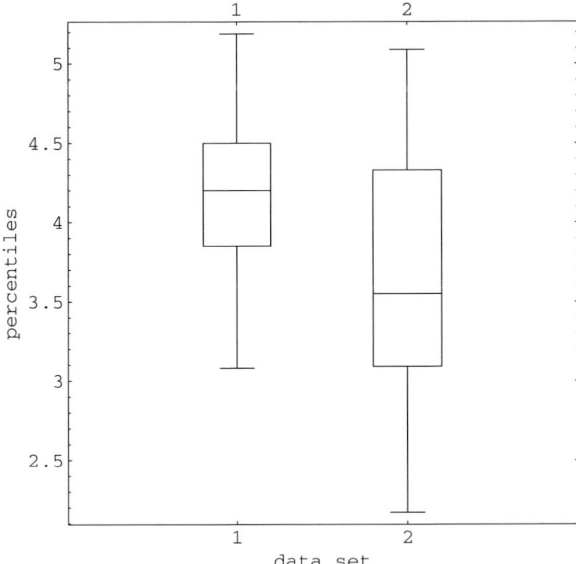

Out[48]= -Graphics-

The minimum and maximum values of each distribution are marked by the horizontal lines at the end of each whisker, whereas the 25th, 50th, and 75th percentiles are indicated by the bottom, middle, and top horizontal lines in the boxes.

> **Computer Note:** Modify **`ListBoxWhiskerPlot`** so that it also plots the mean value of each data set as a dashed line. Make sure that you make a copy of the function before attempting to modify it.

> **Computer Note:** *Mathematica* 5.0 includes the function **`BoxWhiskerPlot`** in the standard package Statistics 'StatisticsPlots', which can be loaded using either **`Needs`** or the **`<<`** operator. See the *Mathematica* documentation for more information about loading packages.

2.9 Well Logs

One of the peculiarities of the geosciences is that they often use data sets in which there is a physically based reason for plotting the independent variable on the vertical axis of a graph, for example when displaying borehole geophysical logs or data collected while measuring a stratigraphic section. The easiest way to plot graphs such as well logs is to switch the usual order of arguments used in *Mathematica*'s **`ListPlot`** function.

To illustrate how vertically oriented data such as well logs can be be imported, manipulated, and plotted using *Mathematica*, we will use a 30 m interval of geophysical log data from an exploratory well drilled to evaluate groundwater resources in clastic sediments comprising the aquifer system beneath Albuquerque, New Mexico. The data were originally supplied by the geophysical logging contractor in a single text file that contained two parts: a header containing information about the well and the logging procedure and a data section containing rows and colums of data. The data section of the file originally contained more than 6300 rows of data collected at 0.5 foot (15 cm) intervals. Although *Mathematica* can easily import and plot files of this size, almost all of the detail is lost when the complete log is displayed at the scale of a computer monitor or book page. Therefore, we will use an abbreviated data set in order to examine the geophysical signatures that distinguish sandy aquifers from clayey aquitards (or, in petroleum industry applications, potential reservoirs from shaley seals or source beds).

The first step in importing the geophysical log data, which has already been accomplished, is to remove the header information and store it in a seperate file (Charles.dat, which is included on the CD accompanying this book). Then, import the remaining data section using **Import**.

```
In[49]:= data = Import[
           "/Users/bill/Mathematica_Book/Charles.dat"];
```

You will have to change the file path as appropriate if you are following this example on your own computer. The first line in the data file is a list of 17 column names:

```
In[50]:= data[[1]]

Out[50]= {DEPT, DT, CGR, NPHI, POTA, SGR, THOR,
           URAN, GR, ILD, ILM, SFLU, SP, CALI, DRHO, PEF, RHOB}
```

From left to right, the column names are DEPT – depth in feet (1 foot = 0.30 m); DT – sonic travel time in microseconds/foot; CGR – gamma ray computed from uranium, thorium, and postassium concentrations, in American Petroleum Institute (API) gamma ray units; NPHI – neutron porosity; POTA – potassium concentration in ppm; SGR – spectroscopic gamma ray in API units; THOR – thorium concentration in ppm; URAN – uranium concentration in ppm; GR – standard gamma ray in API units; ILD and ILM – deep and medium induction resisitivity in ohm-m; SFLU – spherically-focussed resistivity in ohm-m; SP – spontaneous potential in millivolts; CALI – caliper (borehole diameter) in inches (1 inch = 2.54 cm); DRHO – density porosity; PEF – photoelectric factor in barns/electron; and RHOB – bulk density correction in g/cm^3. The numerical data start in the second row.

```
In[51]:= data[[2]]

Out[51]= {1944., 137.625, 60.2598, 0.39575, 0.02481,
           73.7363, 6.56494, 1.94691, 73.8125, 16.6419, 16.2414,
           13.3604, -78.5, 9.17969, -0.01209, 2.21875, 2.10848}
```

During the subsequent steps it will be convenient to have a variable containing the length of the data file.

```
In[52]:= npts = Length[data]

Out[52]= 202
```

To plot the log with the independent variable (depth) as the vertical axis, create a table in which depth is the second of the two variables. Below are tables corresponding to the column names listed above. Notice that the depth has been converted from feet to meters and multiplied by –1 so that depth increases downward on the resulting plots.

```
In[53]:= sflu = Table[{data[[i, 12]], -0.3 data[[i, 1]]},
           {i, 2, npts}];

In[54]:= ild = Table[{data[[i, 10]], -0.3 data[[i, 1]]},
           {i, 2, npts}];

In[55]:= sp = Table[{data[[i, 13]], -0.3 data[[i, 1]]},
           {i, 2, npts}];

In[56]:= nphi = Table[{data[[i, 4]], -0.3 data[[i, 1]]},
           {i, 2, npts}];
```

It is often useful to plot a density porosity curve in addition to a neutron porosity curve. Although the data file does not contain a density porosity column, it does have a bulk density column from which density porosity can be calculated if values for the grain density and fluid density can be inferred. The equation for density porosity, $n_\rho = (\rho_{\text{matrix}} - \rho_{\text{bulk}})/(\rho_{\text{matrix}} - \rho_{\text{fluid}})$, can be incorporated into a function

```
In[57]:= DensityPorosity[ρb_, ρm_, ρf_] := (ρm - ρb)/(ρm - ρf)
```

The example log was obtained from a clastic aquifer system rich in arkosic and volcaniclastic sediments, so reasonable values might be $\rho_{\text{matrix}} = 2.7$ g/cm^{2}and $\rho_{\text{water}} =$ 1.0 g/cm^3. Now that the density porosity function has been defined, it can be used to create a table of porosity values from the bulk density data.

```
In[58]:= dphi = Table[{DensityPorosity[data[[i, 17]], 2.7, 1.],
           -0.3 data[[i, 1]]}, {i, 2, npts}];
```

The next series of statements draws – but, in order to conserve space, does not display – plots of the SP, resistivity, and porosity tables that we have just generated. After all of the plots are drawn, they are shown side-by-side by changing **DisplayFunction → Identity** to **DisplayFunction → $DisplayFunction**.

```
In[59]:= sfluplot = ListPlot[sflu, AspectRatio → 4,
           PlotJoined → True, Frame → True,
           FrameTicks → {{25, 50, 75, 100}, Automatic, {}, {}},
           PlotRange → {All, {-580, -615}}, Axes → None,
           FrameLabel → {"ohm - m", "Depth"},
           DisplayFunction → Identity]

Out[59]= -Graphics-
```

```
In[60]:= ildplot = ListPlot[ild, AspectRatio → 4,
           PlotJoined → True, Frame → True,
           FrameTicks → {{25, 50, 75, 100}, Automatic, {}, {}},
           PlotRange → {All, {-580, -615}}, Axes → None,
           PlotStyle → Dashing[{0.02}],
           FrameLabel → {"ohm - m", "Depth"},
           DisplayFunction → Identity]

Out[60]= -Graphics-

In[61]:= resistivityplot = Show[sfluplot, ildplot,
           DisplayFunction → Identity]

Out[61]= -Graphics-

In[62]:= spplot = ListPlot[sp, AspectRatio → 4,
           PlotJoined → True, Frame → True,
           FrameTicks → {Automatic, Automatic}, Axes → None,
           PlotRange → {{-90, -66}, {-580, -615}},
           FrameLabel → {"mV", "Depth"},
           DisplayFunction → Identity]

Out[62]= -Graphics-

In[63]:= dphiplot = ListPlot[dphi, PlotJoined → True,
           AspectRatio → 4, Frame → True, Axes → None,
           PlotRange → {{0.25, 0.55}, {-580, -615}},
           FrameTicks → {{0.3, 0.4, 0.5}, Automatic, {}, {}},
           FrameLabel → {"porosity", "Depth"},
           DisplayFunction → Identity]

Out[63]= -Graphics-

In[64]:= nphiplot = ListPlot[nphi, PlotJoined → True,
           AspectRatio → 4, Frame → True,
           FrameTicks → {{0.3, 0.4, 0.5}, Automatic, {}, {}},
           Axes → None, PlotStyle → Dashing[{0.02}],
           PlotRange → {{0.25, 0.55}, {-580, -615}},
           FrameLabel → {"porosity", "Depth"},
           DisplayFunction → Identity]

Out[64]= -Graphics-

In[65]:= phiplot = Show[dphiplot, nphiplot,
           DisplayFunction → Identity]

Out[65]= -Graphics-
```

Here are all five geophysical log curves superimposed on three different sets of axes. On the resistivity plot (center), the solid line is the SFLU shallow resistitivity curve whereas the dashed ilne is the ILD deep resistivity curve. On the porosity plot, the solid line is density porosity and the dashed line is neutron porosity.

```
In[66]:= Show[GraphicsArray[{spplot, resistivityplot,
           phiplot}], DisplayFunction → $DisplayFunction]
```

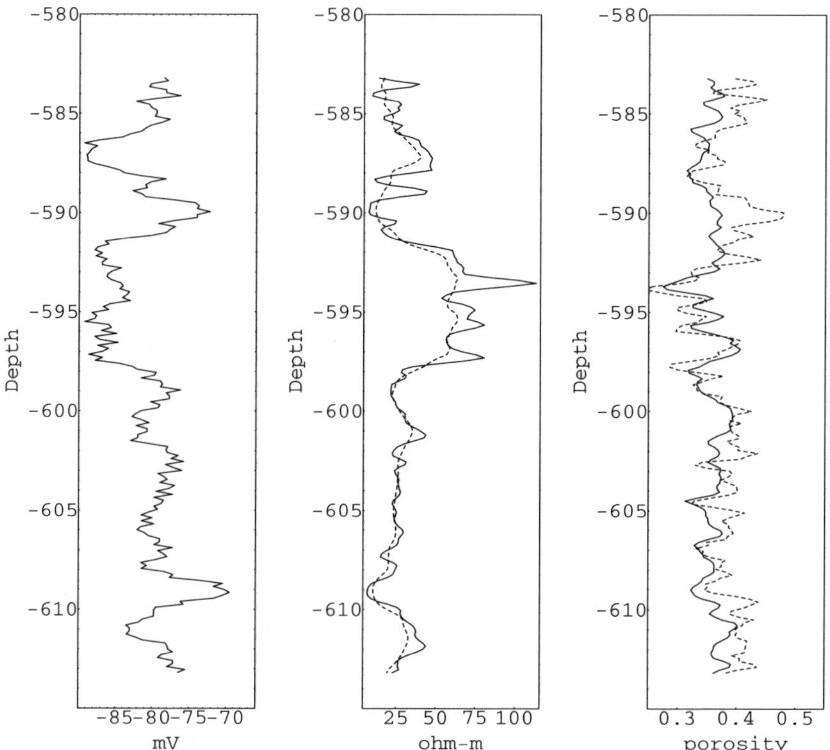


```
Out[66]= -GraphicsArray-
```

Several geophysical attributes can be used to identify sandy zones that have the potential to be productive aquifers or petroleum reservoirs (*e.g.*, Asquith and Gibson, 1982). First, low spontaneous potential (SP) values. In this example, the values range from –90 mV to –70 mV and, without knowing anything more about the specifics of log responses to various sediment types in this basin, –80 mV seems like a good first approximation of a cutoff value. Second, zones in which the spherically focussed resistivity (SFLU: solid resistivity curve) is noticably greater than the deep induction resistivity (ILD: dashed resistivity curve) indicate permeable beds in which the high resistivity drilling mud has been able to displace the low resistivity groundwater. Third, zones in which the neutron porosity log (which responds to either pore water or bound water in clays) shows a higher value than the density porosity log will tend to be clayey intervals that do not make good aquifers or petroleum reservoirs. These three criteria can be combined into a single logical statement that identifies zones with good aquifer or reservoir potential. First, create a table of zeroes.

```
In[67]:= aquifer = Table[{0, sp[[i, 2]]}, {i, npts - 2}];
```

Then, replace 0 with 1 for each depth at which all three of the criteria are satisfied.

```
In[68]:= Do[If[sp[[i, 1]] ≤ -80. && sflu[[i, 1]] > ild[[i, 1]] &&
            nphi[[i, 1]] ≤ dphi[[i, 1]], aquifer[[i, 1]] = 1],
            {i, npts - 2}
         ];
```

Now create, but do not display, a plot of the potential aquifer quality.

```
In[69]:= aquiferplot = ListPlot[aquifer, AspectRatio → 4,
           Frame → True, PlotJoined → True,
           FrameTicks → {{0, 1}, Automatic, {}, {}},
           Axes → None, PlotRange → {{-0.2, 1.2}, {-580, -615}},
           FrameLabel → {"aquifer", "Depth"},
           DisplayFunction → Identity]

Out[69]= -Graphics-
```

Finally, show the aquifer quality plot next to the three geophysical log plots for comparison. As in the previous set of plots, the solid resistivity curve is SFLU shallow resistitivity whereas the dashed ilne is ILD deep resistivity. On the porosity plot, the solid line is density porosity and the dashed line is neutron porosity. The suite of logs suggests that the best potential aquifer or reservoir is the sandy unit from –592 to –597 m.

```
In[70]:= Show[GraphicsArray[{spplot, resistivityplot, phiplot,
           aquiferplot}], DisplayFunction → $DisplayFunction]
```

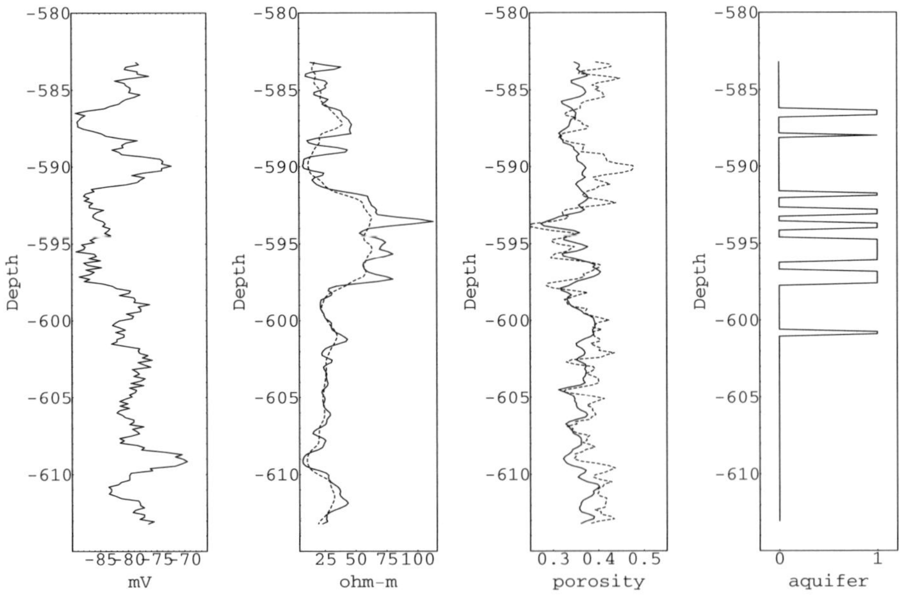


```
Out[70]= -GraphicsArray-
```

2.10 References and Recommended Reading

Aber, J.S., 1988. *Structural Geology Exercises with Glaciotectonic Examples*. Hunter Textbooks, 140 p.

Asquith, G. and Gibson, C.R., 1982, *Basic Well Log Analysis for Geologists*: American Association of Petroleum Geologists.

Fisher, N.I., Lewis, T., and Embleton, B.J.J., 1987, *Statistical Analysis of Spherical Data*: Cambridge University Press.

Glynn, J. and Gray, T., 2000, *The Beginner's Guide to Mathematica Version 4*: Cambridge University Press.

Hobbs, B.E., Means, W.D., and Williams, P.F., 1976, *An Outline of Structural Geology*: New York, Wiley.

Kamb, W.B., 1959, Petrofabric observations from Blue Glacier, Washington: *Journal of Geophysical Research*, v. 64, p. 1908-1909.

Marsaglia, K.M., 2003, Detrital modes of Magdalena fan sandstone support northward displacement of the fan from a sediment source at the mouth of the proto-Gulf of California (abstract): Geological Society of America 2003 Cordilleran Section meeting, http://gsa.confex.com/gsa/2003CD/finalprogram/abstract_50799.htm.

Marshak, S. and Mitra, G., 1998, *Basic Methods of Structural Geology*: Prentice-Hall.

Smith, C. and Blachman, N., 1995, *The Mathematica Graphics Guidebook*: Addison-Wesley.

Twiss, R.J., and Moores, E.M., 1992, *Structural Geology*: W.H. Freeman.

Wickham-Jones, T., 1994, *Mathematica Graphics: Techniques and Applications*: Springer-Verlag.

Wolfram, S., 1999, *The Mathematica Book (4th ed.)*: Cambridge University Press.

3 Manipulating and Solving Equations

3.1 *Mathematica* Packages You Will Need

```
In[1]:= Needs["Graphics`"]
        Needs["Geometry`Rotations`"]
        Needs["CompGeosci`"]
```

Computer Note: The CompGeosci package will load correctly only if it is located in one of the directories in *Mathematica*'s standard file path. Execute the statement **$Path** to see a list of the default paths on your computer and place the file CompGeosci.m in one of those directories. The specific file paths may differ from one operating system to another. See Chapter 1 for more information about installing the CompGeosci package.

3.2 Basic Symbolic Manipulation

Mathematica includes a number of functions that allow equations to be symbolically manipulated. One of the most useful of these is **Simplify**, which simplifies equations with or without conditions imposed by the user. For example, the polynomial shown below can be simplified by factoring to obtain the simplerform

```
In[2]:= Simplify[4x^2 + 4 x + 1]
```

```
Out[2]= (1 + 2 x)^2
```

The same result can in this case be obtained using **Factor**

```
In[3]:= Factor[4x^2 + 4 x + 1]
```

```
Out[3]= (1 + 2 x)^2
```

When **Simplify** was used in this case, *Mathematica* automatically determined that factoring would produce the simplest result. Other techniques might be used for different expressions. The use of **Factor** requires the user to determine that factoring the expression would be a useful thing to do. In some cases it may not be useful. A related function, **FullSimplify**, can do a more thorough job (the *Mathematica* documentation states that it will return a result at least as simple as that returned by simplify), but it can also be extremely slow. Therefore, it is usually best to start with **Simplify** and use **FullSimplify** only if the former does perform well enough. The original polynomial can be recovered using **Expand**

```
In[4]:= Expand[%]

Out[4]= 1 + 4 x + 4 x^2
```

If trigonometric terms are invovled, **TrigExpand** can be used to recast multiple angle expressions in terms of single angles, as shown below.

```
In[5]:= TrigExpand[Sin[α - β]]

Out[5]= Cos[β] Sin[α] - Cos[α] Sin[β]
```

To revert to the original expression, use **TrigFactor** or **TrigReduce**

```
In[6]:= TrigReduce[%]

Out[6]= Sin[α - β]

In[7]:= TrigFactor[%%]

Out[7]= Sin[α - β]
```

In this simple case, both functions return the same result. In general, however, the will produce different results. **TrigReduce** returns a polynomial containing no powers or products, as shown below

```
In[8]:= TrigReduce[Cos[α]^2 + Cos[α]^3]
```

$$\text{Out[8]= } \frac{1}{4}\ (2 + 3\ \text{Cos}[\alpha] + 2\ \text{Cos}[2\,\alpha] + \text{Cos}[3\,\alpha])$$

whereas **TrigFactor** returns a polynomial that can contain products and powers.

```
In[9]:= TrigFactor[Cos[α]^2 + Cos[α]^3]
```

$$\text{Out[9]= } 2\ \text{Cos}\left[\frac{\alpha}{2}\right]^2 \left(\text{Cos}\left[\frac{\alpha}{2}\right] - \text{Sin}\left[\frac{\alpha}{2}\right]\right)^2 \left(\text{Cos}\left[\frac{\alpha}{2}\right] + \text{Sin}\left[\frac{\alpha}{2}\right]\right)^2$$

The two functions **TrigToExp** and **ExpToTrig** allow for easy transformation between trigonometric and exponential expressions

```
In[10]:= TrigToExp[Sin[α]]
```

$$\text{Out[10]= } \frac{1}{2}\ i\ e^{-i\alpha} - \frac{1}{2}\ i\ e^{i\alpha}$$

```
In[11]:= ExpToTrig[%]

Out[11]= Sin[α]
```

Another group of functions is designed to isolate specific parts of symbolic expressions. These include **Part**, **Exponent**, **Coefficient**, **Numerator**, and **Denominator**. **Part** [*expression, n*] gives the *nth* part of an expression. For example, the second term in the polynomial $4x^2 + 4x + 1$ is

```
In[12]:= Part[4x^2 + 4 x + 1, 2]

Out[12]= 4 x
```

Likewise, **Coefficient** [*expression, x*] will return all of the coefficients of x in *expression*.

```
In[13]:= Coefficient[4x^2 + 4 x + 1, x]

Out[13]= 4
```

Notice that **Coefficient** does **not** return the coefficients of all powers of x, just x^1. The coefficients of terms with different powers of x can be isolated by explicitly specifying them as the argument. **Exponent** [*expression, x*] returns the highest power of x in an expression.

```
In[14]:= Exponent[4x^2 + 4 x + 1, x]

Out[14]= 2
```

The functions **Numerator** and **Denominator** are self-explanatory.

```
In[15]:= Numerator[a/b]

Out[15]= a

In[16]:= Denominator[a/b]

Out[16]= b
```

whereas **Together** pulls together expressions using a common denominator

```
In[17]:= Together[a/b + f/g]

Out[17]= (b f + a g)/(b g)
```

and **Collect** [*expression, x*] isolates the coefficients of all powers of x

```
In[18]:= Collect[3 x + 7 x y + 2 y, x]

Out[18]= 2 y + x (3 + 7 y)
```

Collecting the powers of y yields a different result

```
In[19]:= Collect[3 x + 7 x y + 2 y, y]

Out[19]= 3 x + (2 + 7 x) y
```

Another useful form of manipulation is the cancellation of terms

```
In[20]:= Cancel[(3 x + 7 x y)/x]

Out[20]= 3 + 7 y
```

The functions **ExpandNumerator** and **ExpandDenominator** operate on only the numerator and denominator of an expression.

Individual terms can be replaced using **ReplaceAll** or its shorthand equivalent **/.** and a replacement rule. For example,

```
In[21]:= ReplaceAll[3 + x, x → 4]

Out[21]= 7
```

replaces x with 4and evaluates the expression. (The replacement rule arrow → is automatically formed by typing a dash - and greater than > sign in succession on an input line.) The same replacement could have been accomplished using

```
In[22]:= 3 + x/. x → 4

Out[22]= 7
```

Using **ReplaceAll** or **/.** does not permanently change values. To illustrate this, first assign an expression to the variable name **y**

```
In[23]:= y = 3 + x

Out[23]= 3 + x
```

Because an equal sign was used, the value $3 + x$ was permanently assigned to **y**. It will remain so unless it is erased using **Clear[y]**. The following line evaluates **y** using $x = 4$ one time only

```
In[24]:= y /. x → 4

Out[24]= 7
```

and does not permanently change the value of **y**.

```
In[25]:= y

Out[25]= 3 + x
```

To permanently change the value of **y** using a replacement rule, use an equal sign

```
In[26]:= y = y /. x → 4

Out[26]= 7
```

or, if you do not mind permanently changing the value of x, type **x = 4** and the value of **y** will automatically be updated. We will want to use y again, and will clear its value so as not to cause problems later.

```
In[27]:= Clear[y]
```

Replacement rules can also be used to match patterns. Consider this unwieldy expression:

```
In[28]:= expr = Tan[α] + Tan[α]^2 + Tan[α]^3 + Tan[β] + Tan[β]^2 + Tan[β]^3

Out[28]= Tan[α] + Tan[α]^2 + Tan[α]^3 + Tan[β] + Tan[β]^2 + Tan[β]^3
```

It is well known, as illustrated below, that $\tan\theta = \theta$ for small angles ($\theta << 1$ radian).

```
In[29]:= Plot[{Tan[θ], θ}, {θ, 0, 1}, PlotStyle → {Dashing[{0.}],
         Dashing[{0.02}]}, AspectRatio → 1.5,
         AxesLabel → {"θ", " "}, PlotLegend → {"tan θ", "θ"},
         LegendPosition → {-0.4, 0.4}, LegendSize → {0.6, 0.3}]
```

From In[29]:=

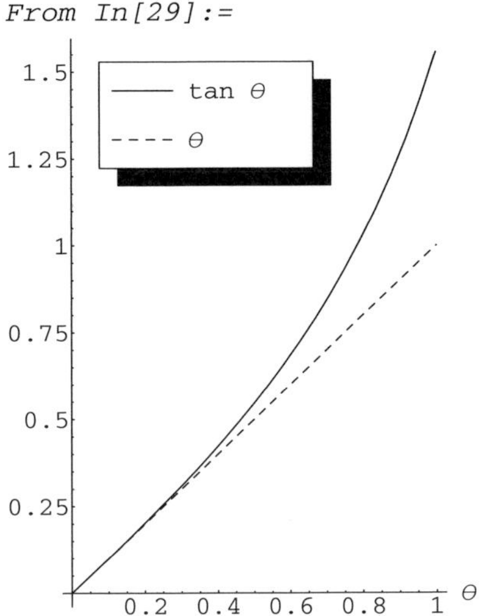

Out[29]= -Graphics-

The two curves are virtually identical for values of $0 \le \theta \le 0.3$ radians. Thus, if α and β are small the expression can be approximated to first order with very little error. First, replace the tangents with the angles

```
In[30]:= expr /. Tan[θ__] → θ
```

Out[30]= $\alpha + \alpha^2 + \alpha^3 + \beta + \beta^2 + \beta^3$

The term **θ__** , in which __ is a double underscore, instructs *Mathematica* to apply the replacement rule to all of the tangent terms regardless of their argument. The next step is to eliminate the powers of the angles, which will be extremely small if $\theta \ll 1$ radian

```
In[31]:= % /. x__^n__ → 0
```

Out[31]= $\alpha + \beta$

In the previous expression, **x__** was used to denote any value of x (in this case, either α or β) and the superscripted **n__** to denote any power (in this case, either 2 or 3). Thus, a single replacement rule was used to eliminate all of the squares and cubes of both α and β. The two replacement rules could also have been combined into a list to perform the entire simplification in one step. (Changing the order of replacement has no effect on the result.)

```
In[32]:= expr /.{ Tan[θ__] → θ, x__^n__ → 0}
```

Out[32]= $\alpha + \beta$

3.3 Matrix and Vector Operations

Mathematica recognizes matrices (and vectors) as such and performs the operations without the user having to invoke any special routines or functions. To illustrate, first define two example matrices.

```
In[33]:= m1 = (a b
               c d)

Out[33]= {{a, b}, {c, d}}

In[34]:= m2 = (h j
               k l)

Out[34]= {{h, j}, {k, l}}
```

Matrices of the same rank can be added or subtracted just like any other variable. Using **//MatrixForm** puts the result into a traditional matrix form.

```
In[35]:= m1 + m2 //MatrixForm

Out[35]= (a + h b + j
          c + k d + l)
```

whereas using **//TraditionalForm** will return the results in the typeset form common to published papers and books.

```
In[36]:= m1 + m2 //TraditionalForm

Out[36]= (a + h b + j
          c + k d + l)
```

The **TableForm** rule will put the results into rows and columns without any surrounding parentheses.

```
In[37]:= m1 + m2 //TableForm

Out[37]= a + h b + j
         c + k d + l
```

Matrix and vector multiplication is accomplished using a dot product. For example,

```
In[38]:= m1.m2 // MatrixForm

Out[38]= (a h + b k a j + b l
          c h + d k c j + d l)
```

is equivalent to

```
In[39]:= Dot[m1, m2] // MatrixForm

Out[39]= (a h + b k a j + b l
          c h + d k c j + d l)
```

Dot works just as well with with combinations of matrices and vectors.

In[40]:= **m1.** $\begin{pmatrix} \mathbf{x} \\ \mathbf{y} \end{pmatrix}$ **//MatrixForm**

Out[40]= $\begin{pmatrix} a\,x + b\,y \\ c\,x + d\,y \end{pmatrix}$

Vector cross products can likewise be obtained using **Cross**, *, or ×.

In[41]:= $\begin{pmatrix} \mathbf{x1} \\ \mathbf{y1} \end{pmatrix} * \begin{pmatrix} \mathbf{x2} \\ \mathbf{y2} \end{pmatrix}$ **// MatrixForm**

Out[41]= $\begin{pmatrix} x1\,x2 \\ y1\,y2 \end{pmatrix}$

Mathematica distinguishes variables that are vectors from those that are not. If the variables are not two vectors, the * and × operators represent standard scalar multiplication. Matrix division is not a traditionally defined operation. If one variable is divided by another and both are matrices, *Mathematica* will return a component-by-component quotient.

In[42]:= **m1/m2 // MatrixForm**

Out[42]= $\begin{pmatrix} \frac{a}{h} & \frac{b}{j} \\ \frac{c}{k} & \frac{d}{l} \end{pmatrix}$

Likewises, multiplying two matrices without the dot results in component-by-component products. The same result can be produced by taking the cross product of two matrices

In[43]:= **m1 m2 // MatrixForm**

Out[43]= $\begin{pmatrix} a\,h & b\,j \\ c\,k & d\,l \end{pmatrix}$

Other standard matrix operations include the trace

In[44]:= **Tr[m1]**

Out[44]= a + d

the determinant

In[45]:= **Det[m1]**

Out[45]= -b c + a d

the transpose

In[46]:= **Transpose[m1] //MatrixForm**

Out[46]= $\begin{pmatrix} a & c \\ b & d \end{pmatrix}$

and the inverse

```
In[47]:= Inverse[m1] // MatrixForm
```

Out[47]= $\begin{pmatrix} \frac{d}{-b\,c+a\,d} & -\frac{b}{-b\,c+a\,d} \\ -\frac{c}{-b\,c+a\,d} & \frac{a}{-b\,c+a\,d} \end{pmatrix}$

For equations that are cast in matrix form, **LinearSolve** can be used to obtain a solution. For example, the equations $3x + 7y = 8$ and $x + 4y = 7$ can be written in matrix form as:

In[48]:= **M** = $\begin{pmatrix} 3 & 7 \\ 1 & 4 \end{pmatrix}$

```
Out[48]= {{3, 7}, {1, 4}}
```

In[49]:= **Y** = $\begin{pmatrix} 8 \\ 7 \end{pmatrix}$

```
Out[49]= {{8}, {7}}
```

Values for x and y are found using **LinearSolve**.

```
In[50]:= LinearSolve[M, Y]
```

Out[50]= $\{\{-\frac{17}{5}\}, \{\frac{13}{5}\}\}$

The same result can be obtained by explicitly taking the inverse of **M** and premultiplying it with **Y**.

```
In[51]:= Inverse[M].Y
```

Out[51]= $\{\{-\frac{17}{5}\}, \{\frac{13}{5}\}\}$

As will be illustrated in the Geoscience Applications section, numerical or symbolic eigenvalues and eigenvectors can also be calculated using **Eigenvalues**, **Eigenvectors**, or **Eigensystem**. The *Mathematica* documentation contains information on additional functions such for matrix decomposition and linear programming.

3.4 Linear Equation Solving

3.4.1 Solve, Roots, and Reduce

The basic **Solve** command tries to obtain algebraic solutions to equations or systems of equations. For example, the following statement solves the equation $3\,x^2 - 4\,x + 7 = y$ for x. The equation to be solved is defined using two equal signs (==), which is a logical operator, rather than a single equal sign (=) which assigns values to variables.

```
In[52]:= Solve[3 x^2 - 4 x + 7 == y, x]
```

Out[52]= $\left\{\left\{x \to \frac{1}{3}\left(2-\sqrt{-17+3y}\right)\right\}, \left\{x \to \frac{1}{3}\left(2+\sqrt{-17+3y}\right)\right\}\right\}$

Mathematica returns solutions as lists of replacement rules, and is capable of obtaining an exact solution to any polynomial equation in one variable in which the highest power is 5. To isolate the first of the two solutions, for example to use in a calculation, simply use the replacement rule to assign the value to a variable.

```
In[53]:= x /. %[[1]]
```

Out[53]= $\frac{1}{3}\left(2-\sqrt{-17+3y}\right)$

or, equivalently,

```
In[54]:= x /. First[%%]
```

Out[54]= $\frac{1}{3}\left(2-\sqrt{-17+3y}\right)$

Another way to obtain the result is to use **Roots**, which determines the roots of polynomial equations and returns them as a logical combination of equations rather than a list of replacement rules. Using the same example equation as above, **Roots** returns

```
In[55]:= Roots[3 x^2 - 4 x + 7 == y, x]
```

Out[55]= $x == \frac{1}{3}\left(2-\sqrt{-17+3y}\right) || x == \frac{1}{3}\left(2+\sqrt{-17+3y}\right)$

in which `||` is the logical "or" operator. A simple rule to follow is to use **Solve** if you want results as a list of replacement rules and **Roots** if you want them as a list of logical expressions. **Solve** and **Roots** can also be used on systems of equations. For example,

```
In[56]:= eq1 = 4 x + 7 == y

         eq2 = 2 x + 3 == y

Out[56]= 7 + 4 x == y

Out[56]= 3 + 2 x == y

In[57]:= Solve[{eq1, eq2}, {x, y}]

Out[57]= {{x → -2, y → -1}}
```

A related function is **Reduce**, which reduces an equation to its simplest form and returns all combinations of variables that satisfy the equation. In the example below, **Reduce** returns the same answer as **Solve** or **Roots**.

```
In[58]:= Reduce[3 x^2 - 4 x + 7 == y, x]
```

Out[58]= $x == \frac{1}{3}\left(2-\sqrt{-17+3y}\right) || x == \frac{1}{3}\left(2+\sqrt{-17+3y}\right)$

In[59]:= **Solve[3 x^2 - 4 x + 7 == y, x]**

Out[59]= $\{\{x \to \frac{1}{3}(2 - \sqrt{-17 + 3 y})\}, \{x \to \frac{1}{3}(2 + \sqrt{-17 + 3 y})\}\}$

To illustrate the difference between **Solve** and **Reduce**, consider a different equation:

In[60]:= **Solve[a x == 0, x]**

Out[60]= {{x → 0}}

Because **Solve** finds generic solutions for the specified variable x, it does not consider the possibility that $a = 0$. It finds the general solution that will satisfy the equation for any value of a. **Reduce**, however, does consider the special case and returns all possible solutions, including $a = 0$.

In[61]:= **Reduce[a x == 0, x]**

Out[61]= a == 0 || x == 0

Here is another example. As illustrated in the *Mathematica* documentation, the generic solution to the quadratic equation is

In[62]:= **Solve[a x^2 + b x + c == 0, x]**

Out[62]= $\{\{x \to \frac{-b - \sqrt{b^2 - 4 a c}}{2 a}\}, \{x \to \frac{-b + \sqrt{b^2 - 4 a c}}{2 a}\}\}$

whereas all of the solutions are:

In[63]:= **Reduce[a x^2 + b x + c == 0, x]**

Out[63]= $a \neq 0 \&\& \left(x == \frac{-b - \sqrt{b^2 - 4 a c}}{2 a} \,||\, x == \frac{-b + \sqrt{b^2 - 4 a c}}{2 a}\right) \,||$
$a == 0 \&\& b \neq 0 \&\& x == -\frac{c}{b} \,||\, c == 0 \&\& b == 0 \&\& a == 0$

in which **&&** is the logical "and" operator. **Reduce** realizes, for example, that the quadratic equation can be satisfied if a, b, and c are zero whereas **Solve** does not. It also realizes that the generic solutions are invalid if $a = 0$.

3.4.2 NSolve and FindRoots

The function **NSolve** returns numerical rather than symbolic solutions. A symbolic solution to the polynomial $3x^2 - 4x + 7 = 9$ is

In[64]:= **Solve[3 x^2 - 4 x + 7 == 9, x]**

Out[64]= $\{\{x \to \frac{1}{3}(2 - \sqrt{10})\}, \{x \to \frac{1}{3}(2 + \sqrt{10})\}\}$

whereas a the numerical solution is

In[65]:= **NSolve[3 x^2 - 4 x + 7 == 9, x]**

Out[65]= {{x → -0.387426}, {x → 1.72076}}

The numerical solution gives the same result as taking the numerical value of the symbolic solution, but with one less step, which can be illustrated by finding the numerical value of the symbolic solution

```
In[66]:= N[%%]
Out[66]= {{x → -0.387426}, {x → 1.72076}}
```

There may also be cases in which **Solve** or **Reduce** cannot obtain a solution, for example to the equation sin$x = x$.

```
In[67]:= Solve[Sin[x] == x/10, x]
```

$$Out[67]= \text{Solve}\left[\text{Sin}[x] == \frac{x}{10}, x\right]$$

A approximate numerical solution for a local region of the function, however, can be estimated using **FindRoot** if an initial guess is specified. Using an initial guess of 2, we find that

```
In[68]:= FindRoot[Sin[x] == x/10., {x, 2}]
Out[68]= {x → 2.85234}
```

Placing the solution into the left-hand side of the equation

```
In[69]:= Sin[x] /. %
Out[69]= 0.285234
```

and then the right-hand side of the equation

```
In[70]:= x/10. /. %%
Out[70]= 0.285234
```

shows that the local numerical solution is correct. As shown in the plot below, however, the equations $y = \sin x$ and $y = x/10$ actually have seven local roots (*i.e.*, the points where the two functions intersect each other).

```
In[71]:= Plot[{Sin[x], 0.1 x}, {x, -15, 15}]
From In[71]:=
```

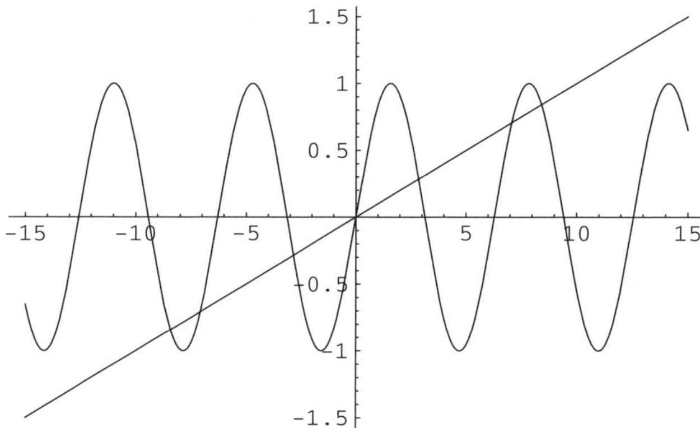


```
Out[71]= -Graphics-
```

Computer Note: Assume that you cannot plot the two functions to determine the number of roots. Write a short *Mathematica* statement to estimate the number of roots to $\sin x = x/10$ that occur within the range $-15 \le x \le 15$.

3.4.3 Geoscience Examples

Three Point Problems in Structural Geology and Hydrogeology

Three point problems arise in both structural geology and hydrogeology. In structural geology, the objective is to calculate the strike and dip (or dipline azimuth and dip angle) of a planar feature such as a dipping bed or fault from the known elevation of the feature at three points. In hydrogeology, the objective is to calculate the magnitude and direction of the hydraulic gradient from the known elevation of the potentiometric surface at three points. Mathematically, however, the two kinds of three point problem are virtually identical.

Although the solution of three point problems is traditionally taught using scaled drawings, it is also possible to solve the problems algebraically.The key is to define a plane containing all three of the points,which has the equation $z = a + b\,x + c\,y$. To find a general solution to the three point problem, start by defining the form of the plane in which all three points lie. Assume that the $+x$ direction is East and the $+y$ direction is North, so the locations of the points can be given in UTM coordinates, state plane coordinates (or similar systems outside the United States), or any other orthogonal coordinate system.

```
In[72]:= plane = a + bx + cy

Out[72]= a + b x + c y
```

Next, determine the slope of the plane in the two coordinate directions, which is done by differentiating with respect to x and y.

```
In[73]:= xslope = ∂x plane

Out[73]= b

In[74]:= yslope = ∂y plane

Out[74]= c
```

Because the slopes are vectors (they have both magnitude and direction), the maximum slope can be found using the Pythagorean theorem.

```
In[75]:= maxslope = √((∂x plane)^2 + (∂y plane)^2)

Out[75]= √(b^2 + c^2)
```

In a hydrogeologic problem, **maxslope** is the maximum hydraulic gradient. In a structural geologic problem, **maxslope** is the tangent of the dip angle. Like other computer programs, *Mathematica* calculates angles in radians. The variable

Degree is a built in conversion factor with a value of $\pi/180$, so dividing an angle measured in radians by **Degree** will convert it to degrees.

```
In[76]:= dipangle = ArcTan[maxslope]/°
```

$$Out[76]= \frac{\text{ArcTan}\left[\sqrt{b^2 + c^2}\right]}{°}$$

The strike of a dipping plane is an ambiguous quantity, because it can be specified in either of two diametrically opposed directions. A strike of 80° is identical to a strike of 80° + 180° = 260°. Dipline azimuths, however, are unambiguous quantities and are therefore easier to calculate than strike directions. The dipline azimuth is defined as the compass direction of the maximum downward inclination of the plane, which is identical to the compass direction of a vector normal to the plane (also known as the *aspect* of the plane). The dipline azimuth is also the direction of the maximum hydraulic gradient in hydrogeologic problems. Slope and aspect will come up again in Chapter 7, where they will be used to help visualize and analyze digital elevation models representing Earth's surface. The azimuth of the dipline is given by

```
In[77]:= azimuth = ArcTan[-yslope, -xslope]/°
```

$$Out[77]= \frac{\text{ArcTan}[-c, -b]}{°}$$

The strike of a dipping plane is found by adding or subtracting 90° to or from the azimuth. Notice that the syntax of the **ArcTan** function used above is different than that used to calculate the dip angle. Most computer languages, including *Mathematica*, make provisions for two kinds of arctangent functions. The first kind requires only one argument and returns a value in the range of $-\pi/2$ to $\pi/2$ radians (–90° to 90°), which is acceptable for dip angle values that can fall only between 0 and $\pi/2$ radians (0° to 90°). The second kind takes two signed arguments and returns a value that can range from $-\pi$ to π radians (–180° to +180°), which is also the range of possible aspect or dipline azimuth values. The *Mathematica* documentation also states that **ArcTan**[*x*,*y*] returns the value of arctan y/x, but this angle is measured from the $+x$ axis. To calculate the dipline azimuth as measured from the $+y$ axis, which corresponds to North, the x and y values are swapped.

Consider the example of a dipping bed that is intersected by three boreholes. The x, y, z coordinates of the borehole-bed intersections are (1000 m, 1000 m, 100 m), (2000 m, 5000 m, 850 m), and (4000 m, 3000 m, 400 m). It is a common mistake to mix units of measurement in three point problems, for example by using x and y coordinates given in kilometers and a z coordinate given in meters. Another common mistake is to use the depths measured in boreholes rather than elevations for the z coordinate. Always check your input to ensure that the units are consistent and correct. Using the coordinates above, we can write three equations describing the plane:

```
In[78]:= eq1 = 100. == a + 1000. b + 1000. c;
         eq2 = 850. == a + 2000. b + 5000. c;
         eq3 = 400. == a + 4000. b + 3000. c;
```

and use **Solve** to find numerical values fora,b,andc.

```
In[79]:= Solve[{eq1, eq2, eq3}, {a, b, c}]

Out[79]= {{a → -65., b → -0.03, c → 0.195}}
```

The dip angle, in degrees, is

```
In[80]:= dipangle /. %[[1]]

Out[80]= 11.1608
```

and the azimuth of the dipline, again in degrees, is

```
In[81]:= azimuth /. %%[[1]]

Out[81]= 171.254
```

The strike of the bed is

```
In[82]:= % + 90

Out[82]= 261.254
```

or

```
In[83]:= %% - 90

Out[83]= 81.2538
```

Computer Note: Write a *Mathematica* function that accepts the x, y, and z coordinates of three co-planar points and calculates the slope and aspect of the plane. How would you specialize the function to calculate strike and dip or hydraulic gradient magnitude and direction?

Stress Tensor Rotation, Principal Stresses, and Eigenvalues

Many quantities of interest to geoscientists are represented by **tensors**, which are groups of numbers that obey certain transformation laws with regard to their coordinate systems. The best known examples of tensor quantities used in the geosciences include **permeability** (or **hydraulic conductivity**), **stress**, and **strain** (*e.g.*, Furbish, 1997; Johnson, 1970; Middleton and Wilcock, 1994; Oertel, 1996). The calculation of principal stress magnitudes and orientations is a classic geoscience problem in which *Mathematica*'s symbolic and matrix manipulation capabilities can be put to good use.

In two dimensions, the state of stress in a continuous medium is represented by a **second-rank tensor** containing two rows and two columns.

$$In[84]:= \sigma = \begin{pmatrix} \sigma xx & \sigma xy \\ \sigma yx & \sigma yy \end{pmatrix}$$

```
Out[84]= {{σxx, σxy}, {σyx, σyy}}
```

In the example above, **σ** refers to the entire tensor whereas the indexed values (*e.g.*, **σxx**) refer to the components of the tensor. The first of the two indices refers to the coordinate direction **in** which that component of stress is acting and the second refers to the orientation of the imaginary surface **on** which it is acting. Thus, **σxx** is the component of the stress tensor acting **in** the x-direction and **on** a surface normal to the x-axis. This is often referred to as the **in-on** notation. Each of the components of the stress tensor is expressed in terms of a force per unit area, which has units identical to those of pressure (*e.g.*, Pascals in SI units).

Computer Note: Although it may seem logical to use *Mathematica*'s typesetting capabilities to more elegantly express the components of the stress tensor using true subscripted indices, for example

$$\sigma = \begin{pmatrix} \sigma_{xx} & \sigma_{xy} \\ \sigma_{yx} & \sigma_{yy} \end{pmatrix}$$

this will not work. If this were entered as input, *Mathematica* would repeatedly assume that each of the components of the tensor σ is the tensor itself with a subscript, and return an error message when its recursion limit of 256 iterations is exceeded. You can try this to see the results if you'd like, but first familiarize yourself with the Abort Evaluation option under the Kernel menu. The easiest way to deal with symbolic subscripts is, therefore, to append them onto the end of the variable name as shown above. Integer indices can be used by defining the tensor using *Mathematica*'s **Array** function, for example **Array[σ, {2, 2}]**. If a tensor is defined in this manner, then one would refer to the second colum of the first row as **σ[1, 2]**. Although it is a matter of personal preference, it seems easier to type **σxy** than **σ[1, 2]**.

The forces acting in the x- and y-directions, respectively, on a plane oriented at an angle θ can be converted to stresses (by dividing each force by the area upon which it acts; see Middleton and Wilcock, 1994, p. 144 for a detailed explanation) and written as two components of a **stress vector**. They are:

```
In[85]:= σx = σxx Cos[θ] + σxy Sin[θ]

Out[85]= σxx Cos[θ] + σxy Sin[θ]
```

and

```
In[86]:= σy = σyx Cos[θ] + σyy Sin[θ]

Out[86]= σyx Cos[θ] + σyy Sin[θ]
```

in which **σx** and **σy** are the components of the stress vector acting in the x- and y-directions, and **θ** is the angle between the positive x-axis and the outward-directed normal to the plane. These two equations are the two dimensional form of **Cauchy's**

formula, which can be used to calculate the stresses acting on a plane of any orientation. They can also be written in matrix form as

```
In[87]:= σ.(Cos[θ]
            Sin[θ]) == (σx
                        σy)

Out[87]= True
```

Mathematica returns a value of **`True`** because the two sides of the equation are separated by a logical operator and are algebraically identical to each other. To determine the normal and shear stresses acting on a plane oriented at angle θ, the stress vectors must be rotated so that they are acting parallel and perpendicular to the plane. This is accomplished using a two dimensional **rotation matrix** that we will call **`R`**

```
In[88]:= R = ( Cos[θ]  Sin[θ]
              -Sin[θ]  Cos[θ])

Out[88]= {{Cos[θ], Sin[θ]}, {-Sin[θ], Cos[θ]}}
```

The rotation matrix can be entered by hand, as shown above. The standard *Mathematica* package Geometry`Rotations` also includes a rotation matrix function

```
In[89]:= RotationMatrix2D[θ] // MatrixForm

Out[89]= ( Cos[θ]  Sin[θ]
          -Sin[θ]  Cos[θ])
```

The normal and shear stresses on a plane with arbitrary orientation θ are the stress vectors premultiplied by the rotation matrix, or

```
In[90]:= Simplify[R.σ.(Cos[θ]
                       Sin[θ])] // MatrixForm

Out[90]= (σxx Cos[θ]^2 + (σxy + σyx) Cos[θ] Sin[θ] + σyy Sin[θ]^2
          σyx Cos[θ]^2 - (σxx - σyy) Cos[θ] Sin[θ] - σxy Sin[θ]^2)
```

Using **`RotationMatrix2D`** will produce the same results:

```
In[91]:= Simplify[RotationMatrix2D[θ].σ.(Cos[θ]
                                         Sin[θ])]// MatrixForm

Out[91]= (σxx Cos[θ]^2 + (σxy + σyx) Cos[θ] Sin[θ] + σyy Sin[θ]^2
          σyx Cos[θ]^2 - (σxx - σyy) Cos[θ] Sin[θ] - σxy Sin[θ]^2)
```

Equilibrium of moments, which is necessary to ensure that the continuous medium is not undergoing rotational acceleration, requires that $\sigma_{xy} = \sigma_{yx}$. The results above can therefore be further simplified to

```
In[92]:= % /. {σyx → σxy} // MatrixForm

Out[92]= (    σxx Cos[θ]^2 + 2 σxy Cos[θ] Sin[θ] + σyy Sin[θ]^2
          σxy Cos[θ]^2 - (σxx - σyy) Cos[θ] Sin[θ] - σxy Sin[θ]^2)
```

which is the same result used in structural geology, geomechanics, and geophysics textbooks. *Mathematica* makes the general assumption that the product of two matrices is another matrix in which each row is a list, even in the special case where one of the matrices (and hence the result) is a vector. Therefore, the result above is in the general matrix form of a list of lists. Each of its two components can be extracted and assigned to variable names for later use.

```
In[93]:= normstress = %[[1, 1]]

Out[93]= σxx Cos[θ]^2 + 2 σxy Cos[θ] Sin[θ] + σyy Sin[θ]^2

In[94]:= shearstress = %%[[2, 1]]

Out[94]= σxy Cos[θ]^2 - (σxx - σyy) Cos[θ] Sin[θ] - σxy Sin[θ]^2
```

To illustrate the use of the results, consider a medium in which the state of stress is given by σ_{xx}= 250 MPa, σ_{yy}= 100 MPa, and σ_{xy}= 75 MPa. What are the normal and shear stresses acting on a plane with an outward directed normal vector that is rotated 18° with respect to the positive *x*-axis?

```
In[95]:= normstress /. {σxx → 250., σyy → 100., σxy → 75., θ → 18.°}

Out[95]= 279.76
```

and the shear stress is given by the second component of **normshear**

```
In[96]:= shearstress/.{σxx → 250., σyy → 100., σxy → 75., θ → 18.°}

Out[96]= 16.5924
```

Another way to visualize the relationship among the components of the stress tensor and the angle θ is to calculate the values of normal and shear stresses as θ varies from 0 to 360°. The table below does that using the state of stress from the example above and angular increments of 2°.

```
In[97]:= pts =
           Table[{normstress, shearstress} /.
             {σxx → 250., σyy → 100., σxy → 75.},
             {θ, 0, 360.°, 2.°}];
```

The pairs of normal and shear stress values can now be plotted to graphically illustrate their relationship, producing a Mohr diagram that will be familiar to students of structural geology, soil and rock mechanics, and geophysics.

```
In[98]:= ListPlot[pts, AspectRatio → 240/300.,
           PlotRange → {{0, 300}, {-120, 120}},
           PlotJoined → True, AxesLabel → {"Normal", "Shear"}]
```

From In[98]:=

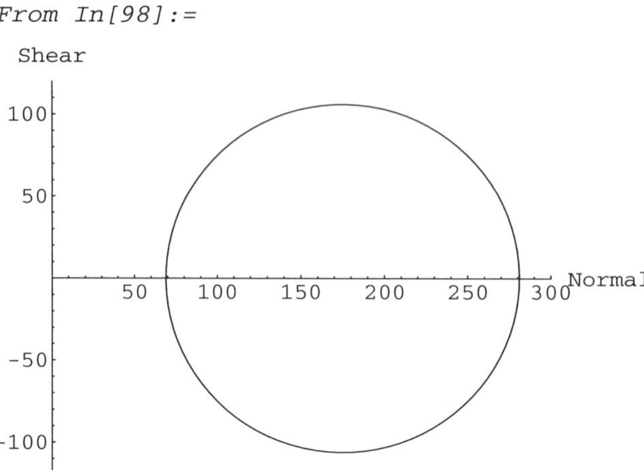

Out[98]= -Graphics-

The two intersections of the Mohr circle with the normal stress axis, for which $\sigma_{xy} = 0$, are the two principal stresses (approximately 69 and 281 MPa). The example state of stress of σ_{xx}= 250 MPa, σ_{yy}= 100 MPa, and σ_{xy}= 75 MPa can be illustrated by plotting the σ_{xx} and σ_{xy} or σ_{yy} and $\sigma_{yx} = -\sigma_{xy}$ values as a point on the Mohr diagram.

```
In[99]:= Show[%,
           Graphics[{PointSize[0.02],Point[{250.,75.}],
             Point[{100.,-75.}],
             Line[{{100.,-75.},{250.,75.}}]}]
         ]
```

From In[99]:=

Out[99]= -Graphics-

Although both the sine and cosine functions in the expressions we have derived can vary over a range of 360°, in physical space the inclination of a plane can vary over only 180°. A plane said to be inclined 200° relative to a horizontal datum, for example, is the same as a plane inclined 200° – 180° = 20°. Therefore, angles on a Mohr diagram must be twice as large as the angles in physical space that the represent (which is why Mohr diagrams are traditionally constructed using the angle 2θ). The angle from the $+x$-axis (the direction in which σ_{xx} acts) to the maximum principal stress, taking into account the double angle relationship between the Mohr diagram and physical space, can be found using some simple trigonometry:

```
In[100]:= 1/2 ArcTan[σxy/(σxx - (σxx+σyy)/2.)]° /.
           {σxx → 250., σyy → 100., σxy → 75.}

Out[100]= 22.5
```

The angle from the y-axis (**not** the x-axis!) to the minimum principal stress direction is, similarly,

```
In[101]:= 1/2 ArcTan[-σxy/(σyy - (σxx+σyy)/2.)]° /.
           {σxx → 250., σyy → 100., σxy → 75.}

Out[101]= 22.5
```

Therefore, the direction from the x-axis to the minimum principal stress is 90° + 22.5° = 112.5°. Both of these results can be confirmed by using a protractor to measure the angles on the Mohr diagram and dividing the results by 2.

The magnitudes and orientations of the two principal stresses can also be found algebraically. One way to accomplish this is to differentiate the expression for the normal stress acting on a plane of arbitrary orientation with respect to θ and set the result equal to zero, in order to find the orientation at which the normal stress is maximized or minimized.

```
In[102]:= ∂θ normstress

Out[102]= 2 σxy Cos[θ]^2 - 2 σxx Cos[θ] Sin[θ]
            +2 σyy Cos[θ] Sin[θ] - 2 σxy Sin[θ]^2
```

It is almost always a good idea to use **Simplify** to see if the result can be put into a simpler form. In this case, **Simplify** converts the angles to double angles because one of its default options is **Trig → True**, which allows the simplification algorithm to perform trigonometric as well as algebraic manipulations.

```
In[103]:= Simplify[%]

Out[103]= 2 σxy Cos[2 θ] + (-σxx + σyy) Sin[2 θ]
```

At this point, it is easy to look at the last equation and see that dividing it through by cos 2θ and rearranging would put the equation into the particularly simple textbook form tan $2\theta = -\sigma_{xy}/(\sigma_{yy} - \sigma_{xx})$. Unfortunately, *Mathematica*'s simplification algorithm does not see any advantage to that form even though it is probably what most humans would prefer. **FullSimplify** returns the same result. Manually dividing through by cos 2θ, however, will accomplish the job.

```
In[104]:= Simplify[%/Cos[2 θ]]

Out[104]= 2 σxy + (-σxx + σyy) Tan[2 θ]
```

> **Computer Note:** The previous step would not have worked if the derivative had already been set to zero, because *Mathematica* would treat the entire equation as a single object divided by cos 2θ. The result would have been
>
> **(2σxyCos[2θ] + (-σxx + σyy) Sin[2θ] == 0) Sec[2θ]**
>
> rather than the desired **2σxy + (-σxx + σyy) Tan[2θ]**. Try each of the two possibilities yourself.

Finally, the result can be set equal to zero and solved for tan 2θ

```
In[105]:= Solve[% == 0, Tan[2 θ]]
```

$$\text{Out[105]= } \left\{\left\{\text{Tan}[2\,\theta] \to \frac{2\,\sigma xy}{\sigma xx - \sigma yy}\right\}\right\}$$

or θ

```
In[106]:= Solve[%% == 0, θ]
```

$$\text{Out[106]= } \left\{\left\{\theta \to \frac{1}{2}\,\text{ArcTan}\left[\frac{2\,\sigma xy}{\sigma xx - \sigma yy}\right]\right\}\right\}$$

which is the same result as we obtained from the Mohr diagram. We will use this result later, so it will be convenient to assign it a name.

```
In[107]:= maxangle = θ /. %[[1]]
```

$$\text{Out[107]= } \frac{1}{2}\,\text{ArcTan}\left[\frac{2\,\sigma xy}{\sigma xx - \sigma yy}\right]$$

The tangent function is periodic and repeats itself every 180°, which can be shown by equating the tangents of angles separated by 180° and applying **Simplify**.

```
In[108]:= Simplify[Tan[α] == Tan[(α + 180 °)]]

Out[108]= True
```

Therefore, minimum and maximum stresses will be oriented at angles of $2\alpha = \pm 180°$, or $\alpha = \pm 90°$, from each other. The directions in which the normal stress is maximized and minimized are also the directions in which shear stress vanishes, which can be illustrated by substituting **maxangle** into the shear stress expression.

```
In[109]:= Simplify[shearstress /. θ → maxangle]

Out[109]= 0
```

Using the same state of stress as in the previous example, the orientation of the maximum normal stress is thus (in degrees)

```
In[110]:= maxangle/° /. {σxx → 250., σyy → 100., σxy → 75.}

Out[110]= 22.5
```

The magnitude of the normal stress acting in this direction is (in MPa)

```
In[111]:= normstress /. {σxx → 250., σyy → 100., σxy → 75.,
            θ → 22.5 °}

Out[111]= 281.066
```

which agrees with the maximum principal stress estimated from the Mohr circle. To check the calculations, we can also calculate the shear stress in this direction. According to the definition of the principal stresses, it should be zero.

```
In[112]:= shearstress /. {σxx → 250., σyy → 100., σxy → 75.,
            θ → 22.5 °}
```

Out[112]= -1.77636×10^{-15}

Small numbers (default < 10^{-10}) can be eliminated using **Chop**.

```
In[113]:= Chop[%]

Out[113]= 0
```

The minimum principal stress is oriented ± 90° from the maximum principal stress, and its value is (in MPa)

```
In[114]:= normstress /. {σxx → 250., σyy → 100., σxy → 75.,
            θ → (22.5 + 90)°}

Out[114]= 68.934
```

which also agrees very well with the estimated Mohr circle value. Its orientation is

```
In[115]:= (maxangle/° + 90) /. {σxx → 250., σyy → 100., σxy → 75.}

Out[115]= 112.5
```

The magnitudes of the two principal stresses can also be calculated as the **eigenvalues** of the stress tensor. Middleton and Wilcock (1994, p. 148-151) provide a very clear explanation of the relationship between the eigenvalues of the stress tensor and principal stresses. We will simply calculate the values to illustrate that they are identical to those calculated using a Mohr diagram or the algebraic method or by algebraically finding the directions of maximum and minimum normal stress. The general forms of the two eigenvalues of the two dimensional stress tensor are

```
In[116]:= Eigenvalues[σ] /. σyx - > σxy // MatrixForm
```

Out[116]=
$$\begin{pmatrix} \frac{1}{2}\left(\sigma xx+\sigma yy-\sqrt{\sigma xx^2+4\,\sigma xy^2-2\,\sigma xx\,\sigma yy+\sigma yy^2}\right) \\ \frac{1}{2}\left(\sigma xx+\sigma yy+\sqrt{\sigma xx^2+4\,\sigma xy^2-2\,\sigma xx\,\sigma yy+\sigma yy^2}\right) \end{pmatrix}$$

Substituting values for the state of stress used in the previous examples, the two principal stresses are calcuated to be (in MPa)

```
In[117]:= %/. {σxx → 250., σyy → 100., σxy → 75.} // MatrixForm
```

Out[117]= $\begin{pmatrix} 68.934 \\ 281.066 \end{pmatrix}$

The orientations of the principal stresses can be found from the **eigenvectors** of the stress tensor. Their symbolic form is

```
In[118]:= Eigenvectors[σ]/. σyx - > σxy // MatrixForm
```

Out[118]= $\begin{pmatrix} -\frac{-\sigma xx + \sigma yy + \sqrt{\sigma xx^2 + 4\,\sigma xy^2 - 2\,\sigma xx\,\sigma yy + \sigma yy^2}}{2\,\sigma xy} & 1 \\ -\frac{-\sigma xx + \sigma yy - \sqrt{\sigma xx^2 + 4\,\sigma xy^2 - 2\,\sigma xx\,\sigma yy + \sigma yy^2}}{2\,\sigma xy} & 1 \end{pmatrix}$

The eigenvectors are the the axes of the stress ellipsoid; therefore, the angles of the two axes can be found by taking the four-quadrants arctangents of the x and y components of each axis. Refer to the *Mathematica* documentation for a discussion of the differences between two- and four-quadrant arctangent functions. Substituting the state of stress from the previous examples, the orientations of the two principal stresses are thus (in degrees)

```
In[119]:= ArcTan[%[[1, 1]], %[[1, 2]]]/° /. {σxx → 250., σyy →
100., σxy → 75.}
```

Out[119]= 112.5

and

```
In[120]:= ArcTan[%%[[2, 1]], %%[[2, 2]]]/° /.
            {σxx → 250., σyy → 100., σxy → 75.}
```

Out[120]= 22.5

Finally, we will use the variable R again in another problem and need to clear its value to avoid problems.

```
In[121]:= Clear[R]
```

3.5 Ordinary Differential Equations

3.5.1 Manual Manipulation and Integration

In some cases, differential equations may be simple enough that their solution is very straightforward and *Mathematica* can be used to implement techniques that are described in calculus and differential equations textbooks. For example, consider the linear first order equation 2 dy/dx + 6 y = 3. According to standard mathematical tables, the solution to an equation of the general form $dy/dx + P(x)\,y - Q(x) = 0$ is

$$y\, e^{\int 6dx} = \int 3e^{\int 6dx} dx + c$$

where c is a constant of integration that is determined from the initial or boundary conditions for a particular problem. In this example, P and Q are constants rather than functions of x. Knowing the form of the general solution, it is possible to type it in as a *Mathematica* statement and use **Solve** to obtain the following solution for y. Notice that the equation must be divided through by 2 in order to put it into the general form of the equation for which the solution is given.

In[122]:= **Solve**$\left[y\, e^{\int 6/2 dx} == \int \frac{3}{2} e^{\int 6/2 dx} dx + c,\ y\right]$

Out[122]= $\left\{\left\{y \to \frac{1}{2} e^{-3x} \left(2c + e^{3x}\right)\right\}\right\}$

Mathematica returns solutions as a list of replacement rules, which is convenient for equations that have more than one root. In this case, though, the list consists of only one element that can be further simplified to arrive at the general solution:

```
In[123]:= GS = Simplify[y /. %[[1]]]
```

Out[123]= $\frac{1}{2} + c\, e^{-3x}$

Next, an initial or boundary condition must be used to obtain a particular solution. In this case, assume that the boundary condition is given by $y = y_0$ at $x = 0$. Evaluate the general solution at $x = 0$ and set the result equal to y_0. The resulting expression is named **BC**, short for boundary condition.

```
In[124]:= BC = y0 == GS /. x → 0
```

Out[124]= $y0 == \frac{1}{2} + c$

Now, solve the boundary condition equation for c.

```
In[125]:= Solve[BC, c]
```

Out[125]= $\left\{\left\{c \to \frac{1}{2} (-1 + 2\, y0)\right\}\right\}$

Finally, substitute the new-found value for the constant into the general solution.

```
In[126]:= PS = GS /. %[[1]]
```

Out[126]= $\frac{1}{2} + \frac{1}{2} e^{-3x} (-1 + 2\, y0)$

Is the solution correct? To find out, substitute the particular solution back into the differential equation and simplify the expression.

In[127]:= **Simplify**$[2\, \partial_x \mathbf{PS} + 6\, \mathbf{PS}] == 3$

```
Out[127]= True
```

As required, the result is 3 and the solution is valid.

This is, in essence, the traditional approach to solving differential equations. The equation is compared to a table of general cases to find an approach that is known to work, the general solution is recast in terms of the specific values of the problem at hand, some algebra and calculus are done, and, assuming no mistakes have been made, a solution is eventually obtained.

3.5.2 Solutions Using DSolve and NDSolve

The traditional approach to solving differential equations of interest to geoscientists rarely involves the development of fundamentally new solutions. Instead, it is simply the application of well-known general rules to specific cases. *Mathematica* is capable of following the same kinds of rules to obtain solutions using the functions **DSolve** and **NDSolve**. The example equation introduced in the previous section, for example, can be solved without reference to tables or handbooks using **DSolve**.

```
In[128]:= DSolve[2 ∂x y[x] + 6 y[x] == 3, y[x], x]
```

Out[128]= $\{\{y[x] \to \frac{1}{2} + e^{-3x}\, C[1]\}\}$

The constant of integration is **C[1]**. In the event that an initial or boundary condition is known, either symbolically or numerically, *Mathematica* can solve a list of equations that includes both the original differential equation and its initial or boundary conditions. For example, if it is known that $y = y_0$at $x = 0$ the particular solution can be obtained directly from **DSolve.**

```
In[129]:= DSolve[{2 ∂x y[x] + 6 y[x] == 3, y[0] == y0}, y[x], x]
```

Out[129]= $\{\{y[x] \to \frac{1}{2} e^{-3x} (-1 + e^{3x} + 2\, y0)\}\}$

Does this solution agree with the one that we found by manually solving the equation?

```
In[130]:= Expand[y[x] /. %[[1]]] == Expand[PS]

Out[130]= True
```

Therefore, the two solutions are algebraically identical.

There may also be cases in which you don't want, or cannot obtain, an analytical solution. In those situations, **NDSolve** can be used to obtain a numerial solution that is represented by something known as an interpolating function. The statement below solves the same differential equation but with a numerical boundary condition of $y = 2$ at $x = 0$. Because this is a numerical solution, it is also necessary to specify the range of x over which a solution is desired, which is arbitrarily set to $0 \leq x \leq 5$.

```
In[131]:= NDSolve[{2 ∂x y[x] + 6 y[x] == 3, y[0]
            == 2.}, y[x], {x, 0, 5}]

Out[131]= {{y[x] → InterpolatingFunction[{{0., 5.}}, <>][x]}}
```

As described in the *Mathematica* documentation, **NDSolve** accepts options that control the accuracy, precision, step size, and other aspects of the numerical solution algorithm. The next statement plots the numerical results by using the replacement rule containing the solution. Interpolating functions will be discussed in more detail in Chapter 6.

```
In[132]:= Plot[y[x] /. %[[1]], {x, 0, 5}, PlotRange → All,
             AxesLabel → {"x", "y"}]
```

From In[132]:=

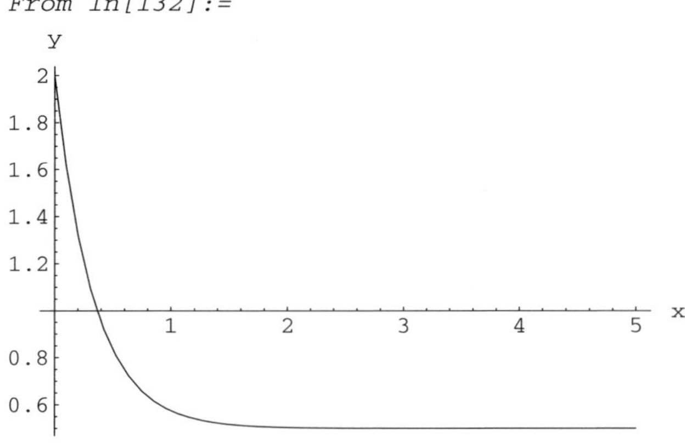

Out[132]= -Graphics-

> **Computer Note:** Set $y = 2$ at $x = 0$ in the analytical solution obtained using **`DSolve`**, then plot the result to verify that it is identical to the numerical solution obtained using **`NDSolve`**.

3.5.3 Geoscience Examples

Velocity and Kinetic Energy of Sliding Boulders

Calculation of the **acceleration**, **velocity**, and **kinetic energy** of a sliding boulder shows how *Mathematica* can be used to integrate simple differential equations. The example below follows the approach of Haneberg and Bauer (1993), who used similar calculations to estimate the velocity and kinematic energy a 2.7×10^5 kg boulder that slid down a steep slope and left a large crater in a road along the Rio Grande gorge of northern New Mexico.

For a rock of mass **m**, the component of weight acting parallel to a slope of angle **β** is

```
In[133]:= m g Sin[β]

Out[133]= g m Sin[β]
```

where **g** is gravitational acceleration (9.81 m/s^2). The resisting force arising from the Mohr-Coulomb shear strength of dry granular soil (which is typical of slopes along the Rio Grande gorge) is

```
In[134]:= m g Cos[β] Tan[φ]

Out[134]= g m Cos[β] Tan[φ]
```

in which the angle of internal friction, ϕ, is a standard soil property reflecting the frictional or non-cohesive component of soil shear strength. Sliding requires a net imbalance of forces, and the net shear force acting parallel to the slope is the difference between the downslope component of weight and the resisting force.

```
In[135]:= %% - %

Out[135]= g m Sin[β] - g m Cos[β] Tan[φ]
```

A force is by definition the product of mass and acceleration, so the net force can also be written as the product of mass and an average slope parallel acceleration yet to be determined.

```
In[136]:= aslope m == %

Out[136]= aslope m == g m Sin[β] - g m Cos[β] Tan[φ]
```

Solving for the slope parallel acceleration,

```
In[137]:= accel = aslope /. Simplify[Solve[%, aslope]][[1]]

Out[137]= g Sec[φ] Sin[β - φ]
```

Computer Note: The previous line combines the solution and simplification of an equation with a replacement rule and variable name assignment. In particular, the expression

```
Simplify[Solve[%, aslope]][[1]]
```

tells *Mathematica* to solve an equation obtained in the previous step, simplify it, and then take the first part of the resulting list of replacement rules. Take a minute or two to carefully read through and understand the combination, then repeat the steps one at a time to reproduce the result on your own computer.

Now that the acceleration has been determined, velocity and distance at time t can be found by integrating the acceleration.

```
In[138]:= velocity = ∫ aslope dt

Out[138]= aslope t

In[139]:= distance = ∫ velocity dt

Out[139]= aslope t^2 / 2
```

Both of the integrals assume that velocity and distance equal zero when $t = 0$. Finally, the kinetic energy of the sliding block is given by

```
In[140]:= energy = 1/2 m velocity^2
```

$$Out[140]= \frac{1}{2}\,aslope^2\,m\,t^2$$

One of the difficult aspects of reconstructing a geologic event such as a rockslide is that not all of the information may be available in the desired form. Because there were no observers (the rockslide occurred while the road was closed because of many landslides and debris flows during a heavy rainstorm), it is impossible to know the duration of sliding. The sliding rock left a visible scar, however, so it was possible to estimate that it moved approximately 500 m down slope. Using representative values of $\phi = 35°$ and $\beta = 37°$, the duration of sliding can therefore be estimated by setting **distance** equal to 500 m and solving for time.

```
In[141]:= Solve[distance == 500, t]
```

$$Out[141]= \left\{\left\{t \to -\frac{10\sqrt{10}}{\sqrt{aslope}}\right\}, \left\{t \to \frac{10\sqrt{10}}{\sqrt{aslope}}\right\}\right\}$$

Both positive and negative roots are returned because the result is found by taking a square root. We will keep only the positive result and assign it to the variable name **totaltime**

```
In[142]:= totaltime = t /. %[[2]]
```

$$Out[142]= \frac{10\sqrt{10}}{\sqrt{aslope}}$$

Using values known or inferred for this particular rockslide, the total time of sliding appears to have been (in seconds)

```
In[143]:= totaltime /. {ϕ → 35. °, β → 37. °, g → 9.81}
```

$$Out[143]= \frac{10\sqrt{10}}{\sqrt{aslope}}$$

The velocity and kinetic energy of the boulder when it struck the road can now be calculated from the total time to be (in m/s and N-m, respectively)

```
In[144]:= velocity/. {ϕ → 35. °, β → 37. °, g → 9.81, t → 49.}
```

$$Out[144]= 49.\,aslope$$

```
In[145]:= energy /. {ϕ → 35. °, β → 37. °, g → 9.81,
            t → 49., m → 2.7 × 10^5}
```

$$Out[145]= 3.24135\times 10^8\,aslope^2$$

Nearby portions of the same highway are protected by energy absorbing rockfall nets. Could similar nets be used to protect the highway from falling or sliding blocks of this size? The capacity of this type of net is on the order of 5×10^5N-m, so the answer is that the this kind of net would not have stopped the boulder if it had been placed just up-slope from the road Haneberg and Bauer (1993) also calculated results for frictionless sliding boulders and rolling boulders, and Wieczorek *et al.* (2000) analyzed a rockfall in which a large slab of rock became airborne and followed a ballistic trajectory. All of these situations are variations on the same simple mechanical problem.

> **Computer Note:** Although a single net placed just above the highway may not be capable of stopping a boulder of the size and velocity described above, it might be possible to use a network of nets placed at intervals upslope from the highway. Using the same values as the example above, calculate 1) the maximum distance that a 2.7×10^5kg boulder can travel before its kinetic energy exceeds the capacity of a rockfall net and 2) the largest boulders (in terms of mass) that could be stopped by nets placed at 100 m intervals up the slope.

Mineral Thermodynamics: Temperature Dependent Enthalpy

The **heat capacity** of a substance is defined as the change in **enthalpy** or **heat content** (H) of the substance relative to a change in its temperature (T), or $dH/dT = C_p$ (Wood and Fraser, 1977). Enthalpy values for minerals are typically tabulated only for the reference temperature of 298°K. Therefore, the differential equation relating enthalpy to heat content must be solved in order to calculate enthalpies at the high temperatures of interest in many geochemical reactions. One way to do this is to isolate H and T on opposite sides of the equation, integrate the results, and solve for H_T(the enthalpy at temperature T). This is particularly simple if the heat capacity is independent of temperature. First, isolate the variables and integrate both sides of the equation

In[146]:= $\int_{H298}^{HT} dH == \int_{298}^{T1} Cp dT$

Out[146]= `-H298 + HT == Cp (-298 + T1)`

then solve for H_T

```
In[147]:= Simplify[Solve[%, HT]]

Out[147]= {{HT → H298 + Cp (-298 + T1)}}
```

Using **`DSolve`** with a specified boundary condition yields the same results in just one step.

```
In[148]:= DSolve[{∂T H[T] == Cp, H[298] == H298}, H[T], T]

Out[148]= {{H[T] → -298 Cp + H298 + Cp T}}
```

Heat capacity is generally not independent of temperature, however, which makes the problem slightly more complicated. Under constant pressure, the relationship between heat content and temperature for values of $T > 298$°K is often written as

```
In[149]:= Clear[a, b, c];
          CpT = a + b T + c/T^2
```

$$Out[149]= a + \frac{c}{T^2} + b\,T$$

The values of a, b, and c were cleared to ensure that they are not set to the values used in the three point problem discussed previously in this chapter. As above, **DSolve** is used to find a solution for H_T

```
In[150]:= Simplify[
            DSolve[{∂_T H[T] == CpT, H[298] == H298}, H[T], T]
          ]
```

$$Out[150]= \left\{\left\{H[T] \to \frac{c}{298} + H298 + a\,(-298 + T) - \frac{c}{T} + \frac{1}{2}\,b\,(-88804 + T^2)\right\}\right\}$$

The value of 88804 is 298^2. It will be convenient to have this result available for future calculations, so we will use the replacement rule to assign the result to the variable **HT**.

```
In[151]:= HT = H[T] /. %[[1]]
```

$$Out[151]= \frac{c}{298} + H298 + a\,(-298 + T) - \frac{c}{T} + \frac{1}{2}\,b\,(-88804 + T^2)$$

For purely cosmetic purposes, this result can be rearranged by collecting terms with the same coefficients.

```
In[152]:= Collect[HT, {a, b, c}]
```

$$Out[152]= H298 + c\left(\frac{1}{298} - \frac{1}{T}\right) + a\,(-298 + T) + \frac{1}{2}\,b\,(-88804 + T^2)$$

You may be wondering why the terms in the rearranged equation are not listed alphabetically. The reason is that the equation is a polynomial in the variable T, and *Mathematica* lists the terms by starting with the constant and then following in ascending powers of T.

As an example of a thermodynamic calculation using the enthalpy equation, consider the formation of jadeite ($NaAlSi_2O_6$) and quartz (SiO_2) from albite ($NaAlSi_3O_8$) at 900°K and 1 bar (100 kPa) as described in Wood and Fraser (1977). The relevant thermodynamic data can be entered in a table containing both text and numerical values. The first element of each row is the mineral name, the second is the enthalpy at 298°K (in cal/mol), and the third through fifth are the coefficients a, b, and c for that mineral. We will perform the calculations using the original units (calories and bars) in order to avoid converting all of the coefficients, and then convert the final result into metric units of kJ/mol.

```
In[153]:= ThermoData =
           {{"albite", -937146., 61.7, 13.9 * 10^-3, -15.01 * 10^5},
            {"jadeite", -719871., 48.16, 11.42 * 10^-3, -11.87 * 10^5},
            {"quartz", -217650., 11.22, 8.2 * 10^-3, -2.7 * 10^5}};
```

These data can be displayed in an orderly fashion using the **TableForm** function, which allows row and column headings to be added. **TableForm** can, like **N** and **MatrixForm**, be appended to an expression using // (two slashes) if options such as **TableHeadings** are not used. The table heading **None** specifies that no row headings are to be shown, because the mineral names are in this case included as the first element of each row. **TableForm** can be wrapped with the related function **PaddedForm** to more precisely control the appearance of the table. See the *Mathematica* documentation for more details. Although it is not necessary to do so, inclusion of the mineral names as an element of each row ensures that each set of thermodynamic data is associated with the name of the corresponding mineral rather than a superficial table heading.

```
In[154]:= TableForm[
            ThermoData,
            TableHeadings→{None,{"Mineral",H298,"a","b","c"}},
            TableSpacing → {1, 1}
          ]
```

Out[154]=

Mineral	H_{298}	a	b	c
albite	-937146.	61.7	0.0139	-1.501×10^{6}
jadeite	-719871.	48.16	0.01142	-1.187×10^{6}
quartz	-217650.	11.22	0.0082	-270000.

Each piece of thermodynamic data can be extracted by referencing the appropriate row and column. The H_{298} value for jadeite, for example, is

```
In[155]:= ThermoData[[2, 2]]

Out[155]= -719871.
```

whereas the c value for quartz is

```
In[156]:= ThermoData[[3, 5]]

Out[156]= -270000.
```

The change in enthalpy in any reaction is the sum of the product enthalpies minus the sum of the reactant enthalpies, or in this case $\Delta H = H_{\text{jadeite}} + H_{\text{quartz}} - H_{\text{albite}}$. The first step in calculating the enthalpy change for the reaction is to calculate the individual enthalpies at 800°K, which is done by substituting values from **ThermoData** into **HT**.

```
In[157]:= Halb800 =
            HT /. {T → 800., H298 → ThermoData[[1, 2]],
              a → ThermoData[[1, 3]], b → ThermoData[[1, 4]],
              c → ThermoData[[1, 5]]}

Out[157]= -905502.
```

```
In[158]:= Hjad800 =
            HT /. {T → 800., H298 → ThermoData[[2, 2]],
              a → ThermoData[[2, 3]], b → ThermoData[[2, 4]],
              c → ThermoData[[2, 5]]}

Out[158]= -695047.

In[159]:= Hqtz800 =
            HT /. {T → 800., H298 → ThermoData[[3, 2]],
              a → ThermoData[[3, 3]], b → ThermoData[[3, 4]],
              c → ThermoData[[3, 5]]}

Out[159]= -210326.
```

The change in enthalpy associated with the decomposition of albite into jadeite and quartz at 800°K and 100 kPa is then found by adding all of the product enthalpies and subtracting all of the reactant enthalpies. The result is (in units of cal/mol)

```
In[160]:= Hjad800 + Hqtz800 - Halb800

Out[160]= 129.432
```

or, in kJ/mol,

```
In[161]:= 4.18 %/1000.

Out[161]= 0.541024
```

Population Growth

The growth of populations, for example those studied in the fossil record and modern ecosystems of interest to many geoscientists, is often described in terms of two end members: **exponential population growth** and **logistic population growth** (Haberman, 1998). Populations that experience exponential growth tend to consist of opportunistic generalists that are able to adapt to unstable environments, for example pioneer species colonizing an intially empty habitat. Those that experience logistic population growth tend to consist of specialists that thrive in stable environments. Exponentially growing populations are further characterized by a constant reproductive rate that is not controlled by population density. Logistically growing populations, in contrast, are characterized by growth rates that decrease as the population density increases, so that the population levels off at a size known as the carrying capacity of the environment.

Exponential population growth is described by the ordinary differential equation $dP/dt = rP$, where P is the population size, t is time, and r is the population growth rate with units of reciprocal time (for example, years^{-1}). It is common to use the variable N to represent population size in population growth models; however, this would conflict with the *Mathematica* **N** function and we will use P instead. The exponential population growth model can easily solved using **DSolve** with the initial condition specified at $P = P_0$ at $t = 0$.

```
In[162]:= DSolve[{∂t P[t] == r P[t], P[0] == P0}, P[t], t]

Out[162]= {{P[t] → ⅇ^(r t) P0}}
```

Assign this solution to the variable **EP**

```
In[163]:= EP = P[t] /. %[[1]]

Out[163]= ⅇ^(r t) P0
```

The results can be visualized by plotting **EP** with several different values of *r* and an initial population of 2. Because exponential population growth occurs so rapidly, we will restrict the plot to 20 time units.

```
In[164]:= Plot[{EP /. {P0 → 2., r → 0.1}, EP /. {P0 → 2., r → 0.15},
             EP /. {P0 → 2., r → 0.2}}, {t, 0, 20},
           PlotStyle → {Dashing[{0.}], Dashing[{0.05}],
             Dashing[{0.01}]},
           AxesLabel → {"t", "P"},
           PlotLegend → {"r = 0.10", "r = 0.15", "r = 0.20"},
           LegendPosition→{-0.75, 0.}, LegendSize→{0.8, 0.5}]
```

From In[164]:=

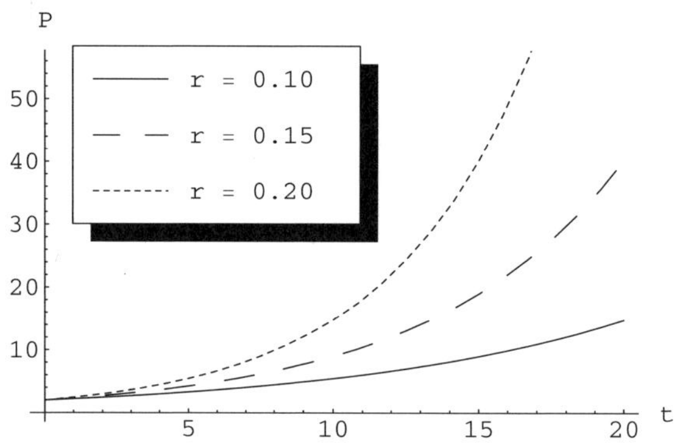


```
Out[164]= -Graphics-
```

Logistic population growth is described by the slightly more complicated differential equation $dP/dt = rP\ (1 - P/K)$, where K is the carrying capacity of the ecosystem. Although the growth rate, r, is shown as a constant the effective growth rate is given by the term $r\ (1 - P/K)$, meaning that population growth will cease when $P = K$. This can be demonstrated by using a substitution rule to evaluate the effective growth rate term.

```
In[165]:= r (1 - P/K) /. P → K

Out[165]= 0
```

As above, the quickest way to solve the equation is to use **DSolve** and specify an initial condition.

```
In[166]:= DSolve[{∂tP[t] == rP[t] (1-P[t]/K), P[0] == P0}, P[t], t]
```

Out[166]= $\left\{\left\{P[t] \to \frac{e^{rt}\,K\,P0}{K - P0 + e^{rt}\,P0}\right\}\right\}$

Now, assign the solution to the variable **LP** (for logistic population)

```
In[167]:= LP = P[t] /. %[[1]]
```

Out[167]= $\frac{e^{rt}\,K\,P0}{K - P0 + e^{rt}\,P0}$

and plot results for the same set of r values as were used in the exponential population growth example and a carrying capacity of 100 organisms. Because logistic population growth is self-regulating, however, we will plot the results over a larger range of 0 to 100 time units to examine what happens as the population approaches the carrying capacity.

```
In[168]:= Plot[{LP /. {P0 → 2., r → 0.1, K → 100.},
            LP /. {P0 → 2., r → 0.15, K → 100.},
            LP /. {P0 → 2., r → 0.2, K → 100.}}, {t, 0, 100},
          PlotStyle → {Dashing[{0.}], Dashing[{0.05}],
            Dashing[{0.01}]},
          AxesLabel → {"t", "P"},
          PlotLegend → {"r = 0.10", "r = 0.15", "r = 0.20"},
          LegendPosition → {0., -0.4}, LegendSize → {0.8, 0.5}]
```

From In[168]:=

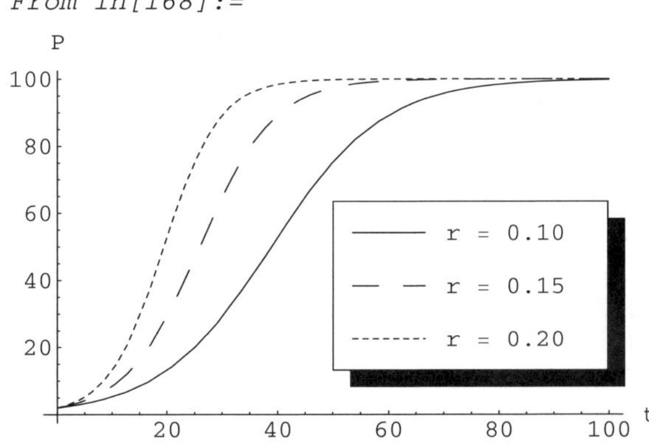

Out[168]= -Graphics-

The logistic population growth curves are similar to the exponential growth curves for the first 20 to 40 time units, depending on the value of r. Beyond that, the growth curves flatten and converge on the specified carrying capacity of 100 organisms.

The equilibrium population is that for which $dP/dt = 0$. Referring to the logistic growth equation, $dP/dt = rP\,(1-P/K)$, the equilibrium population(s) must therefore be given by the roots of $rP\,(1-P/K) = 0$. This can be done using either **Solve** or the related function **Roots** (see the *Mathematica* documentation for a discussion of similarities and differences between the two).

```
In[169]:= Solve[r P (1 - P/K) == 0, P]

Out[169]= {{P → 0}, {P → K}}
```

Thus, a logistic population will either approach the carrying capacity of the ecosystem ($P = K$) or become extinct ($P = 0$). This can be illustrated graphically by plotting the right-hand side of the logistic growth equation (dP/dt) as a function of P, producing a **phase plot**. As above, let $K = 100$ and $r = 0.15$.

```
In[170]:= Plot[r P (1 - P/K) /.{K → 100., r → 0.15}, {P, 0, 110},
            AxesLabel → {"P", "dP/dt"}]

From In[170]:=
dP/dt
```

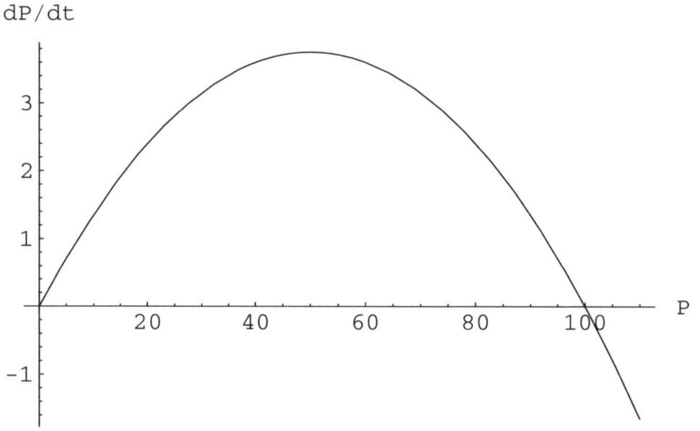

```
Out[170]= -Graphics-
```

The two equilibrium populations are the points at which the dP/dt curve intersects the P axis ($P = 0$ and $P = K$). $P = 0$ represents an unstable state of equilibrium because dP/dt is positive and populations with $P > 0$ can move only away from that state of equilibrium. A population of $P= K$, however, represents a stable state of equilibrium. For values of $P < K$, dP/dt is positive and the population will grow until it achieves the equilibrium state of $P = K$. For values of $P >$ K, though, dP/dt is negative and the population will shrink until it reaches $P = K$.

The nature of logistic growth can also be visualized by superimposing plots showing growth curves for different values of $P_0$for fixed values of r and K. The statement below generates a table filled with plots (with **DisplayFunction → Identity** so the plots are not shown), then shows all of the plots together on the same set of axes (using **DisplayFunction → $DisplayFunction**).

```
In[171]:= Show[
            Table[
              Plot[LP /. {r → 0.15, K → 100.}, {t, 0, 100},
              DisplayFunction → Identity], {P0, 2, 202, 10}
            ], DisplayFunction → $DisplayFunction,
            PlotRange → {0, 200}, AxesLabel → {"t", "P"}
          ]
```

From In[171]:=

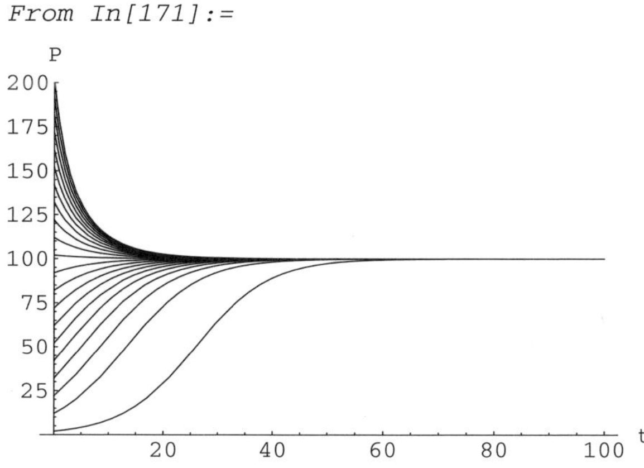

Out[171]= -Graphics-

By plotting the population for different values of P_0, you can show that values of P_0 > K always lead to a population decrease and values of $P_0 < K$ alway lead to a population decrease. Likewise, the effect of changing r values while holding P_0constant can be visualized by copying the previous statement and switching variables.

```
In[172]:= Show[
            Table[
              Plot[LP /. {P0 → 20., K → 100.}, {t, 0, 100},
              DisplayFunction → Identity], {r, -0.2, 0.2, 0.01}
            ], DisplayFunction → $DisplayFunction,
            PlotRange → {0, 100}, AxesLabel → {"t", "P"}
          ]
```

From In[172]:=

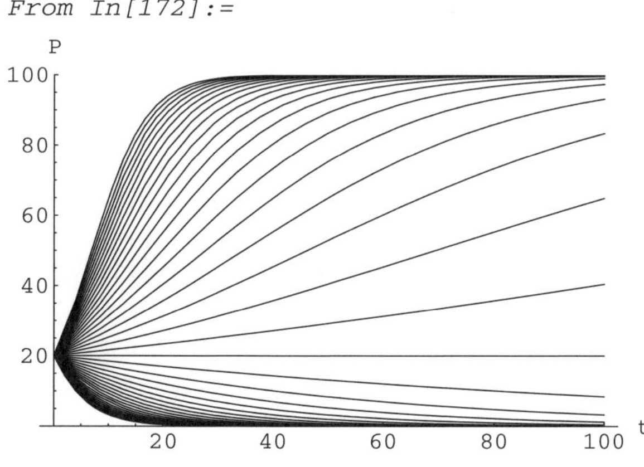

Out[172]= -Graphics-

Computer Note: A variation on the logistic growth model is logistic growth with harvesting, in which harvesting refers to any loss of population due to outside influences (*e.g.*, hunting by humans or other predators). The equation describing logistic growth with a constant rate of harvesting is $dP/dt = rP(1 - P/K) - h$, where h is the number of organisms removed from the population per unit of time. Use **`DSolve`** to obtain a solution and plot results to explore the impact of different h values on the size of populations with different initial populations, growth rates, and carrying capacities. What are the possible equilibrium values of P in a logisitic population with harvesting? (The equilibrium values will depend on h.)

Flexure of Strata Above Laccoliths

Mathematical models of thin elastic plates have been used to simulate the flexure of lithospheric plates as a result of loading by volcanoes or tectonic forces, the folding of layered rocks due to lateral compression or the intrusion of magma bodies, and surface deformation associated with heavy groundwater pumping (Haneberg and Friesen, 1995; Johnson, 1970; Middleton and Wilcock, 1994; Turcotte and Schubert, 2002). This example illustrates how *Mathematica* can be used to solve the differential equation describing the flexure of an elastic plate in order to simulate the deformation of strata above laccoliths, which are igneous intrusions that dome overlying strata to form a distinctive mushroom like shape. Both Johnson (1970) and Turcotte and Schubert (2002) describe the details and limitations of the underlying theory in general and its application to the analysis of laccoliths in particular.

The flexure of a thin elastic plate with no lateral loading is described by the equation $d^4w/dx^4 = p/R$, where w is the vertical deflection of the plate, p is a uniform load applied along the entire length of the plate, and R is the flexural rigidity of the plate. The flexural rigidity can be written in terms of Young's modulus (E), Poisson's ratio (ν), and plate thickness (T) as $R = \frac{e\,T^3}{12(1-\nu^2)}$. Plates are considered to be thin if their thickness is less than about one-tenth of their length. The theory used here is limited to small strains, but this is a reasonable assumption in many geologic problems because the vertical displacement is typically a fraction of the plate thickness and length. In the case of a laccolith, the uniform pressure is applied in an upward direction and is $p = p_m - \rho\, g\, T$, where p_m is the magma pressure, ρ is the density of the overburden, g is gravitational acceleration, and T is the thickness of the overburden. In this example, the overburden is assumed to behave as a single layer. Koch *et al.* (1981) discuss how a composite flexural rigidity can be used to simulate the behavior of overburden consisting of strata with different mechanical properties.

The general solution of the differential equation describing the flexure of strata above laccoliths is

```
In[173]:= DSolve[∂{x,4} w[x] == p/R, w[x], x]
```

$$Out[173]= \left\{\left\{w[x] \to \frac{p\,x^4}{24\,R} + C[1] + x\,C[2] + x^2\,C[3] + x^3\,C[4]\right\}\right\}$$

There is some flexibility in the way that derivatives can be expressed in *Mathematica*. One way, shown above, is to use the traditional-looking partial derivative symbol with subscripts. The forms $\partial_{x,x,x,x}\, w[x]$, **D[w[x], {x, 4}]**, and **D[w[x], {x, x, x, x}]** are all equivalent. The latter two date from early versions of *Mathematica* that did not have advanced typesetting capabilities and required all input and output to be in standard text format.

Four boundary conditions must be specified in order to obtain a particular solution. The first two boundary conditions will specify that there is no deflection at either end of a beam of length L, or $w = 0$ at $x = \pm L/2$.

```
In[174]:= bc1 = w[-L/2] == 0
          bc2 = w[L/2] == 0
```

$$Out[174]= w\left[-\frac{L}{2}\right] == 0$$

$$Out[174]= w\left[\frac{L}{2}\right] == 0$$

The second pair of boundary conditions specify that the plate is horizontal at each end, representing undeformed horizontal strata. This is accomplished by setting the slope of the plate to zero.

```
In[175]:= bc3 = ∂x w[x] == 0 /. x → -L/2
          bc4 = ∂x w[x] == 0 /. x → +L/2
```

$$Out[175]= w'\left[-\frac{L}{2}\right] == 0$$

$$Out[175]= w'\left[\frac{L}{2}\right] == 0$$

Notice that the values of x in the second set of boundary conditions was specified differently than those in the first set. This has to do with the way that *Mathematica* treats derivatives. When the derivatives are specified using the format above or using the notation **D[w[x], x]**, the function must be supplied as the generic $w(x)$ **before** the specific value of x is inserted. Otherwise, *Mathematica* will assume that w is a function of $\pm L/2$ rather than x and, when the derivative with respect to x is evaluated, the result will be zero. If this result forms part of an equation, as above, the result will be

```
In[176]:= ∂x w[L/2] == 0

Out[176]= True
```

because the derivative of $w(L/2)$ with respect to x is indeed equal to zero. If, however, the derivative is written using the notation **w′ [x]**, then the specific value of x can be inserted immediately.

```
In[177]:= w′[-L/2] == 0
          w′[L/2] == 0
```

Out[177]= $w'\left[-\frac{L}{2}\right] == 0$

Out[177]= $w'\left[\frac{L}{2}\right] == 0$

Now that the four boundary conditions have been specified, **DSolve** can be used to obtain the particular solution.

```
In[178]:= Simplify[
            DSolve[{∂{x,4} w[x] == p/R, bc1, bc2, bc3, bc4}, w[x], x]
          ]
```

Out[178]= $\left\{\left\{w[x] \rightarrow \frac{p\,(L^2 - 4\,x^2)^2}{384\,R}\right\}\right\}$

It is often convenient to express solutions in dimensionless forms, which makes them more generally applicable. For example, the vertical deflection w can be normalized relative to the length of the plate L. First, apply a replacement rule to extract the solution from its curly brackets.

```
In[179]:= sltn = w[x] /. %[[1]]
```

Out[179]= $\frac{p\,(L^2 - 4\,x^2)^2}{384\,R}$

Computer Note: Turcotte and Schubert (2002) give the solution to this problem as (taking into account a difference in the sign of w and using the same variables as this example)

$$w = \frac{p}{24R}\left(x^4 - \frac{x^2L^2}{2} + \frac{L^4}{16}\right)$$

Use *Mathematica* to determine whether the two solutions are equal. One way to do this symbolically is to **Expand** both solutions and then equate them using the == operator. It is necessary to expand each solution because *Mathematica* does not recognize a statement of the form $a\,(b + c) == a\,b + b\,c$ as being true because the forms are different. Another way to determine whether the two solutions are identical is to divide one by the other to see if the quotient is 1 or subtract one from the other to see if the result is 0 (in each case using **Simplify** if necessary.

The result can be put into a particularly simple form that is a function of only x/L if it is multiplied by R and divided by pL^4. This means, of course, that this is no longer an expression for w. Instead, it is an expression for the **dimensionless** or **normalized** deflection $wR/(pL^4)$

```
In[180]:= Expand[sltn R/(p L^4)]
```

$$Out[180]= \frac{1}{384} - \frac{x^2}{48\,L^2} + \frac{x^4}{24\,L^4}$$

```
In[181]:= deflect = %/. {(x/L)^2 → X^2, (x/L)^4 → X^4}
```

$$Out[181]= \frac{1}{384} - \frac{X^2}{48} + \frac{X^4}{24}$$

The implication of the dimensionless result is that, although the **magnitude** of the deflection will depend on *p, R*, and L^4 the general **shape** of the laccolith will not. Its general shape will be:

```
In[182]:= Plot[deflect, {X, -1/2, 1/2}, AxesLabel → {"x/L", wR/(pL^4)}]
```

From In[182]:=

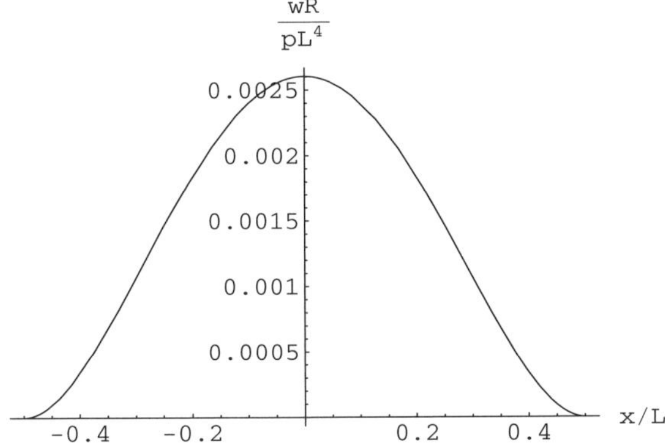

Out[182]= -Graphics-

The maximum deflection, which occurs at $x = 0$, is (in dimensional terms)

```
In[183]:= sltn /. x → 0
```

$$Out[183]= \frac{L^4\,p}{384\,R}$$

The bending moment developed in the plate is given by $M = -R\,d^2w/dx^2$

```
In[184]:= ∂{x,2} sltn
```

$$Out[184]= \frac{p\,(128\,x^2 - 16\,(L^2 - 4\,x^2))}{384\,R}$$

or, in dimensionless form,

```
In[185]:= % R/p/L^2
```

$$Out[185]= \frac{128\,x^2 - 16\,(L^2 - 4\,x^2)}{384\,L^2}$$

As above, the length terms can be grouped together

```
In[186]:= moment = Expand[%] /. (x/L)^2 → X^2
```

$$\text{Out[186]=} -\frac{1}{24} + \frac{X^2}{2}$$

and then plotted in dimensionless form.

```
In[187]:= Plot[moment, {X, -1/2, 1/2}, AxesLabel → {"x/L", M\R/p\L^2}]
```

From In[187]:=

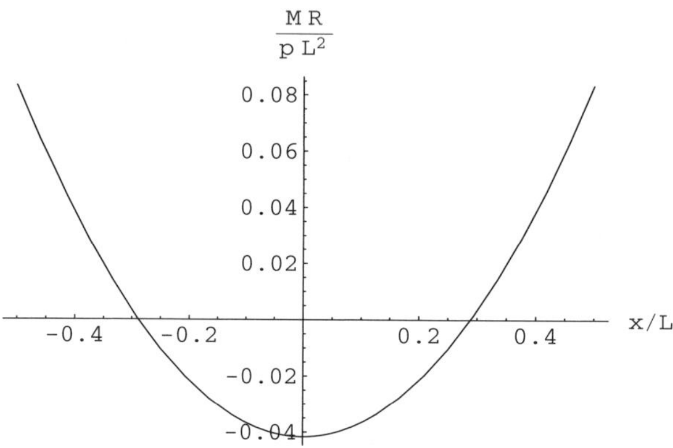

Out[187]= -Graphics-

The bending moment is related to the **curvature** and **fiber strain** developed in the plate (Johnson, 1970; Turcotte and Schubert, 2002). The fiber strain, in particular, is the strain developed in imaginary horizontal fibers located at different distances from the center of the plate as it is bent. For small deflections, the fiber strain is $\epsilon = -y\, d^2w/dx^2$, where y is the distance measured perpendicular to the thickness of the plate. Therefore, there will be tension ($\epsilon > 0$) along the upper edge of the plate ($y > 0$) where the bending moment is negative and along the lower edge of the plate ($y < 0$) where the bending moment is positive. The opposite holds true for compression. The plane defining the center of the plate, $y = 0$, is known as the **neutral surface** because there is neither tension nor compression at $y = 0$ in a thin elastic plate. The locations of the largest fiber strains (at the crest and two edges of the laccolith) are likely to be the locations where joints or dilational fractures form during bending, which is the basis for the curvature mapping techniques employed by structural geologists exploring for productive areas in fractured aquifers or petroleum reservoirs (Fischer and Wilkerson, 2000; Steward and Wynn, 2000). Chapter 7 includes a discussion of curvature mapping using gridded subsurface data.

Groundwater Flow Across Faults

Faults can act either as barriers to or conduits for the flow of groundwater, petroleum, and ore-bearing fluids (Haneberg *et al.*, 1999). As such, it can be useful to have a simple model to make inferences about the hydraulic properties of faults from field data such as hydraulic head measurements from observation wells. This example describes, following the method developed in Haneberg (1995), how steady state groundwater flow across faults can be simulated by simultaneously solving two or three differential equations describing horizontal flow in two aquifers separated by a vertical fault.

Horizontal steady state groundwater flow through a homogeneous and isotropic aquifer with no sinks or sources is described by the differential equation $d^2h/dx^2 = 0$. The hydraulic head, h, is the energy per unit weight of the groundwater, which flows down-gradient from areas in which hydraulic head is high to those in which it is low. This equation has a general solution of the form

```
In[188]:= DSolve[∂x,x h[x] == 0, h[x], x]

Out[188]= {{h[x] → C[1] + x C[2]}}
```

Haneberg (1995) showed how to incorporate recharge or discharge along the fault into the solutions, and Bense *et al.* (2003) used a variation of this method to account for a fault with recharge between two aquifers of differing transmissivity. In order to simulate groundwater flow across two aquifers separated by a fault of finite width, we will write three equations of this form (one for the fault and two for the aquifers) and then solve them simultaneously to ensure that the hydraulic head and flow match at each of the fault-aquifer boundaries. If the width of the fault is zero and it has no hydraulic properties unto itself, then its only effect will be to juxtapose two aquifers of different permeability. In that case, the fault does not have to be explicitly considered and only two equations need be written (one for each aquifer). The general solutions are for the left aquifer (L), fault (F), and right aquifer (R) are (using semi-colons to suppress the output):

```
In[189]:= hL = c1 + c2 x
          hF = c3 + c4 x
          hR = c5 + c6 x

Out[189]= c1 + c2 x

Out[189]= c3 + c4 x

Out[189]= c5 + c6 x
```

In this example, the fault straddles the coordinate system origin and extends over $-w \le x \le +w$. The aquifer to the left of the fault extends over $-L \le x \le -w$ and the aquifer to the right of the fault extends over $w \le x \le L$. This geometry is illustrated below. Most of the graphics commands are self-explanatory, and are given as a list enclosed by curly brackets. A series of replacement rules is used to specify options about the axes and ticks after the closing **`Graphics`** square brace but just inside the closing **`Show`** square brace.

```
In[190]:= Show[
            Graphics[
              {
                Thickness[0.007],
                Line[{{-1., 0}, {-1., 1}}],
                Line[{{1., 0}, {1., 1}}], GrayLevel[0.75],
                Rectangle[{-0.1, 0}, {0.1, 1}], GrayLevel[0.],
                Text["left aquifer", {-0.5, 0.5}, {0, 0}],
                Text["right aquifer", {0.5, 0.5}, {0, 0}],
                Text["fault", {0., 0.9}, {0, 0}]
              }
            ], Axes → True, AxesLabel → {"x", None},
            Ticks → {{{-1, " - L"}, {-0.1, " - w"}, {0.1, "w"},
              {1, " - L"}}, None}
          ]
From In[190]:=
```



```
Out[190]= -Graphics-
```

The next step is to specify six boundary conditions that will allow the six constants to be determined. This can be done in different ways, one of which is illustrated below. We will start by specifying that the hydraulic head is $+\Delta h$ at $x = L$ and $-\Delta h$ at $x = -L$. This gives rise to the following two boundary conditions:

```
In[191]:= bc1 = hL == -Δh /. x → -L
          bc2 = hR == Δh /. x → L

Out[191]= c1 - c2 L == -Δh

Out[191]= c5 + c6 L == Δh
```

The next two boundary conditions apply to the contacts between the fault and the aquifers, where the solutions for hydraulic head will be required to match each other. That is to say, the head in the aquifer must equal the head in the fault along the contact between the two.

```
In[192]:= bc3 = hL == hF /. x → -w
          bc4 = hR == hF /. x → w
```

```
Out[192]= c1 - c2 w == c3 - c4 w

Out[192]= c5 + c6 w == c3 + c4 w
```

The final two boundary conditions relate to the discharge of groundwater across the fault-aquifer contacts. For one-dimensional horizontal flow, the discharge is given by a variation of Darcy's law: $Q = -T\, dh/dx$. Q is the discharge, with units of length^3/time, and T is the aquifer or fault transmissivity, with units of length^2/time. The negative sign is included because groundwater flows down gradient but the discharge must be positive. Transmissivity is the product of the hydraulic conductivity (length/time) and thickness (length) of the aquifer or fault. In the absence of any sources or sinks along the contact, we will require that the volume of water flowing out of one unit be exactly equal to the volume flowing into the adjacent unit. Thus,

```
In[193]:= bc5 = -TL ∂x hL == -TF ∂x hF /. x → -w
          bc6 = -TF ∂x hF == -TR ∂x hR /. x → w

Out[193]= -c2 TL == -c4 TF

Out[193]= -c4 TF == -c6 TR
```

Now that all six boundary conditions have been specified, they can be solved to find algebraic expressions for the six constants

```
In[194]:= constants =
            Simplify[Solve[{bc1, bc2, bc3, bc4, bc5, bc6},
            {c1, c2, c3, c4, c5, c6}]]
```

$$\text{Out[194]= } \Big\{\Big\{c3 \to -\frac{TF\,(TL - TR)\,(L - w)\,\Delta h}{L\,TF\,(TL + TR) - (-2\,TL\,TR + TF\,(TL + TR))\,w},$$

$$c1 \to \left(-1 + \frac{2\,L\,TF\,TR}{L\,TF\,(TL + TR) - (-2\,TL\,TR + TF\,(TL + TR))\,w}\right)\Delta h,$$

$$c5 \to \Delta h - \frac{2\,L\,TF\,TL\,\Delta h}{L\,TF\,(TL + TR) - (-2\,TL\,TR + TF\,(TL + TR))\,w},$$

$$c2 \to \frac{2\,TF\,TR\,\Delta h}{L\,TF\,(TL + TR) - (-2\,TL\,TR + TF\,(TL + TR))\,w},$$

$$c4 \to \frac{2\,TL\,TR\,\Delta h}{L\,TF\,(TL + TR) - (-2\,TL\,TR + TF\,(TL + TR))\,w},$$

$$c6 \to \frac{2\,TF\,TL\,\Delta h}{L\,TF\,(TL + TR) - (-2\,TL\,TR + TF\,(TL + TR))\,w}\Big\}\Big\}$$

Particular solutions for the hydraulic head in the aquifers and the fault can be found by substituting the constants into the general solutions for head.

```
In[195]:= hL = Simplify[hL /. constants]
```

$$\text{Out[195]= } \left\{\frac{(L\,TF\,(-TL + TR) - 2\,TL\,TR\,w + TF\,(TL\,w + TR\,w + 2\,TR\,x))\,\Delta h}{L\,TF\,(TL + TR) - (-2\,TL\,TR + TF\,(TL + TR))\,w}\right\}$$

```
In[196]:= hF = Simplify[hF /. constants]
```

$$\text{Out[196]= } \left\{\frac{(L\,TF\,(-TL + TR) + TF\,(TL - TR)\,w + 2\,TL\,TR\,x)\,\Delta h}{L\,TF\,(TL + TR) - (-2\,TL\,TR + TF\,(TL + TR))\,w}\right\}$$

```
In[197]:= hR = Simplify[hR /. constants]
```

$$Out[197]= \left\{ -\frac{(L\,TF\,(TL - TR) - 2\,TL\,TR\,w + TF\,(TL\,w + TR\,w - 2\,TL\,x))\,\Delta h}{L\,TF\,(TL + TR) - (-2\,TL\,TR + TF\,(TL + TR))\,w} \right\}$$

The Wolfram Research web site shows how this solution can also be obtained using **DSolve** (http://library.wolfram.com/examples/faultflow/)

To illustrate an application of these solutions, consider an example in which both of the aquifer transmissivities are 0.01 m^2/s, the fault transmissivity is 0.001 m^2/s, the head decreases a total of 10 m across a 1 km wide problem domain, and the fault is inferred to be 1 m wide. All of these site-specific values can be put into a list of replacement rules that we will call, for lack of a better name, **sitevals**

```
In[198]:= sitevals = {TL → 0.01, TR → 0.01, TF → 0.0001,
            L → 500., Δh → 5., w → 1.};
```

Each of the three solutions must be plotted separately over its range of validity. One way to accomplish this is to create three plots with **DisplayFunction → Identity** and then combine them using **Show** with **DisplayFunction → $DisplayFunction**

```
In[199]:= pL = Plot[hL /. sitevals, {x, -500., -1},
            PlotStyle → Thickness[0.007],
            DisplayFunction → Identity];
          pF = Plot[hF /. sitevals, {x, -1., 1.},
            PlotStyle → Thickness[0.007],
            DisplayFunction → Identity];
          pR = Plot[hR /. sitevals, {x, 1., 500.},
            PlotStyle → Thickness[0.007],
            DisplayFunction → Identity];
          Show[pL, pF, pR, DisplayFunction → $DisplayFunction,
            Frame → True,
            FrameLabel → {"Horizontal Distance (m)", "Head (m)"}]
```

From In[199]:=

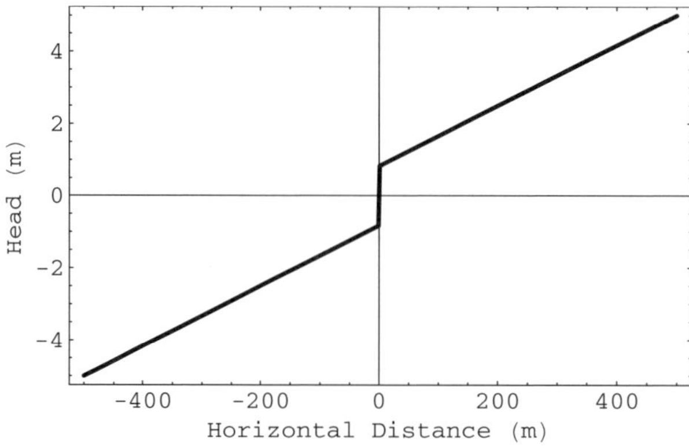

Out[199]= -Graphics-

If the fault is equally as transmissive, or even more so, than the aquifers, it will have no observable effect on the hydraulic gradient. This can be demonstrated by changing **sitevals** so that **TF** is an order of magnitude larger than **TL** and **TR**, and then plotting a new set of head profiles.

```
In[200]:= sitevals = {TL → 0.01, TR → 0.01, TF → 0.1, L → 500.,
             Δh → 5., w → 1.};
```

```
In[201]:= pL = Plot[hL /. sitevals, {x, -500., -1},
             PlotStyle → Thickness[0.007],
             DisplayFunction → Identity];

          pF = Plot[hF /. sitevals, {x, -1., 1.},
             PlotStyle → Thickness[0.007],
             DisplayFunction → Identity];

          pR = Plot[hR /. sitevals, {x, 1., 500.},
             PlotStyle → Thickness[0.007],
             DisplayFunction → Identity];

          Show[pL, pF, pR, DisplayFunction → $DisplayFunction,
             Frame → True,
             FrameLabel → {"Horizontal Distance (m)", "Head (m)"}]
```

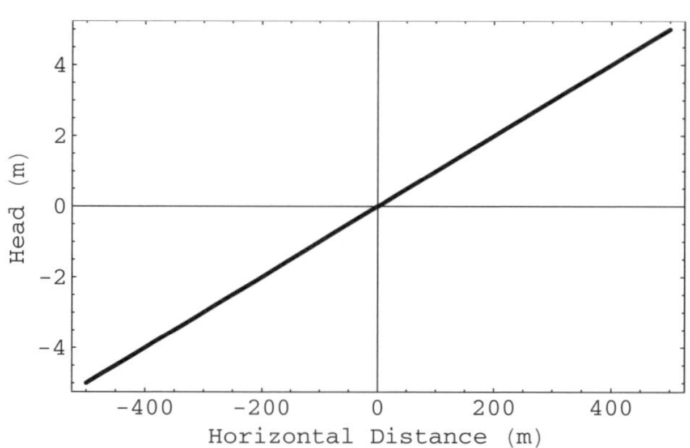


```
Out[201]= -Graphics-
```

What happens if one of the aquifers is more transmissive than the other, for example if the fault juxtaposes highly permeable sands and gravels against lower permeability bedrock? The transmissivity of the aquifers can be changed in **sitevals**

```
In[202]:= sitevals = {TL → 0.01, TR → 0.005, TF → 0.0001,
             L → 500., Δh → 5., w → 1.}
```

```
Out[202]= {TL → 0.01, TR → 0.005, TF → 0.0001, L → 500.,
             Δh → 5., w → 1.}
```

and another set of solutions plotted

```
In[203]:= pL = Plot[hL /. sitevals, {x, -500., -1},
              PlotStyle → Thickness[0.007],
              DisplayFunction → Identity];

           pF = Plot[hF /. sitevals, {x, -1., 1.},
              PlotStyle → Thickness[0.007],
              DisplayFunction → Identity];

           pR = Plot[hR /. sitevals, {x, 1., 500.},
              PlotStyle → Thickness[0.007],
              DisplayFunction → Identity];

           Show[pL, pF, pR, DisplayFunction → $DisplayFunction,
              Frame → True,
              FrameLabel → {"Horizontal Distance (m)", "Head (m)"}]
```

From In[203]:=

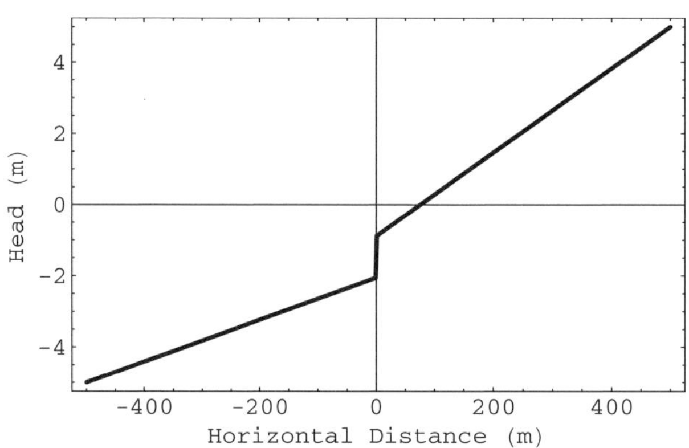


```
Out[203]= -Graphics-
```

Look at the plot carefully and you will see that the ratio of any two of the three transmissivities is the reciprocal of the ratio of the corresponding hydraulic gradients. This kind of irregular stair-step pattern of head changes across faults has been observed in the Albuquerque basin aquifer system, New Mexico, where normal faults bounding the rift basin juxtapose Cenozoic aquifers consisting of poorly lithified sediments against less transmissive Paleozoic bedrock (Titus, 1963; Haneberg, 1995; Reiter, 1999). Bense *et al.* (2003) described and analyzed similar patterns of head changes across large and small faults in the Roer rift of northern Europe.

3.6 Partial Differential Equations

The solution of partial differential equations, in which the dependent variable is a function of two or more independent variables, is considerably more difficult than the solution of ordinary differential equations. Nonetheless, **`DSolve`** can symbolically solve a few types of partial differential equations. **`NDSolve`** can numerically solve a wider range of linear and weakly nonlinear partial differential equations of interest to geoscientists, including diffusion and wave equations. Other kinds of partial differential equations can be solved using numerical methods such as finite differences implemented as *Mathematica* programs.

`NDSolve` solves partial differential equations using the **method of lines**, in which the equations are discretized in one dimension to form a set of ordinary differential equations that can be solved using a variety numerical methods. The particular method chosen will depend on the equation being solved and, according to the *Mathematica* documentation, the computer code to accomplish this is about 500 pages long.

The method of lines requires that an initial condition be specified in terms of one of the independent variables and two boundary conditions be specified in terms of a second independent variable. Therefore, the method is applicable to a broad class of diffusion and wave equations in which the dependent variable is a function of time and one spatial dimension. Problems that **cannot** be solved by this method include those described by the **Laplace** (*e.g.*, steady heat flow, groundwater flow, or chemical diffusion in two spatial dimensions) and **biharmonic** (*e.g.*, deformation of elastic continua in two spatial dimensions) equations, each of which requires that four boundary condtions be specified.

3.6.1 Hillslope Diffusion

Culling (1960) proposed that the geomorphic evolution of landscapes could, under some conditions, be simulated using a linear differential equation analogous to that used to describe unsteady chemical diffusion, groundwater flow, and heat flow. This approach was subsequently adopted by geomorphologists interested in estimating the ages of young fault scarps formed in unlithified surficial deposits such as alluvium (Andrews and Hanks, 1985; Nash, 1980, 1984; Hanks and Wallace, 1985). Others have used it to simulate the growth of prograding deltas and alluvial fans (*e.g.*, Turcotte and Schubert, 2002). This section includes examples illustrating the solution of the linear hillslope diffusion equation to simulate the geomorphic evolution of fault scarps.

The form of the equation originally proposed by Culling is $\partial z/\partial t = K\,\partial^2 z/\partial x^2$, where z is the elevation of Earth's surface, K is the topographic diffusivity (with units of length^2/time), t is time, and x is horizontal distance. Many applications of the diffusion model to simulate fault scarp degradation have assumed that K is linear and not a function of z. Some research, however, suggests that the diffusion constant is related to the slope angle and that the hillslope diffusion equation is therefore nonlinear (Andrews and Hanks, 1985; Roering *et al.*, 1999). This example is limited to the linear case.

For the simple case of a vertical fault displacing a flat horizontal surface, the problem has a well-known analytical solution (Hanks *et al.*, 1984) that can be written as a *Mathematica* function convenient for plotting.

```
In[204]:= z[x_, t_, K_, z0_] := z0/2 (1 + Erf[x/(2. Sqrt[K t])])
```

The implementation of **DSolve** in *Mathematica* 5.0 and earlier cannot obtain symbolic solutions to diffusion equations, so we will use the published solution. In this solution, $z_0$is the initial height of the scarp. The validity of the solution can be demonstrated by differentiating the function as appropriate, expanding the expressions, and equating them.

```
In[205]:= Expand[∂t z[x, t, K, z0]] == Expand[K ∂x,x z[x, t, K, z0] ]

Out[205]= True
```

Estimates of K from different areas suggest that a common value is on the order of $10^{-4} m^2$/yr, so we will use a value of $1 \times 10^{-4} m^2$/yr in this example and plot topographic profiles for different times. The nature of diffusion problems such as this one is that the rate of change is inversely proportional to the square root of time, so the time increments used below increase as the square of the elapsed time.

```
In[206]:= Plot[{z[x, 2, 0.0001, 1], z[x, 4, 0.0001, 1],
              z[x, 8, 0.0001, 1], z[x, 16, 0.0001, 1],
              z[x, 32, 0.0001, 1], z[x, 64, 0.0001, 1],
              z[x, 128, 0.0001, 1], z[x, 256, 0.0001, 1],
              z[x, 512, 0.0001, 1], z[x, 1024, 0.0001, 1],
              z[x, 2048, 0.0001, 1], z[x, 4096, 0.0001, 1]},
           {x, -2, 2}, AspectRatio → 1/4, AxesLabel → {"x", "z"},
           AxesOrigin → {-2, 0}]

From In[206]:=
```

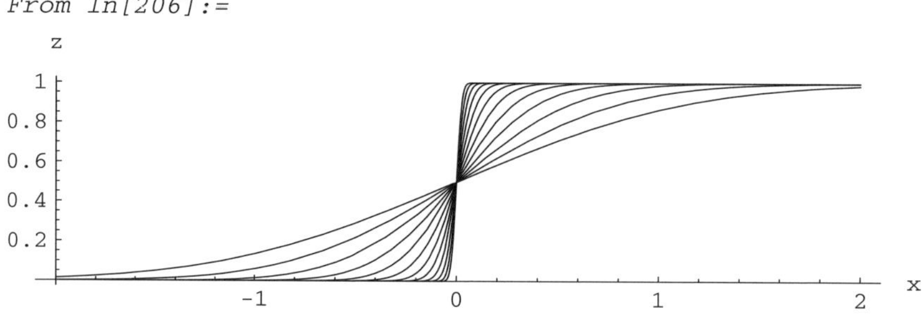


```
Out[206]= -Graphics-
```

Although *Mathematica* cannot obtain an analytical solution for this problem, it can obtain a numerical solution using **NDSolve**. The first step is to specify the initial shape of the topography to be simulated. In the case of a simple vertical fault scarp, this can be accomplished using the *Mathematica* **UnitStep** function (also known as a Heaviside step function). The plot below shows the **UnitStep** representation of a fault scarp with a height of 1 m.

```
In[207]:= Plot[UnitStep[x], {x, -2, 2}, AspectRatio → 1/4,
            PlotStyle → Thickness[0.007], AxesLabel → {"x","z"},
            AxesOrigin → {-2, 0}]
```

From In[207]:=

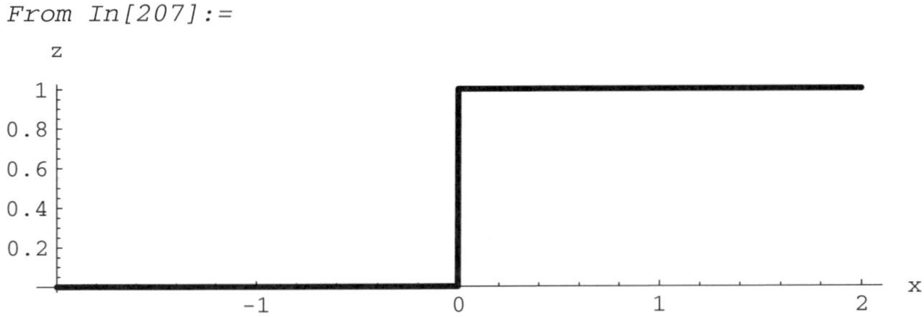

Out[207]= -Graphics-

Two spatial boundary conditions must also be specified, so we will hold the elevation constant at some finite distance from the fault ($z = 0$ at $x = -2$ m and $z = 1$ at $x = +2$ m). The analytical solution is for an infinite space, but numerical solutions are limited to finite problem domains. Next, define the equation to be solved using a value of $K = 0.001\ m^2$/yr.

```
In[208]:= lineareqn = ∂t z[x, t] == 0.0001 ∂x,x z[x, t]
```

Out[208]= $z^{(0,1)}[x, t] == 0.0001\, z^{(2,0)}[x, t]$

and the initial and boundary conditions

```
In[209]:= ic = z[x, 0] == UnitStep[x]

          bc1 = z[-2., t] == 0.

          bc2 = z[2., t] == 1.
```

Out[209]= z[x, 0] == UnitStep[x]

Out[209]= z[-2., t] == 0.

Out[209]= z[2., t] == 1.

Then, use **NDSolve** to obtain a numerical solution by simultaneously solving the diffusion equation, the initial condition, and the two boundary conditions over the interval -2 m $\leq x \leq +2$ m and 1 year $\leq t \leq$ 5000 years. **NDSolve** also requires the ranges of the two variables to be specified, and they must agree with those used in the boundary and intial conditions if a solution is to be obtained.

```
In[210]:= NDSolve[{lineareqn, ic, bc1, bc2}, z, {x, -2, 2},
           {t, 0, 5000}]

Out[210]= {{z → InterpolatingFunction[{{-2., 2.},
           {0., 5000.}}, <>]}}
```

The result of **NDSolve** is returned as an interpolating function that can be plotted just like any other *Mathematica* function (see Chapter 6 for a discussion of interpolation). Like those of **Solve** and **DSolve**, the results of **NDSolve** are returned as a list of replacement rules that can be assigned to a variable name.

```
In[211]:= z2 = z /. %[[1]]

Out[211]= InterpolatingFunction[{{-2., 2.}, {0., 5000.}}, <>]
```

This numerical solution can be compared by plotting it on the same set of axes as the analytical solution. For $t = 500$ years, the two curves are:

```
In[212]:= Plot[{z2[x, 500]}, {x, -2, 2},
           PlotStyle → {Dashing[{0.}], Dashing[{0.01}]},
           AspectRatio → 1/4, AxesLabel → {"x", "z"},
           AxesOrigin → {-2, 0}]

From In[212]:=
```

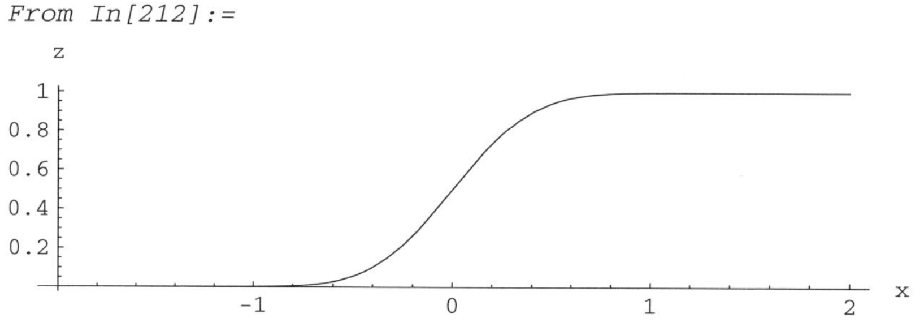


```
Out[212]= -Graphics-
```

> **Computer Note:** Numerical solutions can contain artifacts related to the way in which the problem was formulated and the method chosen for its solution. The implementation of **NDSolve** used in *Mathematica* 4.2 and earlier produces oscillations, known as **Gibbs oscillations**, for small values of t in this example. The oscillations occur because the infinitely steep fault scarp is approximated by a Fourier series of sine waves, and very short wavelength components must be used to approximate the vertical step. The implementation of **NDSolve** in *Mathematica* 5.0, however, eliminates the oscillations. If you are using *Mathematica* 4.2 or earlier, using the option **DifferenceOrder → 12** in **NDSolve** will greatly reduce, but not completely eliminate, the Gibbs oscillations. They die off rapidly and do not affect the solution for $t > 50$ years.

Here is a plot of the numerical solution for various times.

```
In[213]:= Plot[{z2[x, 2], z2[x, 4], z2[x, 8], z2[x, 16], z2[x, 32],
            z2[x, 64], z2[x, 128], z2[x, 256], z2[x, 512],
            z2[x, 1024], z2[x, 2048], z2[x, 4096]}, {x, -2, 2},
           AspectRatio → 1/4, AxesLabel → {"x", "z"},
           AxesOrigin → {-2, 0}]
```

From In[213]:=

Out[213]= -Graphics-

> **Computer Note:** Use **Table** to generate a series of fault scarp profiles for different times, then animate them. This can be done by selecting all of the plots and choosing Animate Selected Graphics from the Cell menu. Consult the *Mathematica* documentation for your particular front end to learn more about animating graphics.

One of the advantages of numerical solutions is that they can be easily adapted to boundary conditions more complicated than a simple vertical fault. For example, consider a listric normal fault along which the hangingwall was been rotated as it slipped downward.

```
In[214]:= topography =
           Interpolation[{{-2., 0.1}, {-1.5, 0.08}, {-1, 0.05},
            {-0.6, -0.02}, {-0.3, -0.1}, {0.3, 1.}, {2., 1.}},
            InterpolationOrder → 1]

Out[214]= InterpolatingFunction[{{-2., 2.}}, <>]
```

The statement above interpolates a series of first order (straight-line) polynomials between each of the points in the list. The **Interpolation** function will be discussed in more detail in Chapter 6 and is also described in the *Mathematica* documentation. The interpolated topography is shown below.

```
In[215]:= Plot[topography[x], {x, -2, 2},
           PlotStyle → Thickness[0.005],
           AspectRatio → 1/4, AxesLabel → {"x", "z"},
           AxesOrigin → {-2, 0}]
```

From In[215]:=

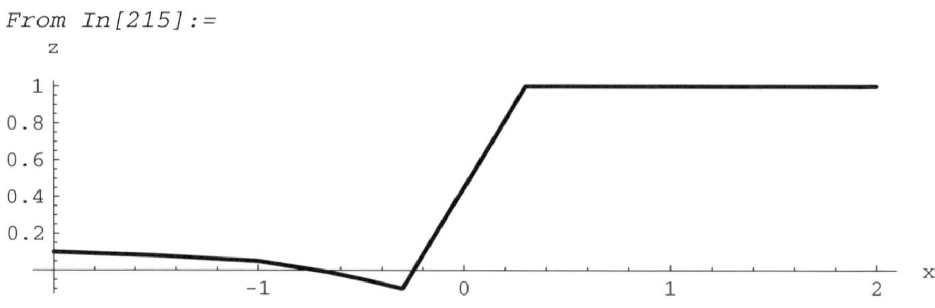

Out[215]= -Graphics-

Now that the topography is represented by a function, it can be used to specify the initial and boundary conditions.

```
In[216]:= ic = z[x, 0] == topography[x]
          bc1 = z[-2, t] == topography[-2.]
          bc2 = z[2, t] == topography[2.]
```

```
Out[216]= z[x, 0] == InterpolatingFunction[{{-2., 2.}}, <>][x]
Out[216]= z[-2, t] == 0.1
Out[216]= z[2, t] == 1.
```

The diffusion equation is solved in the same way as before, again using $K = 10^{-4}\ m^2/\text{yr}$.:

```
In[217]:= NDSolve[{lineareqn , ic, bc1, bc2}, z, {x, -2, 2},
            {t, 0, 5000}]
```

```
Out[217]= {{z → InterpolatingFunction[{{-2., 2.},
            {0., 5000.}}, <>]}}
```

```
In[218]:= z3 = z/.%[[1]]
```

```
Out[218]= InterpolatingFunction[{{-2., 2.}, {0., 5000.}}, <>]
```

```
In[219]:= Plot[{z3[x, 2], z3[x, 4], z3[x, 8], z3[x, 16], z3[x, 32],
              z3[x, 64], z3[x, 128], z3[x, 256], z3[x, 512],
              z3[x, 1024], z3[x, 2048], z3[x, 4096]},
            {x, -2, 2}, AspectRatio → 1/4, AxesLabel → {"x", "z"},
            AxesOrigin → {-2, 0}]
```

From In[219]:=

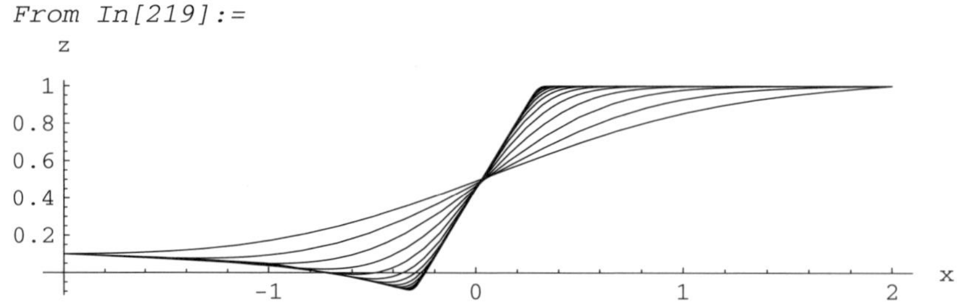

Out[219]= -Graphics-

Another possibility would be to use a detailed topographic profile of a fault scarp measured in the field shortly after an earthquake.

3.6.2 Periodic Heat Flow

Another intesting application of the diffusion equation involves a class of problems in which one of the spatial boundary conditions is periodic in time. Geoscientific examples include subsurface temperature fluctuations as a result of daily and seasonal surface temperature variations (Turcotte and Schubert, 2002) and subsurface hydraulic head or pore water pressure fluctuations as a result of periodic rainstorms (Baum and Reid, 1995; Haneberg, 1991 a, b; Iverson and Major, 1987; Keller *et al*, 1989; Reid, 1995). This example problem illustrates the use of the linear heat diffusion equation with a periodic boundary condition to simulate the annual heating and cooling of Earth's shallow subsurface.

Unsteady heat flow in one spatial dimension is described by a diffusion equation similar to the hillslope diffusion equation, except that the dependent variable (T) is temperature, z is depth rather than horizontal distance, and the diffusivity term refers to the thermal diffusivity of soil or rock. Although the value will vary according to mineralogy, and is especially sensitive to the amount of quartz (Haneberg *et al.*, 1994), the thermal diffusivity of most rocks is on the order of 10^{-4} m^2/s. Because this example simulates annual heating, it is convenient to recast the typical thermal diffusivity of rock in units of area per year.

The linear diffusion equation is defined as in the example above, but with different variables and K given in units of m^2/a.

```
In[220]:= eqn = ∂t T[z, t] == K∂z,z T[z, t] /. K → 10^-4 365 * 24 * 60 * 60.

Out[220]= T^(0,1)[z, t] == 3153.6 T^(2,0)[z, t]
```

The initial boundary conditions for this example, however, are noticably different. In this example, we will assume that the intial temperature is 15° throughout the subsurface. The temperature along Earth's surface (z = 0) is given in terms of a mean annual temperature of 15° minus a sinusoidal seasonal component with an amplitude of 10°.

```
In[221]:= Plot[15. - 10. Sin[2 π t], {t, 0, 1},
            AxesLabel → {"t (yr)", "T (°)"},
            AxesOrigin → {0, 15}]
```

From In[221]:=

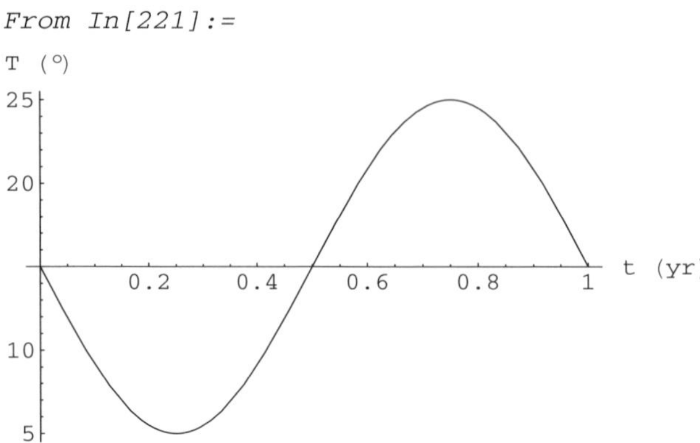

Out[221]= -Graphics-

The second boundary condition will state that the thermal gradient, $\partial T/\partial z = 0$ at great depth. In analytic solutions to the problem (*e.g.*, Carslaw and Jaeger, 1959; Turcotte and Schubert, 2002), one of the constants of integration can be heuristically eliminated by assuming that $\partial T/\partial z = 0$ at $z = -\infty$. In numerical solutions, however, the depth must be finite and we will use an arbitrarily chosen value of $z = 500$ m. In *Mathematica* input format, then, the initial and boundary conditions are:

```
In[222]:= ic = T[z, 0] == 15.
          bc1 = T[0, t] == 15. - 10. Sin[2 π t]
          bc2 = ∂z T[z, t] == 0. /. z → -500.

Out[222]= T[z, 0] == 15.

Out[222]= T[0, t] == 15. - 10. Sin[2 π t]

Out[222]= T^(1,0) [-500., t] == 0.
```

Solve the equation and assign the result to the variable **Temp**, so as not to overwrite **T**(in case the equation is to be solved again, for example with a different amplitude or wavelength temperature fluctuation).

```
In[223]:= NDSolve[{eqn , ic, bc1, bc2}, T, {z, -500, 0}, {t, 0, 5}]

Out[223]= {{T → InterpolatingFunction[{{-500., 0.},
            {0., 5.}}, <>]}}

In[224]:= Temp = T /. %[[1]]

Out[224]= InterpolatingFunction[{{-500., 0.}, {0., 5.}}, <>]
```

Carslaw and Jaeger (1959) contains an analytical solution to this problem. The results of the numerical solution can be visualized in several different ways. One approach is to use superimposed plots of the temperature fluctuations at different depths, as shown below for the first five years of the solution.

```
In[225]:= Plot[{Temp[0,t],Temp[-25,t],Temp[-50,t],
            Temp[-100,t]},{t,0,5},
          PlotStyle→{GrayLevel[0.6],GrayLevel[0.4],
            GrayLevel[0.2],GrayLevel[0.]},
          AxesLabel→{"t", "T"},
          PlotLegend→{"0", "25", "50", "100"},
          LegendLabel- > "Depth (m)",
          LegendPosition→{0.9, 0.1},
          LegendSize→{0.4, 0.5},AxesOrigin→{0, 15}]
```

From In[225]:=

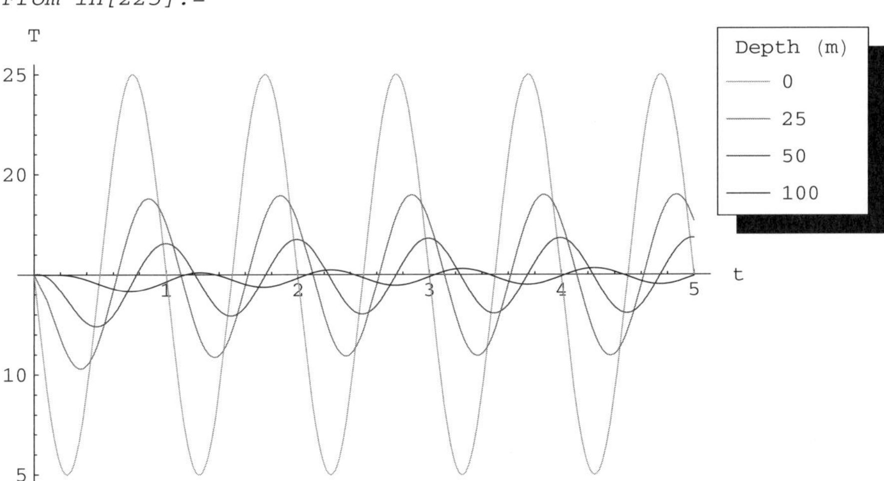

Out[225]= -Graphics-

This plot shows several important characteristics of periodic diffusion problems. First, the dependent variable (temperature in this case) attenuates exponentially with depth, and there is virtually no change in subsurface temperature at a depth of –100 m. Second, as shown by the offet peaks and troughs of the waveforms above, there is a lag between temperature changes on the surface and those at depth. Third, the solution consists of two components: a gradual adjustment of the initial conditions (we specified uniform temperature throughout) to the temperature oscillation at the surface and a so-called steady-state periodic component that will persist once the temperature is equilibrated to the surface fluctuations. The adjustment to initial conditions is most obvious in the $z = -100$ m curve, which increases in value from left to right.

Superimposing several curves on top of each other can create a confusing graph. Another way to visualize the periodic solution is to make a contour plot of T as a function of z and t. *Mathematica* uses 10 contour intervals by default, and the example below creates a table of contours in 2.5° increments to override the default. Very little temperature change occurs at depths below –100 m, so the plot is truncated there (recall that the lower boundary of $z = -500$ m was chosen only as a finite approximation of infinity, and has no physical significance).

```
In[226]:= tempcontourplot = ContourPlot[Temp[z, t], {t, 0, 5},
            {z, -100, 0}, PlotRange → All, PlotPoints → 100,
            AspectRatio→1/2, FrameLabel→{"years","depth (m)"},
            ColorFunction → (GrayLevel[0.2 + 0.8 #]&),
            Contours → Table[c, {c, 6, 24, 2.}], PlotRange → All]
```

From In[226]:=

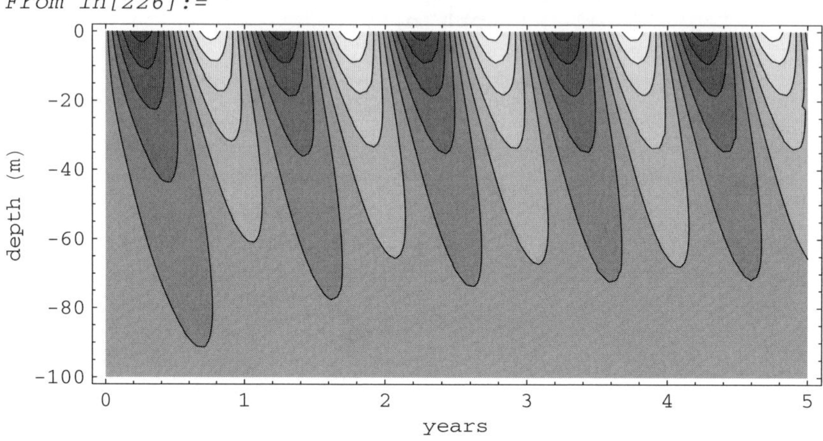

Out[226]= -ContourGraphics-

The contour plot more clearly shows the adjustment from the initial conditions to the periodic steady state that occurs over the first 2 years or so, after which the temperature oscillations appear to be identical. A legend can be added using the **ShowLegend** function contained in the Graphics`Legend` standard add-on package. The legend below specifies 8 increments ranging from 5° to 25°.

```
In[227]:= ShowLegend[tempcontourplot,
            {GrayLevel, 8, "5°", "25°", LegendShadow → None,
            LegendPosition→{1.1, -0.5}, LegendSize→{0.2, 1.}}]
```

From In[227]:=

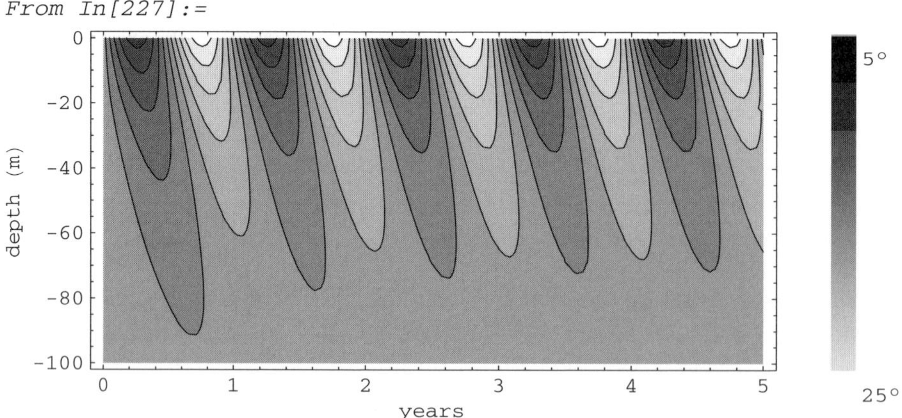

Out[227]= -Graphics-

A third way to visualize the solution is to use a 3D plot. To create a wire mesh version without a shaded surface, use the option **Lighting → False**.

```
In[228]:= Plot3D[Temp[z, t], {t, 0, 5}, {z, -200, 0},
            PlotRange → All, PlotPoints → 50, Lighting → False,
            AxesLabel → {"time", "depth", "T"},
            BoxRatios → {1, 0.6, 0.23}]
```

From In[228]:=

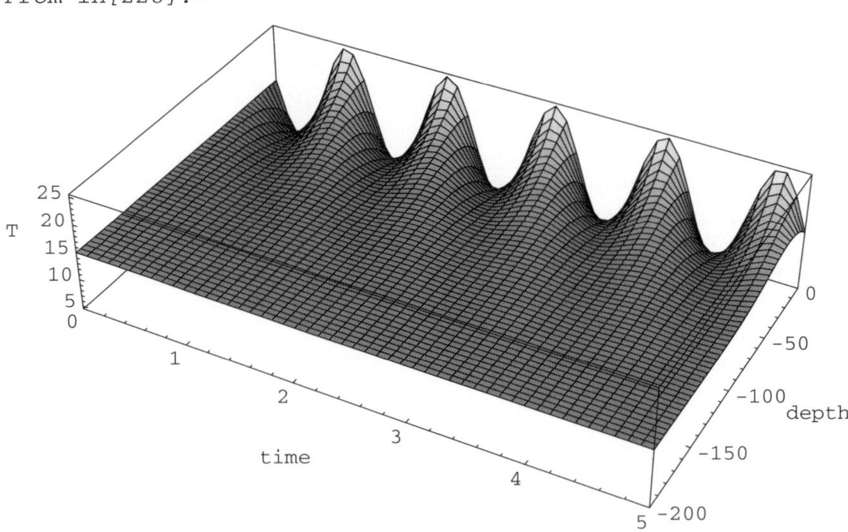

Out[228]= -SurfaceGraphics-

> **Computer Note:** Solve the periodic heat flow problem for diurnal temperature fluctuations that occur as the result of daily heating and cooling. What is the relationship between the frequency of temperature fluctuations and the depth to which they propagate? How deep would the temperature change associated with a 1,000,000 year long ice age propagate into the Earth?

3.6.3 Topographic Loading of Earth's Crust

Perturbations in the subsurface state of stress arising from surface topography have been proposed as an explanation for a number of hydrogeologic and geomorphologic phenomena. These include the origin of valley anticlines with fold axes parallel to deeply incised canyons; enhanced permeability and aquifer productivity beneath valleys; decreased rock strength, P-wave velocities, and coal mine roof stability beneath valleys; ridgetop depressions known as *sackungen*; and the overconsolidation of poorly lithified aquifer systems (Haneberg, 1999 and references therein). One particularly simple way to analyze the influence of topography on the subsurface state of stress is to assume that the effects of topography are approximately equal to those of a sinusoidally-varying load superimposed on a flat surface (Jeffreys, 1976). While this idealization introduces some errors by not explicitly accounting for the material within the hills or removed from the valleys, it is very straightforward and provides an order-of-magnitude estimate of the effects of topography on the state of stress at depth. Haneberg (1999) shows how to solve a more complicated version of this problem in which the upper surface is an arbitrary waveform or combination of waveforms.

Under conditions of plane strain, the distribution of stress in an elastic material can be described using a **biharmonic equation** written in terms of an **Airy stress function**, ϕ (Davis and Selvadurai, 1996; Timoshenko and Goodier, 1970):

$$\frac{\partial^4\phi}{\partial x^4} + 2\frac{\partial^4\phi}{\partial x^2\partial z^2} + \frac{\partial^4\phi}{\partial z^4} = 0$$

in which x and z are the two spatial coordinates. The Airy stress function is in turn related to the components of the 2-D stress tensor by the derivatives $\sigma_{xx} = \partial^2\phi/\partial z^2$, $\sigma_{zz} = \partial^2\phi\,/\,\partial x^2$, and $\sigma_{xz} = -\partial^2\phi\,/\,(\partial x\partial z)$. If the stresses along the top and bottom edges of a 2-D beam can be expressed in terms of sine or cosine curves, then the biharmonic equation has a general solution of

In[229]:= **ϕ = ((c1 + z c2) $e^{n\pi z/L}$ + (c3 + z c4) $e^{-n\pi z/L}$) Cos[n π x/L]**

Out[229]= $\left(e^{\frac{n\pi z}{L}}\ (c1 + c2\ z) + e^{-\frac{n\pi z}{L}}\ (c3 + c4\ z)\right)\ \text{Cos}\left[\frac{n\pi x}{L}\right]$

L is the wavelength of the topography, for example the crest-to-crest or trough-to-trough distance in a series of valleys and ridges. Is this a valid solution to the biharmonic equation?

In[230]:= **Expand[$\partial_{x,x,x,x}\phi$ + 2 $\partial_{x,x,z,z}\phi$ + $\partial_{z,z,z,z}\phi$] == 0**

Out[230]= True

Now that the validity of the general solution has been established, we can move on to the boundary conditions and a particular solution. To do this, first define the three components of the 2-D stress tensor in terms of ϕ

In[231]:= **σxx = Simplify[$\partial_{z,z}\phi$]**

Out[231]= $\frac{1}{L^2}\left(e^{-\frac{n\pi z}{L}}\ n\pi \left(\left(c3 + c1\ e^{\frac{2n\pi z}{L}}\right) n\pi + c4\ (-2\,L + n\pi z) + c2\ e^{\frac{2n\pi z}{L}}\ (2\,L + n\pi z)\right) \text{Cos}\left[\frac{n\pi x}{L}\right]\right)$

In[232]:= **σzz = Simplify[$\partial_{x,x}\phi$]**

Out[232]= $-\frac{n^2\pi^2\left(e^{\frac{n\pi z}{L}}\ (c1 + c2\ z) + e^{-\frac{n\pi z}{L}}\ (c3 + c4\ z)\right)\ \text{Cos}\left[\frac{n\pi x}{L}\right]}{L^2}$

```
In[233]:= σxz = Simplify[-∂x,z ϕ]
```

$$\text{Out[233]= } \frac{1}{L^2}\left(e^{-\frac{n\pi z}{L}}\, n\pi \left(\left(-c3 + c1\, e^{\frac{2n\pi z}{L}}\right) n\pi + c4\,(L - n\pi z) + c2\, e^{\frac{2n\pi z}{L}}\,(L + n\pi z)\right) \text{Sin}\left[\frac{n\pi x}{L}\right]\right)$$

Next, specify the boundary conditions along the surface ($z = 0$). The first boundary condition will represent the load imposed on the flat surface by topography. A positive value of A will indicate vertical compression along the surface as a consequence of a mountain, whereas a negative value will indicate tension because of a valley.

```
In[234]:= bc1 = σzz == A Cos[n π x/L] /. z → 0
```

$$\text{Out[234]= } -\frac{(c1 + c3)\, n^2 \pi^2 \,\text{Cos}\left[\frac{n\pi x}{L}\right]}{L^2} == A\ \text{Cos}\left[\frac{n\pi x}{L}\right]$$

The second boundary condition assumes that the surface is frictionless. This will introduce some error into the solutions, but it is not an unreasonable first approximation of a complicated problem.

```
In[235]:= bc2 = σxz == 0 /. z → 0
```

$$\text{Out[235]= } \frac{n\pi\,(c2\,L + c4\,L + (c1 - c3)\,n\pi)\ \text{Sin}\left[\frac{n\pi x}{L}\right]}{L^2} == 0$$

The next two boundary conditions represent the state of stress at great depth, which can be finite or infinite. In the previous example of periodic heat flow, we used a finite boundary condition to obtain a numerical solution. This time, we will assume that the effects of topography die off and have no effect an infinite distance from the surface. This is not a problem that *Mathematica* can handle symbolically, but it can be solved using some human input. Recall the definition of ϕ. If both $c_3$and $c_4$are **not** zero, the magnitude of the stress function would become infinitely large with depth. This is exactly the opposite of what we would like to occur. Therefore, we can declare that $c_3= c_4= 0$ for this particular problem. This would not be the case if the lower boundary were located at a finite depth.

```
In[236]:= c3 = 0
          c4 = 0

Out[236]= 0

Out[236]= 0
```

The value of **bc2** will be automatically updated to incorporate these definitions.

```
In[237]:= bc2
```

$$\text{Out[237]= } \frac{n\pi\,(c2\,L + c1\,n\pi)\ \text{Sin}\left[\frac{n\pi x}{L}\right]}{L^2} == 0$$

Values for the two remaining constants can be found using **Solve**

```
In[238]:= constants = Simplify[Solve[{bc1, bc2}, {c1, c2}]]
```

Out[238]= $\left\{\left\{c2 \to \frac{A\,L}{n\,\pi},\ c1 \to -\frac{A\,L^2}{n^2\,\pi^2}\right\}\right\}$

and then substituted into the stress expressions.

```
In[239]:= σzz = Simplify[σzz /. constants[[1]]]
```

Out[239]= $\frac{A\,e^{\frac{n\pi z}{L}}\,(L - n\,\pi\,z)\,\mathrm{Cos}\left[\frac{n\pi x}{L}\right]}{L}$

```
In[240]:= σxx = Simplify[ σxx /. constants[[1]]]
```

Out[240]= $\frac{A\,e^{\frac{n\pi z}{L}}\,(L + n\,\pi\,z)\,\mathrm{Cos}\left[\frac{n\pi x}{L}\right]}{L}$

```
In[241]:= σxz = Simplify[σxz /. constants[[1]]]
```

Out[241]= $\frac{A\,e^{\frac{n\pi z}{L}}\,n\,\pi\,z\,\mathrm{Sin}\left[\frac{n\pi x}{L}\right]}{L}$

One way to check the solutions is to plot the stresses along a boundary, which must agree with the distribution specified in the boundary condition. For example, the vertical normal stress along the surface is

```
In[242]:= Plot[σzz/. {n → 2, L → 1, A → 1, z → 0}, {x, -1/2, 1/2},
            PlotRange → All]
```

From In[242]:=

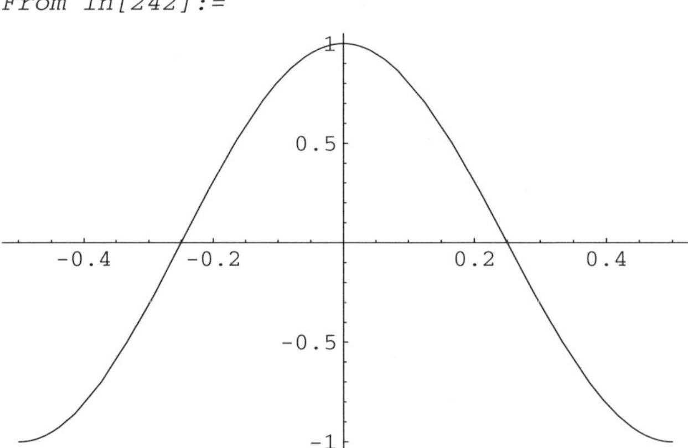

Out[242]= -Graphics-

This agrees with the boundary condition specified while solving the problem. Each of the stress components can be plotted individually, for example as below. Light areas indicate large positive values (compressive stress), whereas dark areas indicate large negative values (tensile stress)

```
In[243]:= ContourPlot[σzz/. {n → 2, L → 1, A → 1}, {x, -1/2, 1/2},
            {z, -1, 0}, PlotRange → All, PlotPoints → 25,
            FrameLabel → {"x", "z"}]
```

From In[243]:=

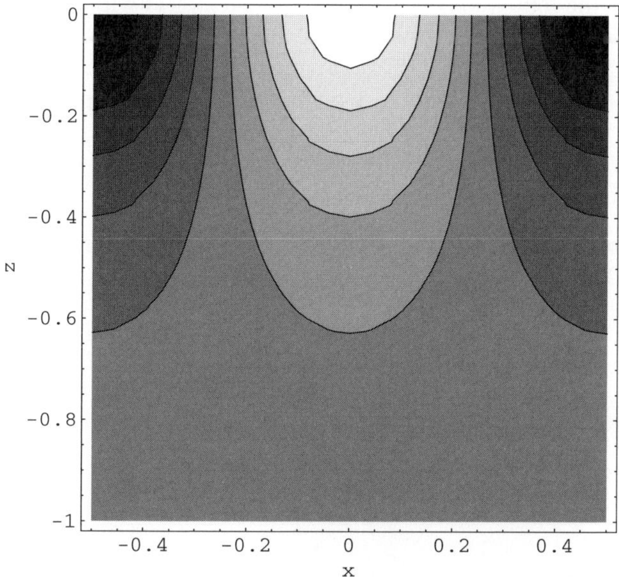

Out[243]= -ContourGraphics-

The effect of a topography with a mountain centered at $x = 0$ (note that $A = +1$) is a bulb of compression that dies off exponentially with depth beneath the mountain, flanked by two bulbs of tension beneath the adjacent valleys. It is also easy to plot the variation of σ_{zz} with depth beneath the mountain.

```
In[244]:= Plot[σzz/. {n → 2, L → 1, A → 1, x → 0}, {z, -1, 0},
            PlotRange → All, AxesLabel → {"z", σzz}]
```

From In[244]:=

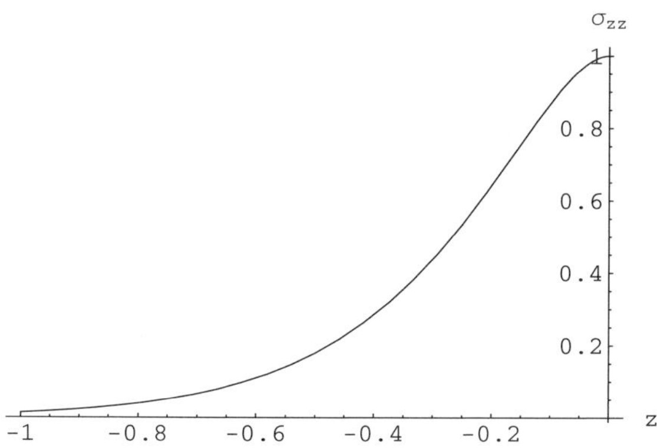

Out[244]= -Graphics-

As shown in both of the plots above, the effect of topography vanishes at a depth equal to the wavelength of the topography. Therefore, we can expect wide valleys or mountains to have a greater effect on the subsurface state of stress than narrow canyons or peaks. Although the relief (A) of the topography will affect the magnitude of stress very near the surface, the perturbation will still die off at depths less than $z = -L$.

The distribution of shear stress with depth follows a different pattern with twin bulbs of equal magnitude but opposite sign centered beneath the mountain.

```
In[245]:= ContourPlot[σxz/. {n → 2, L → 1, A → 1}, {x, -1/2, 1/2},
            {z, -1, 0}, PlotRange → All, PlotPoints → 25,
            FrameLabel → {"x", "z"}]
```

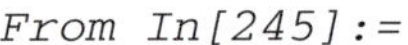

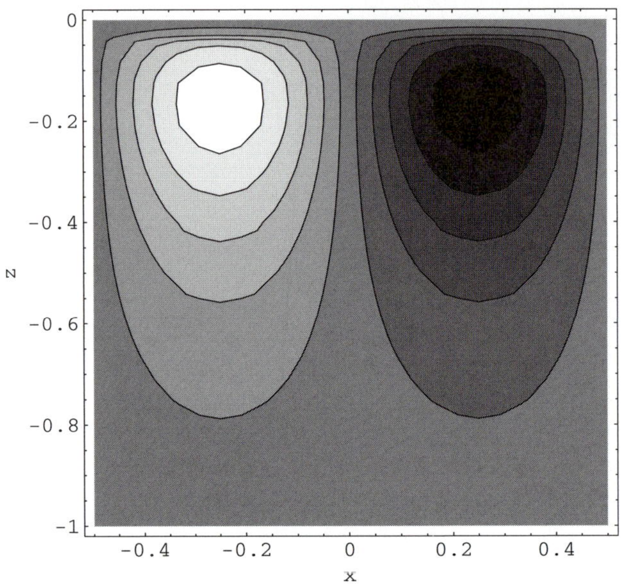


```
Out[245]= -ContourGraphics-
```

Mathematica makes it easy to combine results, for example to plot the mean normal stress.

```
In[246]:= ContourPlot[(σzz + σxx)/2./. {n → 2, L → 1, A → 1},
            {x, -1/2, 1/2}, {z, -1, 0}, PlotRange → All,
            PlotPoints → 25, FrameLabel → {"x", "z"}]
```

From In[246]:=

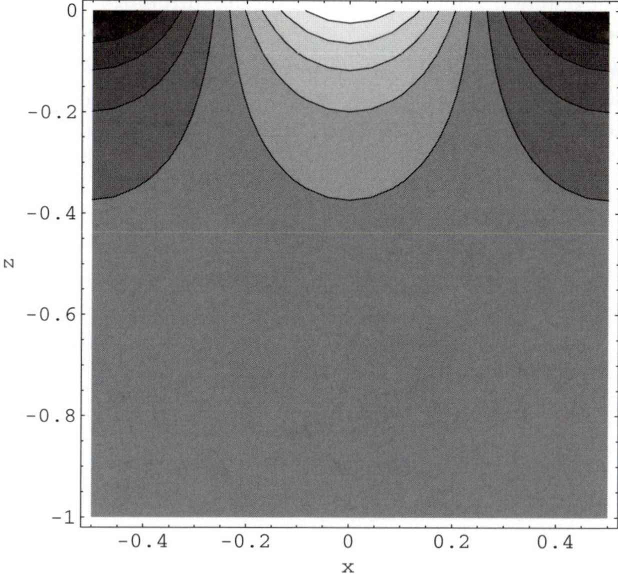

Out[246]= -ContourGraphics-

In terms of mean normal stress, therefore, we can tentatively conclude that the mechanical effects of topography along Earth's surface will persist to a depth of about 1/2 the wavelength of the topography.

> **Computer Note:** Assume that the topography across an area representative of the Basin and Range province of the southwestern United States has a wavelength of 100 km and an amplitude of 2000 m. Might the topography have any influence on the location of magma bodies at depths of 10 to 15 km?

3.6.4 Two Dimensional Steady Groundwater Flow

Steady groundwater flow through a two dimensional homogeneous and isotropic aquifer is described by the Laplace equation, $\partial^2 h/\partial x^2 + \partial^2 h/\partial y^2 = 0$, which is the same equation describing steady two dimensional heat flow and chemical diffusion. Because the right-hand side of the Laplace equation is zero, the conductivity or **diffusivity** term in unsteady flow or diffusion equations (as in the previous two examples) vanishes. A few analytical solutions of the Laplace equation exist, and one of the best known is the solution for topographically driven groundwater flow developed by Toth (1962). This example shows how **finite difference approximations** can be used to obtain a numerical solution to the problem topographically drived groundwater flow problem. This class of problems is one that cannot be solved by *Mathematica*'s current numerical differential equation solver, but a program can be written to obtain a solution by iteration. Smith (1985), Press *et al.* (1992), and Wang and Anderson (1982) provide detailed discussions of other finite difference methods applicable to diffusion-type problems.

Finite difference solutions are based on numerical approximations of the derivatives in differential equations. The derivatives are approximated using the differences in values between adjacent points spaced finite distances apart on a regular grid, hence then name finite difference. A simple one dimensional finite difference approximation of a first derivative at point i on a grid is

$$\frac{\partial f}{\partial x} = \frac{f_{i+1} - f_{i-1}}{2\,\Delta x}$$

where Δx is the distance between adjacent grid points. For example, consider the following list representing the values of some dependent variable f on a finite difference gird, with each value separated by distance $\Delta x = 0.1$.

```
In[247]:= f = {3., 2.6, 2.9};

In[248]:= Δx = 0.1;
```

Using the equation above, a finite difference approximation of the first derivative is:

```
In[249]:= (f[[3]] - f[[1]]) / (2. Δx)

Out[249]= -0.5
```

A second derivative is the first derivative of a first derivative so, using the same kind of reasoning, the second derivative of f at point i can be approximated as

$$\frac{\partial^2 f}{\partial x^2} = \frac{\frac{(f_{i+1} - f_i)}{\Delta x} - \frac{(f_i - f_{i-1})}{\Delta x}}{\Delta x}$$

In the expression above, the first derivatives are calculated for the imaginary points $i + 1/2$ and $i - 1/2$, and the second derivative is calculated for grid point i by taking the the difference between the two. A finite difference approximation of the second derivative is

```
In[250]:= 1/Δx ((f[[3]] - f[[2]])/Δx - (f[[2]] - f[[1]])/Δx)

Out[250]= 70.
```

The finite difference approximation of the one dimensional Laplace equation, $\partial^2 f/\partial x^2 = 0$ can also be solved for f to yield $f_i = (f_{i-1} + f_{i+1})/2$. Extending this logic to two spatial dimensions, the value of f at grid point (r, c), where r represents the row and c the column of the finite difference grid, is $f_{r,c} = (f_{r+1,c} + f_{r-1,c} + f_{r,c+1} + f_{r,c-1})/4$. This is the equation that will form the basis of our finite difference solution. The approach is to fill the grid with arbitrarily selected values, apply some boundary conditions, and then repeatedly apply the finite difference approximation until there is only a negligible change in results for each grid point between iterations.

Two types of boundary conditions can be specified. First, the value of the dependent variable can be constant, in this case specified hydraulic head along a boundary. Second, the flux across a boundary can be specified. The simplest form of flux boundary condition is a no-flow boundary. The flow of groundwater per unit cross-sectional area (*i.e.*, specific discharge) is given by Darcy's law, $q = -K\,dh/dx$, so a no flow condition requires $dh/dx = 0$. Recalling our earlier finite difference approximation of a first derivative, we can write $f_{i-1} = f_{i+1}$ to specify a no flow condition at grid point i. Because point i lies along a boundary, this means that we will either have to use an imaginary grid point outside of the problem domain or write a different finite difference expression for points along the boundary. We will choose the first option in the examples below.

Toth (1962) studied groundwater flow in small drainage basins in Alberta, and inferred from field observations that groundwater flowed downward beneath ridges separating basins and upward toward the axes of the basins, where it could be discharged into a stream. He assumed that seasonal water level fluctuations were small, so that the problem could be simulated using a steady state approach, and that the flow occurred in a homogeneous and isotropic rectangular aquifer. He also assumed that there was no groundwater flow across drainage divides imposed by ridges and streams ($\partial h/\partial x = 0$) and into the bedrock beneath the aquifer system ($\partial h/\partial z = 0$), and that head varied as a linear function of distance along the top of the aquifer. The geometry of the problem is illustrated below using a series of *Mathematica* graphics functions.

```
In[251]:= Show[
            Graphics[{Text["h = b + mx", {0.5, 0.95}],
                Text["∂h/∂x = 0", {0.97, 0.5}],
                Text["∂h/∂x = 0", {-0.02, 0.5}],
                Text["∂h/∂z = 0", {0.5, 0.05}],
                Text["stream", {0., 1.1}],
                Text["ridge", {1., 1.1}],
                Line[{{0, 0}, {0, 1}, {1, 1}, {1, 0}, {0, 0}}],
                Dashing[{0.01}], Line[{{0, 1}, {1, 1.2}}]}],
            PlotRange → {{-0.2, 1.2}, {-0.2, 1.2}}
          ]

From In[251]:=
```



```
Out[251]= -Graphics-
```

The right-hand side of the problem domain illustrated above represents a topographic drainage divide such as a ridge crest, whereas the left-hand side represents a drainage divide in the form of a stream to which the groundwaer is discharged. Following the general nature of the topography, hydraulic head along the upper boundary increases from the stream along the basin axis to the ridge along the basin margin.

First, define the number of rows and columns in the finite difference grid. We will use ten rows and ten columns but, in order to deal with the no-flow boundary conditions, we will have to include two extra columns and one extra row.

```
In[252]:= nr = 11;
          nc = 12;
```

and then create two tables, **old** and **new** to store the estimates.

```
In[253]:= old = Table[0., {r, nr}, {c, nc}];
          new = Table[0., {r, nr}, {c, nc}];
```

The next step is to establish the hydraulic head along the upper boundary. In this example, we will use a simple linear function so that head ranges from 0 to 1.

```
In[254]:= Do[old[[nr, c]] = N[(c - 2)/(nc - 3)], {c, 2, nc - 1}];
```

Once the boundary values have been specified, the actual finite difference approximation is calculated for each of the non-boundary points in **old** and the result is put into **new**. The equation in the expression below is the solution of the finite difference approximation of the Laplace equation for $h_{r,c}$.

```
In[255]:= Do[
            new[[r, c]] =
              (old[[r + 1, c]] + old[[r - 1, c]] + old[[r, c + 1]]
              +old[[r, c - 1]])/4.,
            {r, 2, nr - 1}, {c, 2, nc - 1}
          ];
```

At this point, **new**consists of mostly zeroes because successive changes to the solution will propagate away from the boundary along which head is specified. Row **nr** appears at the **bottom** of the matrix below because that is the standard convention for matrices. When the solution is plotted using **ListContourPlot** or **ListDensityPlot**, however, row **nr** will be at the top. The statement **Round[100new]/100** truncates the results to two decimal places.

```
In[256]:= Round[100 new]/100. // MatrixForm
```

Out[256]=

$$\begin{pmatrix} 0 & 0 & 0 & 0 & 0 & 0 & 0 & 0 & 0 & 0 & 0 & 0 \\ 0 & 0 & 0 & 0 & 0 & 0 & 0 & 0 & 0 & 0 & 0 & 0 \\ 0 & 0 & 0 & 0 & 0 & 0 & 0 & 0 & 0 & 0 & 0 & 0 \\ 0 & 0 & 0 & 0 & 0 & 0 & 0 & 0 & 0 & 0 & 0 & 0 \\ 0 & 0 & 0 & 0 & 0 & 0 & 0 & 0 & 0 & 0 & 0 & 0 \\ 0 & 0 & 0 & 0 & 0 & 0 & 0 & 0 & 0 & 0 & 0 & 0 \\ 0 & 0 & 0 & 0 & 0 & 0 & 0 & 0 & 0 & 0 & 0 & 0 \\ 0 & 0 & 0 & 0 & 0 & 0 & 0 & 0 & 0 & 0 & 0 & 0 \\ 0 & 0 & 0 & 0 & 0 & 0 & 0 & 0 & 0 & 0 & 0 & 0 \\ 0 & 0 & 0.03 & 0.06 & 0.08 & 0.11 & 0.14 & 0.17 & 0.19 & 0.22 & 0.25 & 0 \\ 0 & 0 & 0 & 0 & 0 & 0 & 0 & 0 & 0 & 0 & 0 & 0 \end{pmatrix}$$

The finite difference solution continues by repeating the previous computational step until the difference between **old** and **new** falls below some specified tolerance. The maximum difference between **old** and **new** after the first iteration is found by taking the maximum of the absolute values of the differences between **old** and **new**, excluding the boundary rows and columns.

```
In[257]:= maxerr = Max[
            Abs[
              Table[old[[r, c]] - new[[r, c]], {r, 2, nr - 1},
              {c, 2, nc - 1}]
            ]
          ]

Out[257]= 0.25
```

Next, put the values held in **new** into **old** in order to prepare for the next iteration, taking care not to overwrite the upper boundary head values in **old[[nr]]**.

```
In[258]:= Do[
            old[[r, c]] = new[[r, c]], {r, 2, nr - 1}, {c, 2, nc - 1}
          ]
```

The no-flow boundaries must now be reset to that $h_{1,c}= h_{3,c}$, $h_{r,3}= h_{r,1}$, and $h_{\mathrm{nc},r}= h_{\mathrm{nc}-1,r}$. Because *Mathematica* stores tables as lists of lists, assigning one row to another is easy. Assigning one column to another,is more complicated and requires a **Do** loop.

```
In[259]:= old[[1]] = old[[3]];

In[260]:= Do[
            Module[{},
              old[[r, 1]] = old[[r, 3]];
              old[[r, nc]] = old[[r, nc - 2]];
            ], {r, nr}
          ]
```

> **Computer Note:** Simplify the **Do** loop above using two **All** statements to replace the values in all rows without iterating.

The table **old** now looks like (again recalling that the rows are reversed):

```
In[261]:= Round[100 old]/100. // MatrixForm
```

Out[261]=

0	0	0	0	0	0	0	0	0	0	0	0
0	0	0	0	0	0	0	0	0	0	0	0
0	0	0	0	0	0	0	0	0	0	0	0
0	0	0	0	0	0	0	0	0	0	0	0
0	0	0	0	0	0	0	0	0	0	0	0
0	0	0	0	0	0	0	0	0	0	0	0
0	0	0	0	0	0	0	0	0	0	0	0
0	0	0	0	0	0	0	0	0	0	0	0
0	0	0	0	0	0	0	0	0	0	0	0
0.03	0	0.03	0.06	0.08	0.11	0.14	0.17	0.19	0.22	0.25	0.22
0.11	0	0.11	0.22	0.33	0.44	0.56	0.67	0.78	0.89	1.	0.89

This process is repeated until **maxerr** falls below a specified tolerance, which should be very small compaed to the magnitude of the head values. The following *Mathematica* program implements a finite difference solution of Toth's problem by combining the individual steps above. The number of rows and columns specified, **nr** and **nc**, should not include the extra rows and columns necessary for the no-flow boundary conditions. Values of **tolerance** should be very small compared to the magnitude of the head values. How small is small enough? One way to find out is to experiment with increasingly smaller values until the solution stabilizes from trial to trial.

```
In[262]:= Toth[Δh_, rows_, cols_, tolerance_] :=
            Module[{old, new, maxerr},

              (*
                Set number of rows and columns,
                create and initialize the necessary tables.
                  Make maxerr large to allow entry into
                  the While loop.
                *)

              nr = rows + 1;
              nc = cols + 2;

              old = Table[0., {r, nr}, {c, nc}];
              new = Table[0., {r, nr}, {c, nc}];
              maxerr = 1000;
```

```
  (* Initialize the upper boundary head values *)

  Do[old[[nr, c]] = N[Δh (c - 2) / (nc - 3)], {c, 2, nc - 1}];

  While[maxerr > tolerance,
    Module[{},

      maxerr = 0.;

      (* Apply the finite difference approximation for h *)

      Do[
        new[[r, c]] =
          (old[[r + 1, c]] + old[[r - 1, c]] + old[[r, c + 1]]
            +old[[r, c - 1]]) /4., {r, 2, nr - 1}, {c, 2, nc - 1}
      ];

      (* Determine the maximum error in this iteration *)

      maxerr = Max[
          Abs[
            Table[old[[r, c]] - new[[r, c]], {r, 2, nr - 1},
            {c, 2, nc - 1}]
          ]
        ];

      (* Swap the new and old values for interior grid points *)

      Do[
        old[[r, c]] = new[[r, c]], {r, 2, nr - 1}, {c, 2, nc - 1}
      ];

      (* Reset the three no - flow boundary nodes *)

      old[[1]] = old[[3]];

      Do[
        Module[{},
          old[[r, 1]] = old[[r, 3]];
          old[[r, nc]] = old[[r, nc - 2]];
        ], {r, nr}
      ];

    ];
  ];
  Return[Table[old[[r, c]], {r, 2, nr}, {c, 2, nc - 1}]]
]
```

Now that the finite difference function has been written, calculate the solution to the Toth problem on a 20 × 20 grid using a tolerance of 10^{-6}.

```
In[263]:= h = Toth[1, 20, 20, 1. 10^-6] ;
```

Here is a contour plot of the solution, with large values of head represented by light gray and small values represented by dark gray. Groundwater will flow perpendicular to the head contours.

```
In[264]:= headcontourplot = ListContourPlot[h, PlotRange → All,
            Contours → Table[c, {c, 0., 1., 0.05}]]
```

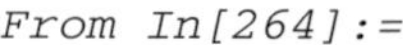

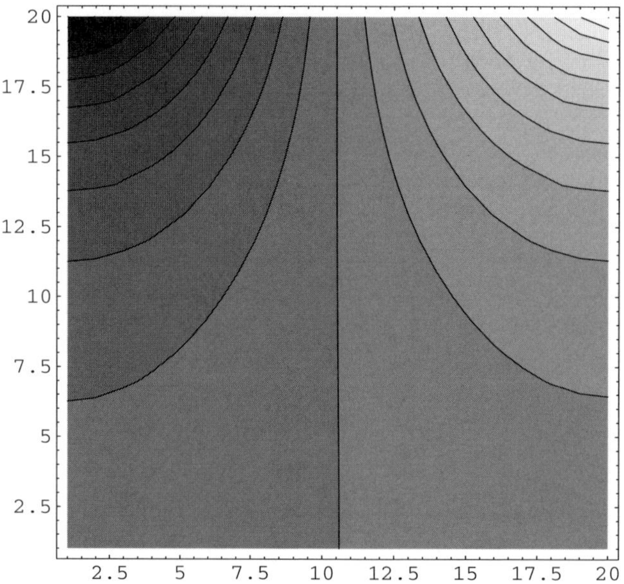


```
Out[264]= -ContourGraphics-
```

The contour plot can be embelished with vectors showing the specific discharge, or magnitude and direction of groundwater flow. *Mathematica*'s standard packages include two functions for plotting vectors. **ListPlotVectorField**, which works with tables such as **h**, takes as its argument a table of vectors. Such a table could be calculated using finite difference approximations of the first derivative, but there is an easier way. The first step is to interpolate a 2-D polynomial function that passes through each of the calculated finite difference results. Chapter 6 contains detailed information about the **ListInterpolation** function.

```
In[265]:= ListInterpolation[h]

Out[265]= InterpolatingFunction[{{1., 20.}, {1., 20.}}, <>]
```

The next step is to use the function **PlotGradientField**, which calculates the gradient of any scalar field (such as hydraulic head) and then plots the corresponding vector field. The length of the vectors is proportional to the magnitude of the hydraulic gradient, and **ScaleFactor** sets the relative length of the longest vector. Because the plot has spatial coordinates in terms of the finite difference row and column numbers, **ScaleFactor → 1** means that the longest vector will be 1/20 of the plot height and width. **PlotPoints → 10** plots a vector at every other point on the finite difference grid so that the vectors will not be too crowded. Also, notice that the row and column indices are reversed, because x values correspond to columns and y values to rows, and the negative of the gradient field is plotted because groundwater flows down, not up, the hydraulic gradient. See the *Mathematica* documentation for other options.

```
In[266]:= vectorplot = PlotGradientField[-%[c, r], {r, 1, 20},
            {c, 1, 20}, PlotPoints → 10, ScaleFactor → 3]

From In[266]:=
```

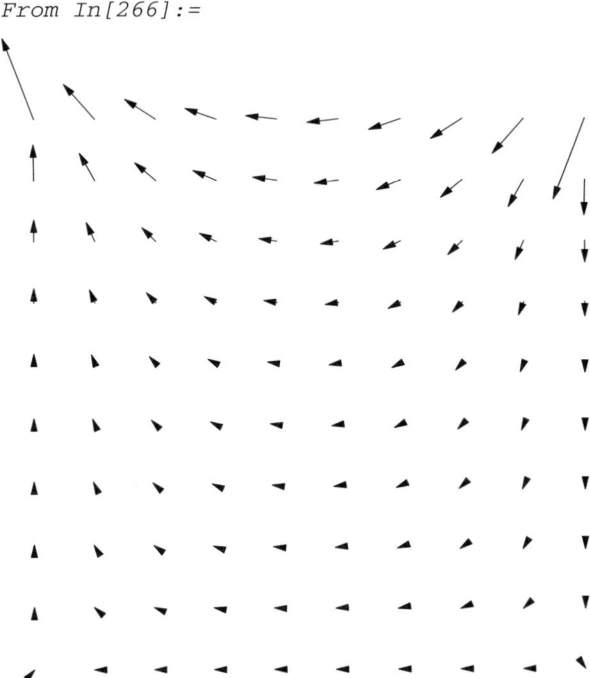

```
Out[266]= -Graphics-
```

One of the difficulties associated with vector plots of this type is that it can be difficult to scale the arrows. In the plot above, chosing a scale factor that makes the longest vectors a reasonable length also makes the shortest vectors so small that their shafts are not plotted. To make all of the vectors the same length regardless of their magnitude, list the option **ScaleFunction → (1&)** before **ScaleFactor** when plotting a gradient field. The two plots can now be superimposed to show both the contours and the vectors. **AspectRatio**, which applies to the entire plot, must be increased to keep the aquifer square when the large upward directed vectors are added to the plot.

```
In[267]:= Show[headcontourplot, vectorplot, Frame → False,
              AspectRatio → 22/20.]
```

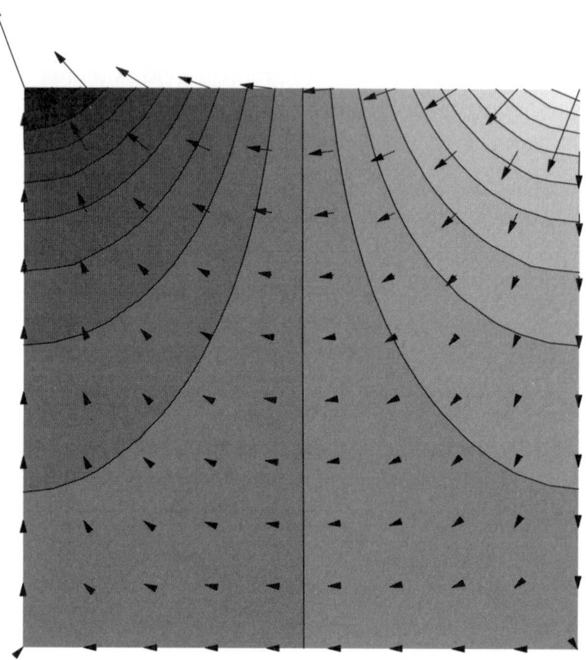

```
Out[267]= -Graphics-
```

> **Computer Note:** In a subsequent paper, Toth (1963) used a combined linear and sinusoidal upper boundary head distribution to investigate the effect of different scales of topography on groundwater flow systems. Modify the finite difference routine above so that $h = \Delta h\ x/L + A\ \sin(2\ n\pi x/L)$, where $\Delta h/L$ is the basin-scale slope of the water table, n is the periodicty of the local topography, A is the amplitude of the local topography, and L is the width of the aquifer being modeled. Use **`ContourPlot`** and **`PlotGradientField`** to visualize your results. Experiment with different values of n to determine how deep the effect of localized topography of different scales persists.

3.7 References and Recommended Reading

Andrews, D.J. and Hanks, T.C., 1985, Scarps degraded by linear diffusion: Inverse solution for age: *Journal of Geophysical Research 90*, p. 10193–10208.

Baum., R.L. and Reid, M.E., 1995, Geology, hydrology, and mechanics of a slow-moving clay-rich landslide, *in* W.C. Haneberg and S.A. Anderson, editors, *Clay and Shale Slope Instability*: Geological Society of America Reviews in Engineering Geology, v. 10, p. 79–104.

Bense, V.F., R.T. Van Balen, and J.J. De Vries, 2003, The impact of faults on the hydrogeological conditions in the Roer Valley Rift System: an overview,*Netherlands Journal of Geosciences/Geologie en Mijnbouw*, v. 82, p. 41–53.

Carslaw, H.S. and Jaeger, J.C., 1959, *Conduction of Heat in Solids (2d ed.)*: Cambridge University Press.

Culling, W.E.H., 1960, Analytical theory of erosion: *Journal of Geology*, v. 68, p. 336–344.

Davis, R.O. and Selvadurai, A.P.S., 1996, *Elasticity and Geomechanics*: Cambridge University Press.

Fischer, M.P., and M.S. Wilkerson, 2000, Predicting the orientation of joints from fold shape: Results of pseudo-three-dimensional modeling and curvature analysis: *Geology*, v. 28, p. 15–18.

Furbish D.J., 1997, *Fluid Physics in Geology*: Oxford University Press.

Haberman, R., 1998, *Mathematical Models: Mechanical Vibrations, Population Dynamics, & Traffic Flow*: Society for Industrial and Applied Mathematics.

Haneberg, W.C., 1991a, Observation and analysis of short-term pore pressure fluctuations in a thin colluvium landslide complex near Cincinnati, Ohio: *Engineering Geology*, v. 31, p. 159–184.

Haneberg, W.C., 1991b, Pore pressure diffusion and the hydrologic response of nearly-saturated, thin landslide deposits to rainfall: *Journal of Geology*, v. 99, p. 886–892

Haneberg, W.C., 1995, Steady state groundwater flow across idealized faults: *Water Resources Research*, v. 31, p. 1815–1820.

Haneberg, W.C., 1999, Effects of valley incision on the subsurface state of stress – theory and application to the Rio Grande valley near Albuquerque, New Mexico: *Environmental & Engineering Geoscience*, v. 5, p. 117–131.

Haneberg, W.C. and Bauer, P.W., 1993, Geologic setting and dynamics of a rockslide along NM 68, Rio Grande gorge, northern New Mexico: *Bulletin of the Association of Engineering Geologists*, v. 30, p. 7–16.

Haneberg, W.C. and Friesen, R.L., 1995, Tilts, strains, and ground-water levels near an earth fissure in the Mimbres Basin, New Mexico: *Geological Society of America Bulletin*, v. 107, p. 316–326.

Haneberg, W.C., Goodwin, L. B., and Ferranti, C. J., 1994, Pseudotachylyte in a metamorphic core complex – analytical modeling of the effect of compositional variation on frictional melting: *Geological Society of America, 1994 Annual Meeting Abstracts with Programs*, v. 26, n. 7, p. 269.

Hanks, T.C. and Wallace, R.E., Morphological analysis of the Lake Lahontan shoreline and beachfront fault scarps, Pershing County, Nevada: *Bulletin of the Seismological Society of America*, v. 75, p. 835–846.

Iverson, R.M. and Major, J.J., 1987, Rainfall, ground-water flow, and seasonal movement at Minor Creek landslide, northwestern California: Physical interpretation of empirical relations: *Geological Society of America Bulletin*, v. 99, p. 579–594.

Jeffreys, H., 1976, *The Earth: Its Origin, History, and Physical Constitution (6th ed.)*: Cambridge University Press.

Johnson, A.M., 1970, *Physical Processes in Geology*: Freeman-Cooper.

Keller, C.K., van der Kamp, G., and Cherry, J.A., 1989, A multiscale study of the permeability of a thick clayey till: *Water Resources Research*, v. 25, p. 2299–2317.

Middleton, G.V. and Wilcock, P.R., 1994, *Mechanics in the Earth and Environmental Sciences*: Cambridge University Press.

Nash, D.B., 1980, Morphologic dating of degraded normal fault scarps: *Journal of Geology*, v. 88, p. 353–360.

Nash, D.B., 1984, Morphologic data of fluvial terrace scarps and fault scarps near West Yellowstone, Montana: *Geological Society of America Bulletin*, v. 95, p. 1413–1424.

Oertel, G., 1996, *Stress and Deformation: A Handbook on Tensors in Geology*: Oxford University Press.

Press, W.H., Teukolsky, S.A., Vetterling, W.T., and Flannery, B.P., 1992, *Numerical Recipes in FORTRAN (2d ed.)*: Cambridge University Press.

Reid, M.E., 1995, A pore-pressure diffusion model for estimating landslide-inducing rainfall: *Journal of Geology*, v. 102, p. 709-717.

Reiter, M., 1999, Hydrogeothermal studies on the southern part of Sandia National Laboratories/Kirtland Air Force Base – Data regarding ground-water flow across the boundary of an intermontane basin, *in* W.C. Haneberg, P.S. Mozley, J.C. Moore, and L.B. Goodwin, editors, *Faults and Subsurface Fluid Flow in the Shallow Crust*: American Geophysical Union, Geophysical Monograph 113, p. 207–222.

Roering, J.J., Kirchner, J.W., and Dietrich, W.E., 1999, Evidence for nonlinear, diffusive sediment transport on hillslopes and implications for landscape morphology: *Water Resources Research*, v. 35, p. 853–870.

Smith, G.D., 1982, *Numerical Solution of Partial Differential Equations: Finite Difference Methods (3d ed.)*: Oxford University Press.

Stewart, S.A. and T.J. Wynn, 2000, Mapping spatial variation in rock properties in relationship to scale-dependent structure using spectral curvature: *Geology*, v. 28, p. 691–694.

Timoshenko, S.P. and Goodier, J.N., 1970, *Theory of Elasticity (3d ed.)*: McGraw-Hill.

Titus, F.B., Jr., 1963, *Geology and ground-water conditions in eastern Valencia County, New Mexico*: New Mexico Bureau of Mines & Mineral Resources Ground-Water Report 7.

Toth, J., 1962, A theory of groundwater motion in small drainage basins in central Alberta, Canada: *Journal of Geophysical Research*, v. 67, p. 4375–4387.

Toth, J., 1963, A theoretical analysis of groundwater flow in small drainage basins: *Journal of Geophysical Research*, v. 68, p. 4795–4812

Turcotte, D.L. and Schubert, G., *Geodynamics* (2d. ed.): Cambridge University Press.

Wieczorek, G.F., Snyder, J.B., Waitt, R.B., Morrissey, M.M., Uhrhammer, R.A., Harp, E.L., Norris, R.D., Bursik, M.I., and Finewood, L.G., 2000, Unusual July 10, 1996, rock fall at Happy Isles, Yosemite National Park, California, *Geological Society of America Bulletin*, v. 112, p. 75–85.

Wang, H.F. and Anderson, M.P., 1982, *Introduction to Groundwater Modeling*: W.H. Freeman.

Wolfram, S., 1999, *The Mathematica Book (4th ed.)*: Cambridge University Press.

Wood, B.J. and Fraser, D.G., 1977, *Elementary Thermodynamics for Geologists*: Oxford University Press.

4 Random Variables and Univariate Probability Distributions

4.1 *Mathematica* Packages You Will Need

Be sure to execute the following statement to ensure that you will have available all of the add-on and book-specific *Mathematica* functions used in this chapter.

```
In[1]:= Needs["Statistics`ContinuousDistributions`"]
        Needs["Statistics`DiscreteDistributions`"]
        Needs["Statistics`DescriptiveStatistics`"]
        Needs["Statistics`DataManipulation`"]
        Needs["Statistics`ConfidenceIntervals`"]
        Needs["Statistics`HypothesisTests`"]
        Needs["Statistics`NonlinearFit`"]
        Needs["Graphics`Graphics`"]
        Needs["CompGeosci`"]
```

Computer Note: The CompGeosci package will load correctly only if it is located in one of the directories in *Mathematica*'s standard file path. Execute the statement **`$Path`** to see a list of the default paths on your computer and place the file CompGeosci.m in one of those directories. The specific file paths may differ from one operating system to another. See Chapter 1 for more information about installing the CompGeosci package.

4.2 The Concept of Random Variables

A **random variable** is the outcome of an experiment, test, or process, that cannot be known in advance. Classic examples of random variables include the result of a coin flip or the sum of two fair dice rolls. Geologic properties such as porosity can also be viewed as random variables. It is impossible to predict or infer every detail in the sequence of erosion, sediment transport, deposition, compaction, diagenesis, and measurement imprecision that combine to produce an observed value of porosity at a certain point within a given rock. Therefore, although it may be theoretically possible to use basic physics to predict the result of a coin flip or complicated string of events that produces porosity in a rock, it is practically impossible to do so. A random variable can also be the product of several other random variables, for example the likelihood of a landslide that is the result of interaction between other random variables such as pore water pressure or soil shear strength.

The processes of interest to geologists are commonly so complicated that they can be treated as if they were the results of random processes. Treating a variable as if it were the result of a random process is not the same as arguing that geologic processes are random rather than mechanistic. Rather, it is a pragmatic concession that is made in order to obtain useful, affordable, and timely answers to pressing problems. It is, in essence, an admission that our knowledge of the world and computational capabilities are not yet advanced enough to correctly formulate and solve mathematical models of complicated geological processes. In the meantime, the concept of random variables gives us a tool with which to quantify geologic uncertainty and make useful predictions of the likelihood of events such as floods, earthquakes, and landslides.

Random variables can be either **continuous**, meaning that there are an infinite number of possible values, or **discrete**, meaning that there are a limited number of possible values. Porosity can be considered to be a continuous random variable because it can assume any value within the range $0 \le n \le 1$. The number of floods or earthquakes that likely to occur in an area over the next decade can be considered a discrete random variable because the result must be an integer.

Random variables are represented by **probability density functions (PDFs)** that give the likelihood that any given value of the variable will occur. The most widely known theoretical PDF is the bell-shaped curve of the normal or Gausssian distribution, although in many geologic problems it is the logarithms of the variables, rather than the variables themselves, that seem to be normally distributed. The peak of a normal distribution simply indicates that values selected at random from it are more likely to lie near the peak than the two extremes. There are, however, many other probability distributions useful to geologists and geological engineers. We'll sample a few of them in this chapter. An excellent and free compilation of probability distributions is *A Compendium of Common Probability Distributions*, which can be downloaded as a pdf file from www.causascientia.org/math_stat/Dists/Compendium.pdf. The probability that a random variable X is less than or equal to some value x, or Prob$\{X \le x\}$, is given by the **cumulative distribution functionor CDF**. The CDF is the integral of the PDF from its lower limit to x.

An important use of random variables is to develop **probabilistic** or **stochastic** models of geologic processes or events. For example, instead of stating with certainty that the porosity of a certain formation is 0.30 a geologist using a probabilistic model of porosity might state that there is a 75% probability that the average porosity of the formation is between 0.20 and 0.40. Or, he or she could say that the porosity of the formation follows a lognormal distribution with a certain mean and variance. The apparent randomness of geological variables such as porosity can be the result of spatial or temporal **variability**, meaning that the property does indeed vary in space or time, or **uncertainty** arising from measurement errors. The knowledge of variability and uncertainty encapsulated in probabilistic models can be incorporated into calculations that make use of porosity. A good example might be estimates of permeability that are essential for groundwater flow models, contaminant transport models, and petroleum reservoir simulation models. One way to perform probabilistic simulations, which will be explored in this chapter, is to use **Monte Carlo** methods in which a calculation or calculations are repeated using different values for each random variable. The result is thus a random variable that is represented by its own PDF.

4.3 Some Continuous Distributions

4.3.1 Normal Distribution

Normal distributions are symmetric and characterized by two parameters: the mean (μ), which reflects the location of the center of the distribution, and the standard deviation (σ), which reflects the dispersion or width of the distribution. In practice, those two Greek letters are commonly used to indicate the values of an underlying distribution from which random values are drawn, whereas $\bar{x}$ and s are commonly used to indicate estimates calculated from samples (such as field or experimental data). The PDF of a normal distribution for the random variable X is

```
In[2]:= PDF[NormalDistribution[μ, σ], X]
```

Out[2]= $\frac{e^{-\frac{(X-\mu)^2}{2\sigma^2}}}{\sqrt{2\pi}\,\sigma}$

and the *Mathematica* command to plot the PDF of a normal distribution with a mean of 0 and a standard deviation of 1, often referred to as the **standard normal distribution**, is

```
In[3]:= Plot[PDF[NormalDistribution[0, 1], X], {X, -5, 5},
          AxesLabel → {"X", "PDF(X)"}]
```

From In[3]:=

PDF(X)
0.4
0.3
0.2
0.1
-4 -2 2 4 X

Out[3]= -Graphics-

The probability that X will be less than or equal to x is given by the CDF, which is the integral of the PDF from its lower limit to x. For example, the probability of drawing a value ≤ 1 from the normal distribution plotted above is

```
In[4]:= ∫_{-∞}^{1.} PDF[NormalDistribution[0, 1], X] ⅆX
Out[4]= 0.841345
```

The decimal point after the upper limit of integration forces *Mathematica* to return numerical, rather than symbolic, output. It can sometimes be handy to have a symbolic expression rather than a numerical one. For example, it is easy to show that the integral of the PDF is indeed equal to the CDF evaluated at X.

```
In[5]:= ∫_{-∞}^{1} PDF[NormalDistribution[0, 1], X] ⅆX
```

Out[5]= $\frac{1}{2}\left(1 + \text{Erf}\left[\frac{1}{\sqrt{2}}\right]\right)$

```
In[6]:= CDF[NormalDistribution[0, 1], 1]
```

Out[6]= $\frac{1}{2}\left(1 + \text{Erf}\left[\frac{1}{\sqrt{2}}\right]\right)$

Here is a plot of the normal CDF:

```
In[7]:= Plot[CDF[NormalDistribution[0, 1], X], {X, -5, 5},
          AxesLabel → {"X", "CDF(X)"}]

From In[7]:=
```

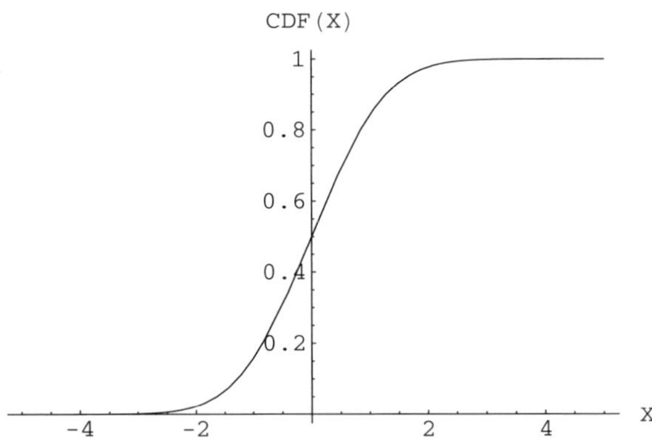


```
Out[7]= -Graphics-
```

The probability that a variable drawn from the same normal distribution will fall between, say, −0.7 and +1.4, can be found by subtracting CDFs.

```
In[8]:= CDF[NormalDistribution[0, 1], 1.4]-
          CDF[NormalDistribution[0, 1], -0.7]

Out[8]= 0.67728
```

The total area beneath **any** PDF equals 1, as shown for a normal distribution by

```
In[9]:= CDF[NormalDistribution[0, 1], ∞]

Out[9]= 1
```

It means that there is a 100% chance that any randomly selected value will fall between the minimum and maximum values of the distribution (in this case $-\infty$ and ∞).

4.3.2 Log-Normal Distribution

Log-normal distributions are those in which log X, rather than X itself, is normally distributed. As such, they range over $0 \leq \log X \leq \infty$. There are two ways to work with log-normal distributions in *Mathematica*. First, take the logarithms of the random variables and proceed to treat them as if they were normally distributed. Second, use the built-in log-normal distribution. Either way, remember that *Mathematica* uses natural base e logs by default and you'll need to specify the base if you want to use something else. The common logarithm of x, for example, would be obtained using **Log[10,x]**

The lognormal PDF is

```
In[10]:= PDF[LogNormalDistribution[μL, σL], X]
```

$$\text{Out[10]=} \frac{e^{-\frac{(\text{Log}[X]-\mu_L)^2}{2\sigma_L^2}}}{\sqrt{2\pi}\, X\, \sigma_L}$$

In this case, μ_L and σ_L are the mean and standard deviation of the logarithms, not the original arithmetic variables. A graphic example of a log-normal PDF is

```
In[11]:= Plot[PDF[LogNormalDistribution[3, 1], X], {X, 0, 100},
           AxesLabel → {"X", "PDF(X)"}]

From In[11]:=
```

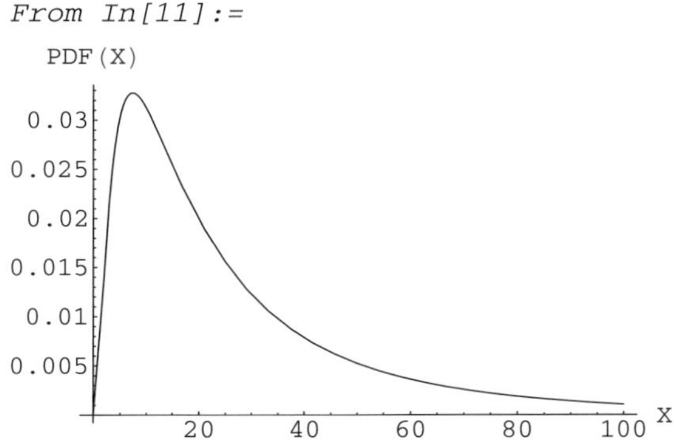


```
Out[11]= -Graphics-
```

4.3.3 Uniform Distribution

Uniform distributions are those, unlike distributions with central tendencies, in which all values of X have an equal probability of occuring. The uniform distribution returns a value that is what many people think of as a random number, such as that returned by the random number generators in spreadsheets and programming languages. It is defined by lower and upper limits. Its PDF is

```
In[12]:= PDF[UniformDistribution[minval, maxval], X]
```

```
          -Sign[-maxval + X] + Sign[-minval + X]
Out[12]= ----------------------------------------
                   2 (maxval - minval)
```

Sign[x] is a standard *Mathematica* function that takes on a value of −1, 0, or 1 depending on whether x is negative, zero, or positive. Below is a plot of the PDF for a uniform distribution ranging over $-3 \le X \le 4$.

```
In[13]:= Plot[PDF[UniformDistribution[-3, 4], X],
           {X, -5, 5}, AxesLabel → {"X", "PDF(X)"}]
```

```
From In[13]:=
```

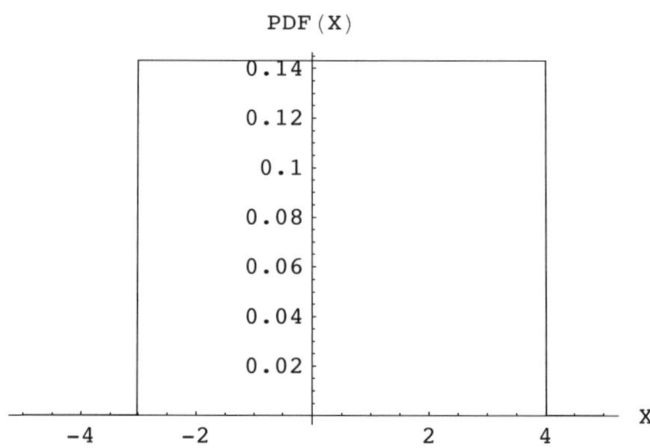


```
Out[13]= -Graphics-
```

Notice that the PDF is constant between the lower and upper limits and zero elsewhere. Just as with the other distributions, the area underneath the PDF is equal to 1.

Uniform distributions can be useful in cases where the available data are uniformly distributed or data are sparse and, although there might be reasonable minima and maxima, there is no compelling reason to assume that there is a central tendency. For example, you might have a good idea that the porosity of a certain formation or facies ranges from, say, 0.3 to 0.5 but don't have enough information to know whether there is a central tendency. The result of using a uniform distribution instead of one with a central tendency (e.g., a normal distribution) is to increase the uncertainty of any results calculated using that distribution.

4.3.4 Extreme Value Distribution

The extreme value distribution is of interest to geologists because many of the events that we study are extreme events, for example the largest floods or highest pore water pressures that are recorded each year. It turns out that even though data representing a process, say flooding, may follow one distribution, the extreme events from that distribution tend to follow a skewed distribution. The first parameter in the extreme value distribution specifies location, much like a mean value. The second parameter specifies the scale, similar to a standard deviation. The extreme value distribution PDF is:

```
In[14]:= PDF[ExtremeValueDistribution[α, β], X]
```

$$\text{Out[14]=}\ \frac{e^{-e^{\frac{-X+\alpha}{\beta}}+\frac{-X+\alpha}{\beta}}}{\beta}$$

A graphical example of an extreme value distribution PDF for $\alpha = 0$ and $\beta = 3$ is

```
In[15]:= Plot[PDF[ExtremeValueDistribution[0, 3], X],
           {X, -10, 20}, PlotRange → All,
           AxesLabel → {"X", "PDF(X)"}]
```

From In[15]:=

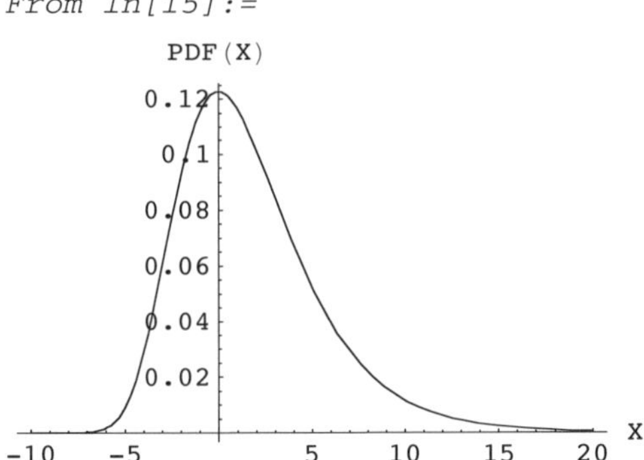

Out[15]= -Graphics-

4.3.5 Beta Distribution

The beta distribution is defined by either two or four parameters (depending on its implementation), and can take on a variety of shapes. In *Mathematica*, the beta distribution ranges from 0 to 1 and its shape is controlled by two parameters, P and Q.

```
In[16]:= PDF[BetaDistribution[P, Q], X]
```

$$\text{Out[16]=}\ \frac{(1-X)^{-1+Q}\,X^{-1+P}}{\text{Beta}[P, Q]}$$

in which Beta[P, Q] is the Euler beta function. It's built into *Mathematica*, so there's no need to take any extra steps to calculate it.

The beta distribution PDF is symmetric when P and Q are equal, with the magnitude of both controlling the peakedness of the PDF. Unequal values of P and Q cause the PDF to be skewed to the right or left. Here is a plot of the beta distribution PDF for P = 7.5 and Q = 3.1. Substitute your own P and Q values and replot the PDF to see what shape it will take (P and Q **must** be positive or you will receive an error message!).

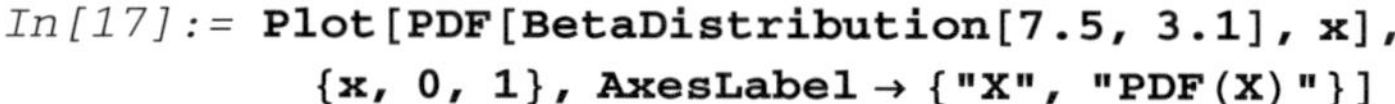

```
In[17]:= Plot[PDF[BetaDistribution[7.5, 3.1], x],
           {x, 0, 1}, AxesLabel → {"X", "PDF(X)"}]
```

From In[17]:=

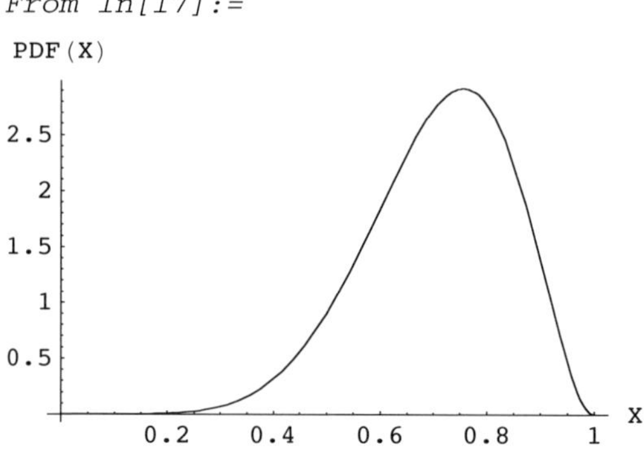

Out[17]= -Graphics-

It is easy to visualize the effect of changing P and Q by creating an array of PDFs in which the two are systematically varied. The process is to first generate a *Mathematica* data structure called a graphics array, which is simply an array or matrix in which each of the elements is a graphic instead of a number or symbol. All of the options control the details of the plots. For example, **PlotRange** is used to ensure that all of the vertical axes will have the same range (otherwise, *Mathematica* would scale each plot individually). **DisplayFunction → Identity** suppresses output of the individual graphs as they are created (you can remove this option to see what happens otherwise). After the GraphicsArray is filled, it is displayed using the Show command with the option **DisplayFunction → $DisplayFunction**.

```
In[18]:= GraphicsArray[
           Table[Plot[BetaDistribution[P, Q], x], {x, 0, 1},
             PlotRange → {{-0.05, 1.05}, {-0.5, 10.5}},
             Frame → True, FrameTickes → {{0, 1}, {0, 5, 10}, {}, {}},
             Epilog → {Text["P = ", {0.6, 9}], Text[P, {0.8, 6}],
               Text["Q = ", {0.6, 6}], Text[Q, {0.8, 6}]},
             DisplayFunction → Identity], {P, 1, 10, 2},
               {Q, 1, 10, 2}]];
         Show[%, DisplayFunction → $DisplayFunction]
```

From In[18]:=

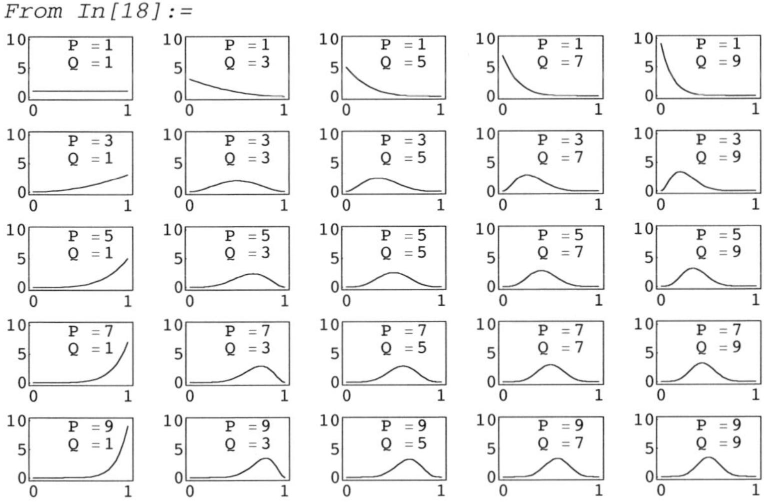


```
Out[18]= -GraphicsArray-
```

4.3.6 Pareto Distribution

The Pareto distribution is useful when there are many small values and an exponentially decreasing number of larger values. It is often used to describe the distribution of incomes, wind speeds, and the distribution of rock fragment sizes such as cataclastic grain size distributions. It is defined by a minimum value $k > 0$ and a shape parameter, $\alpha > 0$. Its PDF is

```
In[19]:= PDF[ParetoDistribution[k, α], X]
```

Out[19]= $k^{\alpha}\, X^{-1-\alpha}\, \alpha$

Here is a plot of the Pareto distribution PDF for $k = 3$ and $\alpha = 5$:

```
In[20]:= Plot[PDF[Paretodistribution[3., 5.], X], {X, 3, 10},
            PlotRange → All, AxesLabel → {"X", "PDF(X)"}]
```

From In[20]:=

PDF (X)

1.5
1.25
1
0.75
0.5
0.25

4 5 6 7 8 9 10 X

```
Out[20]= -Graphics-
```

Should you ever find yourself working with *Mathematica*'s built-in Pareto distribution, be careful not to plot or evaluate it for values of $X < k$. You'll calculate results, but they're nonsensical because the distribution is defined only for $X \geq k$. That's why the plot above starts at $X = k = 3$.

4.4 Some Discrete Distributions

4.4.1 Poisson Distribution

The Poisson distribution is a discrete distribution, characterized by a mean value μ, that can be used to simulate the number of times an event is likely to occur over a given time span. In geologic applications, the events might be earthquakes or landslides. An important limitation of the Poisson distribution, however, is that the events being simulated are mutually independent. As implemented in *Mathematica*, the Poisson distribution is the function of only one parameter, a so-called average. In geologic applications you are more likely, however, to see it formulated in terms of an **average recurrence interval**, r, and an **increment of time**, t.

```
In[21]:= PDF[PoissonDistribution[t/r], n]
```

$$Out[21]= \frac{e^{-\frac{t}{r}} \left(\frac{t}{r}\right)^n}{n!}$$

The probability that two or fewer (n =2) events with an averge recurrence interval of r = 100 years will occur in any given century (t = 100 years) is

```
In[22]:= CDF[PoissonDistribution[100/100.], 2]

Out[22]= 0.919699
```

Thus, there is a 0.92 probability that 0, 1, or 2 events will occur over a century. The probability of exactly two events occuring is given by the difference between the probability of two or fewer events and the probability of one or no events. To wit,

```
In[23]:= CDF[PoissonDistribution[100/100.], 2] -
           CDF[PoissonDistribution[100/100.], 1]

Out[23]= 0.18394
```

which is, for a discrete probability distribution, the same as

```
In[24]:= PDF[PoissonDistribution[100/100.], 2]

Out[24]= 0.18394
```

Note that this equivalency **does not** apply to continuous distributions. Because continuous random variables can take on an infinite number of values, the PDF cannot be used to calculate the probability at a point. Instead, the probability that X falls between two bounds must be calculated.

The Poisson PDF cannot be graphed using `Plot[]` because it is a series of discrete values, between which it is undefined, rather than a continuous function.

But, it can be plotted by first creating a table of discrete values and then using **ListPlot[]**. Here is a plot of Poisson probabilities for various values of n using the same $r = t = 100$ values as above.

```
In[25]:= ListPlot[
           Table[{n, PDF[PoissonDistribution[100/100.],n]},
             {n, 0, 5}], PlotRange → All,
           PlotStyle → PointSize[0.02],
           AxesLabel → {"X", "PDF(X)"}
         ]
```

From In[25]:=

PDF (X)
0.35
0.3
0.25
0.2
0.15
0.1
0.05
1 2 3 4 5 X

Out[25]= -Graphics-

Similarly, the CDF of a discrete distribution such as the Poisson distribution is a step function rather than a smooth curve.

```
In[26]:= Plot[CDF[PoissonDistribution[100/100.],n], {n, 0, 5},
           PlotRange → {0, 1}, AxesLabel → {"X", "CDF(X)"}]
```

From In[26]:=

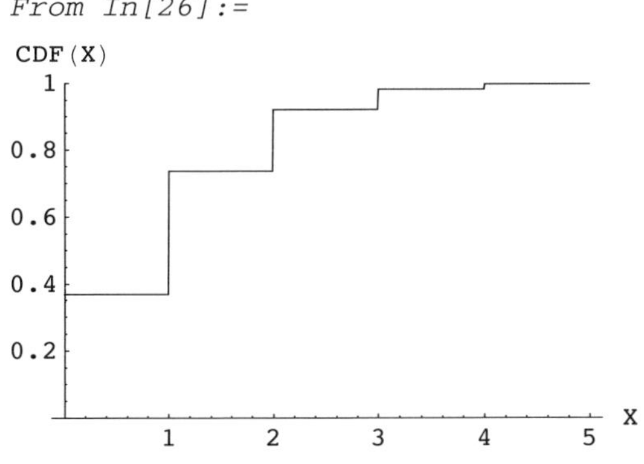

Out[26]= -Graphics-

4.4.2 Binomial Distribution

Another discrete distribution that is similar to the Poisson distribution is the binomial distribution, which gives the probability of n occurrences in t trials, each of which has a probability p.

```
In[27]:= PDF[BinomialDistribution[t,p],n]

Out[27]= (1 - p)^(-n+t) p^n Binomial[t, n]
```

in which **Binomial[t,n]** is a standard *Mathematica* function for the binomial coefficient. Drawing a parallel with the Poisson example above, we can recast that problem in terms of the probability of n occurrences in t years, each of which has a probability $p = 1/r$ where r is the average recurrence interval. To wit, the probability of exactly $n = 2$ floods with a probability of $p = 0.01$ (i.e., 100 year recurrence interval) occurring over $t = 100$ years is

```
In[28]:= PDF[BinomialDistribution[100, 0.01], 2]

Out[28]= 0.184865
```

The result is very close to that predicted by the Poisson distribution. What is the difference between the two? Crovelli (2000) shows that the binomial distribution is actually a discrete time approximation to the continuous time Poisson distribution. Note that we're using discrete and continuous in a different sense here. Both are discrete probability distributions in the sense that the random variable, n, is discrete in both whereas the way that t is treated differs between the two. Crovelli shows that using the binomial distribution introduces noticable errors for events with short recurrence intervals over short times.

4.5 Relating Distributions to Data: Method of Moments

If probabilistic modeling is to be of much use in the real world, there has to be a way to estimate distribution parameters such as means and standard deviations from field or laboratory data. The **method of moments,** proposed by the statistician Carl Pearson in the 1920s, is based on the premise that the moments (mean, variance, skewness, kurtosis) of a sample are reasonably good estimates of those describing the distribution from which they were drawn. To illustrate the method of moments, we will first define a list of values that follow a normal distribution (the generation of random values from probability distributions will be covered further on in this chapter). The values are:

```
In[29]:= data = {0.091, -0.143, -0.655, -1.69, 0.49, 0.678, 1.55,
           -0.353, -0.0644, -0.141};
```

The distribution of the values can be visualized using a simple histogram, in this case scaled so that the area of the bars sums to 1. This will allow the histogram to be plotted on the same axes as a PDF of the distribution from which the values were drawn.

```
In[30]:= plot1 = Histogram[data, HistogramScale → 1,
           BarStyle → GrayLevel[0.6]]
```

From In[30]:=

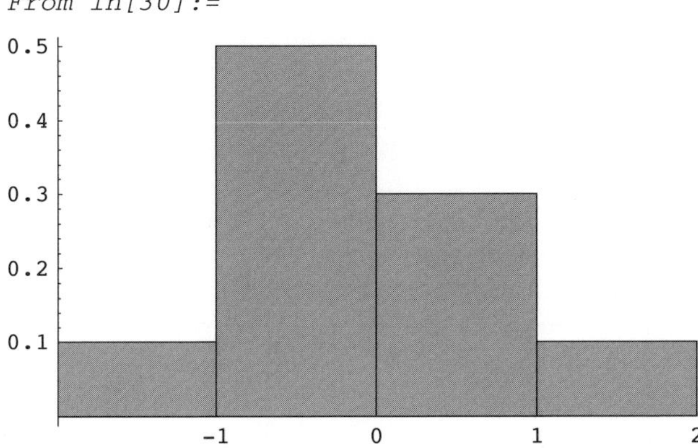


```
Out[30]= -Graphics-
```

The operational definition of the mean of N values is $\overline{x} = \frac{1}{N}\sum_{i=1}^{N} x_i$, and can be calculated using the standard *Mathematica* function **Mean**.

```
In[31]:= meanval = Mean[data]

Out[31]= -0.02374
```

The standard deviation is the square root of the variance, which is $\sigma^2 = \frac{1}{N-1}$ $\sum_{i=1}^{N}(x_i - \overline{x})^2$, and is likewise most easily calculated using the built in functions **Variance** or **StandardDeviation**

```
In[32]:= Variance[data]

Out[32]= 0.728793

In[33]:= dev = StandardDeviation[data]

Out[33]= 0.853694
```

The variance and standard deviation calculated above are sometimes referred to as the **sample variance** and **sample standard deviation** because they are based upon an incomplete sample of the population. The sum of squared deviations is divided by $N - 1$ rather than N, which has the effect of increasing the variance by a small amount in order to reflect the uncertainty associated with the use of an estimated value. If the data represent the entire population, then the **population variance** and **population standard deviation** are calculated by

```
In[34]:= VarianceMLE[data]

Out[34]= 0.655914
```

```
In[35]:= StandardDeviationMLE[data]

Out[35]= 0.809885
```

For large sample sizes, say tens of numbers or more, the difference is generally small enough to ingore. For small sample sizes, such as the 10 values in **data**, the difference can be more noticable.

Mathematica also includes the functions **SampleRange**, **GeometricMean**, **HarmonicMean**, **Median**, **Skewness**, **Kurtosis**, **LocationReport**, **DispersionReport**, and other descriptive statistics that we will not use. They are all, however, described in the paper and electronic documentation accompanying the program.

What about other kinds of distributions? For example, say that we have a data set describing the maximum annual flood discharge from a stream gauging station and would like to fit an extreme value distribution to the data. There are no built-in functions to calculate the α and β parameters that define the extreme value distribution PDF, but it is easy to calculate the sample mean and standard deviation of the data set. These two statistics can in turn be related to the parameters of interest using *Mathematica*'s symbolic manipulation capabilities. The symbolic expression for the mean value of an extreme value distribution with parameters α and β, for example, is

```
In[36]:= Mean[ExtremeValueDistribution[α, β]]

Out[36]= α + EulerGamma β
```

Therefore, if the sample mean and standard deviation are known it is a simple matter to set the symbolic expression for the mean equal to the sample mean and the symbolic expression for the standard deviation equal to the sample standard deviation, forming two equations in two variables.

```
In[37]:= eq1 = 0. == Mean[ExtremeValueDistribution[α, β]]

Out[37]= 0. == α + EulerGamma β

In[38]:= eq2 = 1.
            == StandardDeviation[ExtremeValueDistribution[α, β]]
```

Out[38]= $1. == \frac{\pi\beta}{\sqrt{6}}$

Solving the equations for α and β yields two numerical values.

```
In[39]:= Solve[{eq1, eq2}, {α, β}]

Out[39]= {{α → -0.450053, β → 0.779697}}
```

Below is a plot of the extreme value distribution PDF along with the normal distribution PDF using the mean and standard deviation of 0 and 1.

```
In[40]:= p1 = Plot[PDF[ExtremeValueDistribution[α, β], x]
            /.%[[1]], {x, -5, 5}, PlotRange → All,
            DisplayFunction → Indentity];
         p2 = Plot[PDF[NormalDistribution[0, 1], x], {x, -5, 5},
            PlotRange → All, PlotStyle → Dashing[{0.01}],
            DisplayFunction → Indentity];
         Show[p1, p2, DisplayFunction → $DisplayFunction]
```

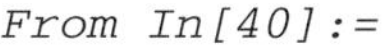
From In[40]:=

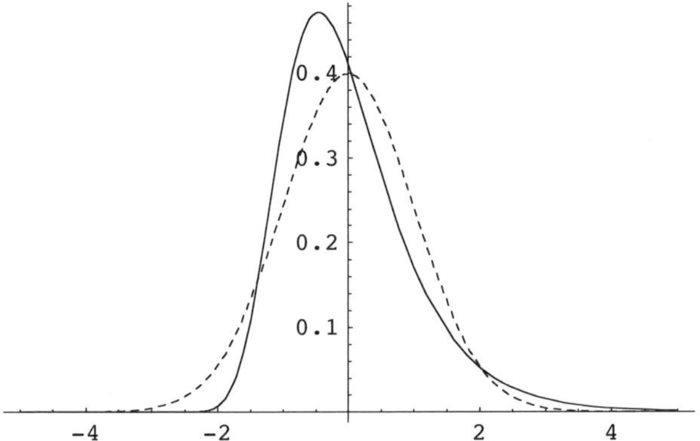

Out[40]= -Graphics-

There are some limitations to this method. For example, the beta distribution built into *Mathematica* has a range of $0 \le x \le 1$ and it will be impossible to calculate valid P and Q parameters from data sets that fall outside of that range. Data with a sample mean and standard deviation of 0 and 1, as used in the previous example, will yield $P = 0$ and $Q = -1$. Because P and Q must both be positive in the *Mathematica* implementation of the beta distribution, these results are nonsensical and any attempts to use them will produce an error. There are, however, two solutions to this problem. First, it is possible to write a custom *Mathematica* function for the beta distribution that ranges between user-specified minimum and maximum values. The necessary PDF equation is available in many probability and statistics textbooks. Second, we can shift the mean and re-scale the data set so that ranges from 0 to 1.

Similar problems arise if the **`Solve`** function is used in an attempt to calculate values of μ and σ for a lognormal distribution from an arithmetic mean and variance. The easiest solution to this problem is to calculate the mean and standard deviation of the logarithms of the data.

4.5.1 How Good Are Those Estimates?

How good is the fit between **`data`** and a normal distribution with its sample mean and standard deviation? Or, between the derived distributions and the underlying distribution? One way to examine the fit is graphically, by plotting a normal distribution PDF using the mean and standard deviation found using the method of moments, then superimpose it with the histogram of the random values and the normal distribution from which the values were drawn.

Here is a PDF generated using the method of moments mean and standard deviation:

```
In[41]:= plot2 = Plot[PDF[NormalDistribution[meanval, dev], x],
           {x, -4, 4}, PlotRange → All,
           PlotStyle → {Thickness[0.006], Dashing[{0.02}]}]
```

From In[41]:=

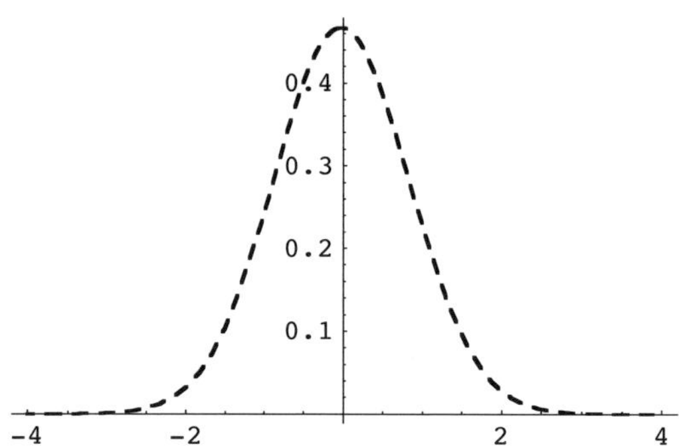

Out[41]= -Graphics-

And here is the underlying normal distribution:

```
In[42]:= plot3 = Plot[PDF[NormalDistribution[0, 1], x],
           {x, -4, 4}, PlotRange → All,
           PlotStyle → {Thickness[0.005]}]
```

From In[42]:=

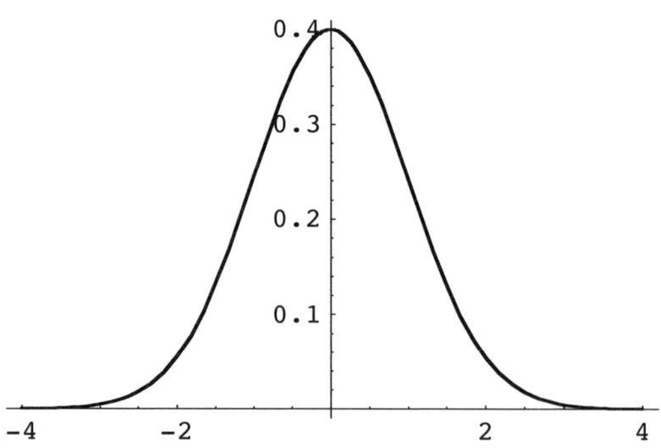

Out[42]= -Graphics-

Combining all three plots into one,

```
In[43]:= Show[plot1, plot2, plot3, PlotRange→{{-4, 4}, {0, 0.5}},
           Ticks → {{-4, -2, 0, 2, 4}, Automatic},
           AxesOrigin → {-4, 0}]
```

```
From In[43]:=
```

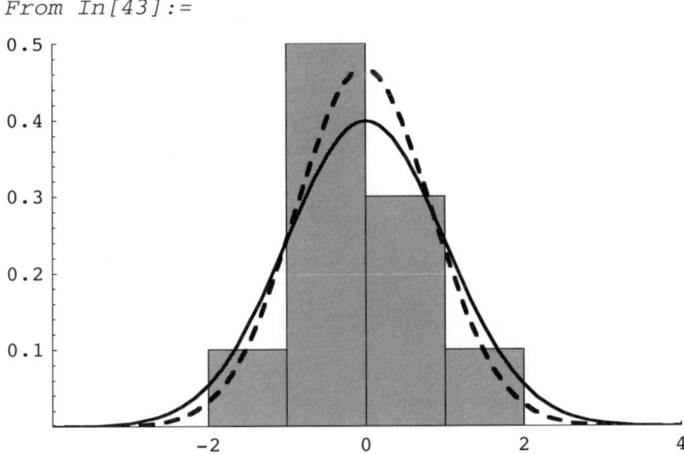


```
Out[43]= -Graphics-
```

Notice that the histogram was specified first in the **Show** command so that it is in the background and doesn't obscure the PDFs. Try reversing the order to see what happens otherwise.

4.6 Parametric Hypothesis Testing: *t* and *F* tests

Having obtained a set of measurements in the laboratory or from the field, and perhaps having calculated some summary statistics such as the sample mean and variance, the next step may be to ask whether the data are likely to have been drawn from a population with some specific mean or variance. Another question might be whether one data set is likely to have been drawn from a population with a mean or variance different from that for a second data set. Performing a simple visual calculation as described above is the first step. Beyond that, there are several techniques for testing hypotheses about samples that are believed to have been drawn from underlying normal distributions. The tests described below are limited to normally distributed data, although the Central Limit Theorem (described in a subsequent section) suggests that they can be used on unimodal data sets that are not highly skewed and that are composed of large numbers of samples. Tests that are based upon an assumption of the underlying distribution are known as **parametric** tests.

Below is a series of soil lead measurements (in ppm) from a 100 x 100 foot survey square at a heavily contaminated smelter site, along with a histogram and its calculated sample mean and sample variance.

```
In[44]:= PbData = {1300., 1200., 500., 3500., 4200., 1700.,
           3300., 2000., 800., 1000., 1300., 2000., 2400.}

Out[44]= {1300., 1200., 500., 3500., 4200., 1700., 3300.,
           2000., 800., 1000., 1300., 2000., 2400.}
```

```
In[45]:= Histogram[PbData, AxesLabel→{"Pb\n(ppm)", "Frequency"},
          BarStyle → GrayLevel[0.6]]
```

From In[45]:=

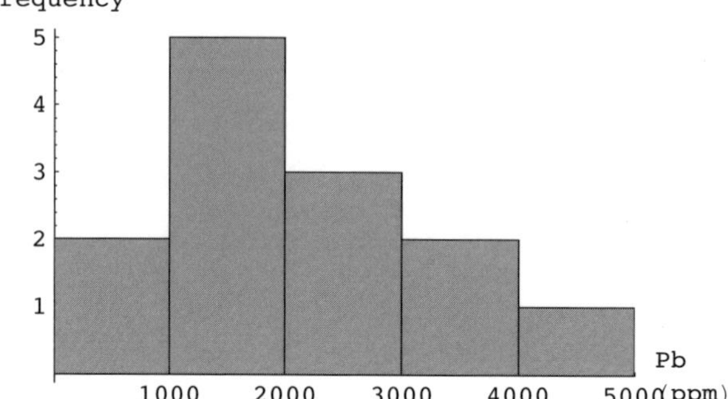


```
Out[45]= -Graphics-

In[46]:= PbMean = Mean[PbData]

Out[46]= 1938.46

In[47]:= PbVariance = Variance[PbData]
```

Out[47]= 1.27423×10^6

Although the data are skewed, they do have a strong central tendency and the sample size is small. Therefore, it is not unreasonable to conclude that they may follow something reasonably close to a normal distribution. We'll use this data set to show how hypotheses about the mean and variance of the population from which the data were drawn can be tested.

4.6.1 The *t* Statistic

The question that we seek to answer is whether the soil lead data were drawn from a normal distribution with a population mean of 1938 ppm and an unknown population variance. Like all statistical tests, the question is framed in terms of the **null hypothesis** that there is no difference between the underlying population mean and the calculated sample mean. The alternative hypothesis is that the population mean is different than the calculated sample mean. The null hypothesis is evaluated by calculating the *t* statistic, which takes into account the calculated sample mean, the postulated population mean, the calculated sample variance, and the number of data used to estimate the mean and variance. Using the sample statistics and the postulated population mean of 1938 ppm, the *t* statistic for this example is

In[48]:= $\texttt{tvalue} = \dfrac{\texttt{PbMean - 1938.}}{\sqrt{\texttt{PbVariance/Length[PbData]}}}$

```
Out[48]= 0.0014742
```

A slightly different test is used if the underlying population variance is known and does not have to be calculated from the data. This situation, however, rarely occurs in practical applications. Consult the *Mathematica* documentation for more details.

The t statistic follows the Student's t distribution, which is similar to a normal distribution. Its exact shape, however, depends on the number of degrees of freedom (typically one less than the number of samples). The composite plot below shows Student's t distribution PDFs for 1 (short dashes), 5 (long dashes), and 10 (solid line) degrees of freedom. Student's t distribution is indistinguishable from the normal distribution for large numbers of samples, which is typically taken to mean 30 or so.

```
In[49]:= p1 = Plot[PDF[StudentTDistribution[1],x], {x, -8, 8},
           PlotStyle → Dashing[{0.01}],
           DisplayFunction → Identity];
         p2 = Plot[PDF[StudentTDistribution[5],x], {x, -8, 8},
           PlotStyle → Dashing[{0.02}],
           DisplayFunction → Identity];
         p3 = Plot[PDF[StudentTDistribution[10],x], {x, -8, 8},
           DisplayFunction → Identity];
         Show[p1, p2, p3, DisplayFunction → $DisplayFunction,
           AxesLabel → {"DOF", "PDF"}]

From In[49]:=
```

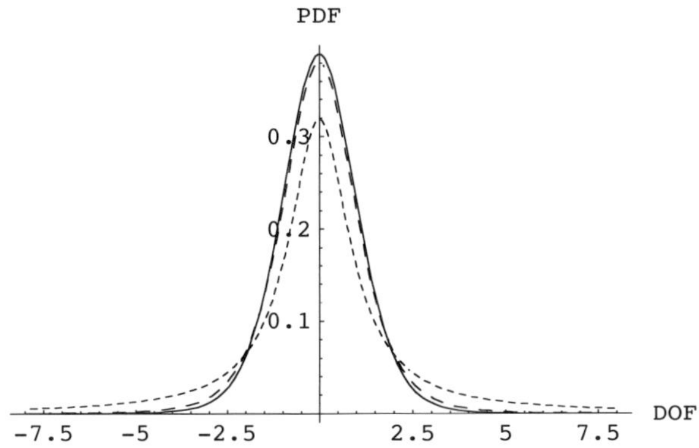


```
Out[49]= -Graphics-
```

4.6.2 Critical t Values

What does the t value mean with regard to the null hypothesis? The traditional way to determine the significance of t is to find a statistics manual and look up the critical value of Student's t distribution for a specified **level of significance**, say 0.05, and a specified number of degrees of freedom. The level of significance is the probability of committing a **Type I error**, meaning that the null hypothesis is rejected even though it is correct. Thus, specifying a significance level of 0.05 means that we are willing to accept a 1/20 chance of incorrectly rejecting the null hypothesis. A Type I error can only occur in situations where the null hypothesis is rejected. It is possible to make a **Type II** error, meaning that the null hypothesis is incorrect even though it is accepted, but it is generally not possible to calculate the probability of committing a Type II error. The null hypothesis is rejected if the calculated t value exceeds the tabulated critical value of t. In this case, the degrees of freedom would be one less than the number of data used in the calculations (in order to account for the fact that the variance had to be estimated from the data). The critical t value for $13 - 1 = 12$ degrees of freedom and an 0.05 level of significance is 1.782, which is much larger than the calculated $t = 0.0014742$. Therefore, the null hypothesis cannot be rejected in this case.

We can accomplish the same thing using a series of steps in *Mathematica*. First, the we calculate the probability of obtaining a t value smaller than or equal to $t = 0.0014742$ from a Student's t distribution with 12 degrees of freedom. Because the calculated sample mean is larger than the postulated population mean, the probability is

```
In[50]:= 1 - CDF[StudentTDistribution[12], tvalue]

Out[50]= 0.499424
```

This result is known as a one-sided P-value because it gives only the probability of obtaining a t value less than or equal to our calculated t. Had the sample mean been smaller than the population mean, the t statistic would have been negative and the P-value would have been given by **`CDF[StudentTDistribution[12], tvalue]`**. Next, we need to determine the critical t value against which the calculated t value is to be compared. Because we have selected a 0.05 level of significance, this will be the value of t for which the CDF is $1 - 0.05 = 0.95$. It is, unfortunately, not possible to do this using *Mathematica*'s **`Solve`** function. One option is to plot the Student's t CDF and visually interpolate the critical value.

```
In[51]:= Graphics[{Dashing[{0.01}],
           Line[{{0, 0.95}, {1.78, 0.95}, {1.78, 0}}]}];

         Plot[CDF[StudentTDistribution[12], t], {t, -5, 5},
           DisplayFunction → Identity,
           AxesLabel → {"t", "CDF(t)"}];

         Show[%, %%, Displayfunction → $DisplayFunction]
```

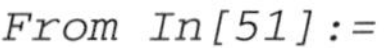

```
From In[51]:=
```

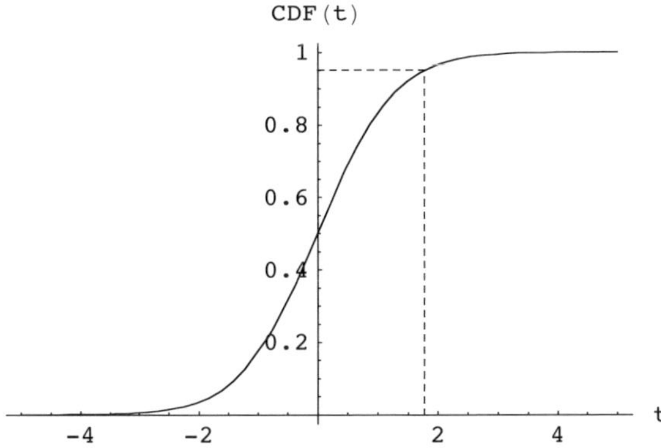


```
Out[51]= -Graphics-
```

Reading across from 0.95 on the vertical axis to the CDF curve and then down to the horizontal axis, it is easy to see that the critical *t* value must be about 1.79.

At this point it may seem easier to look up a critical value in a book rather than plotting and interpolating by eye. The same results, though, can be obtained in one step using the function **MeanTest**, which minimally takes as its arguments the data set and the mean against which the data are to be tested. **MeanTest** returns a one-sided P-value by default.

```
In[52]:= MeanTest[PbData, 1938]

Out[52]= OneSidedPValue → 0.499424
```

The option **FullReport→True** will force **MeanTest** to return the calculated sample mean, the calculated *t* statistic, and the distribution used to calculate the P-value. The **SignificanceLevel** option will return an an answer with regard to the null hypothesis, as shown below.

```
In[53]:= MeanTest[PbData, 1938, FullReport → True,
           SignificanceLevel → 0.05]

Out[53]=
    {FullReport → Mean      TestStat   Distribution
                  1938.46   0.0014742  StudentTDistribution[12]',
    OneSidedPValue → 0.499424,
    Fail to reject null hypothesis at significance level → 0.05}
```

Another use of **MeanTest** is to determine whether the mean value of a data set is significantly different than some critical threshold. For example, the action level that triggered remediation of the smelter site from which the data set came was a mean lead concentration of 500 ppm. What is the probability that the lead data were drawn from a normal distribution with a mean of 500 ppm?

```
In[54]:= MeanTest[PbData, 500., FullReport → True,
           SignificanceLevel → 0.05]

Out[54]=
    {FullReport → Mean     TestStat Distribution
                  1938.46  4.59458  StudentTDistribution[12] ,
    OneSidedPValue → 0.00030834,
    Reject null hypothesis at significance level → 0.05}
```

The answer is virtually none.

4.6.3 Comparing Two Means or Variances

`MeanDifferenceTest` is a variation that can be used to test whether the means of the populations from which two samples were drawn are equal. To illustrate, consider a second set of soil lead values obtained from the same survey square by a different group of geologists:

```
In[55]:= PbData2 = {2700., 1900., 1800., 1900., 1300., 1700.,
           2000., 1700., 1700., 1900., 2000., 2000., 1000.}
Out[55]= {2700., 1900., 1800., 1900., 1300., 1700., 2000.,
           1700., 1700., 1900., 2000., 2000., 1000.}
```

Before conducting a mean difference test, we can plot the PDFs of normal distributions calculated using the sample means and variances from each data set.

```
In[56]:= Plot[PDF[NormalDistribution[Mean[PbData],
             StandardDeviation[PbData]], x], {x, -5000, 5000.},
           PlotRange → All, DisplayFunction → Indentity];
         Plot[PDF[NormalDistribution[Mean[PbData2],
             StandardDeviation[PbData2]], x], {x, -5000, 5000.},
           PlotRange → All, DisplayFunction → Indentity];
         Show[%, %%, DisplayFunction → $DisplayFunction]

From In[56]:=
```

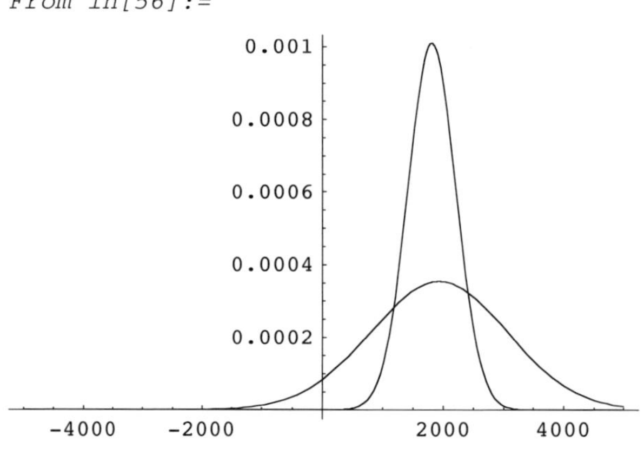


```
Out[56]= -Graphics-
```

The distribution inferred from first data set has considerably more variability than that inferred from the second set, although their mean values do not appear to be significantly different. The plot above also illustrates one of the drawbacks to using normal distributions: they can include negative values that may be physically meaningless. Although it doesn't do anything more than suggest that a normal distribution may not be an appropriate one in this case, thornier problems can arise when physically unrealistic values are generated in more complicated simulations. This issue will be addressed in Chapter 5. The mean values can be rigorously compared using the function **MeanDifferenceTest** from Statistics `HypothesisTests`, as shown below.

```
In[57]:= MeanDifferenceTest[PbData, PbData2, 0.,
           FullReport- > True, SignificanceLevel → 0.05]

Out[57]=
    {FullReport →
      MeanDiff  TestStat  Distribution
      123.077   0.371008  StudentTDistribution[14.9022] ,
    OneSidedPValue → 0.357926,
    Fail to reject null hypothesis at significance level → 0.05}
```

The two variances can be compared using an analogous test, often referred to in statistics texts as an *F* test because ratios of variances follow what is known as an *F* distribution. Thus, it is the ratio of variances rather than their difference that is tested. The null hypothesis is that the two samples were drawn from populations with the same variances, which have a ratio of 1.

```
In[58]:= VarianceRatioTest[PbData, PbData2, 1.,
           FullReport- > True, SignificanceLevel → 0.05]

Out[58]=
                     Ratio     TestStat Distribution
    {FullReport →    8.14672   8.14672  FRatioDistribution[12, 12] ,
      OneSidedPValue → 0.00048347,
      Reject null hypothesis at significance level → 0.05}
```

As suggested by the preliminary plot of PDFs, the variances are different enough that the null hypothesis must be rejected. What is the explanation for our conclusions that there is no significant difference between population means but that there **is** a significant difference between the population variances? One possible explanation lies in the techniques used by the two groups of geologists. The group that produced the first data set used a portable x-ray fluoresence unit to obtain lead concentrations in the field whereas the second group took soil samples to the laboratory for atomic absorption analysis. Another possibility is that we have committed a Type I error by incorrectly rejecting the null hypothesis, although this is unlikely because the results do not change even with **SignificanceLevel→0.001**. Thus, there is a less than 1/1000 chance that a Type I error has occurred. Before drawing conclusions about the precision of the methods used by the two groups of geologists, however, consider the fact that there was no significant difference in variances for data sets from two other survey squares at the same site.

4.7 Nonparametric Hypothesis Testing: K-S Tests

All of the tests above assumed that there was a normal distribution, or something close to it, behind the set of randomly selected values. What happens if the data are clearly not normally distributed? One way to compare a sample to a hypothesized underlying distribution or to compare two sample distributions is to use a Kolmogorov-Smirnov (K-S) test, which is a measurement of the maximum different between two cumulative distribution curves. K-S tests and plots aren't included in *Mathematica*'s standard packages, but there are several K-S routines in the *Mathematica* package that accompanies this book. **KSOneList** calculates the K-S statistic between a list of data and a normal distribution having the sample mean and variance. **KSTwoList** calculates the K-S statistic between two data lists. **KSProb** calculates the K-S probability, which is equivalent to the significance level used in the previous section, from a K-S statistic and the number of data points. Two additional functions, **KSOneListPlot** and **KSTwoListPlot**, produce cumulative frequency plots corresponding to each of the two K-S statistic functions. Both of the two-list functions require that both lists be of equal length, although this restriction was incorporated to make programming simpler and is by no means necessary. The K-S functions also make use of the function **CumFreqs**, which calculates the cumulative frequencies of a list of data. **CumFreqPlot** produces a cumulative frequency plot of a data set. Each of these functions is demonstrated below.

Here is a cumulative frequency plot of PbData. The last two arguments of the function are the minimum and maximum values of the plot.

```
In[59]:= CumFreqPlot[PbData, 0, 5000,
            AxesLabel → {"Pb", "Cum\nFreq"}]
```

```
From In[59]:=
```

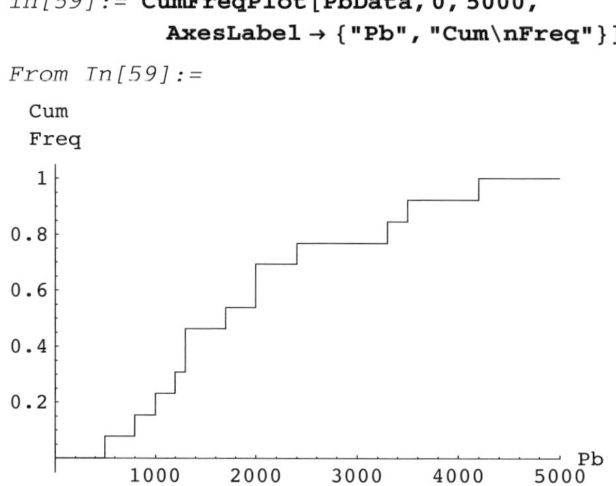


```
Out[59]= -Graphics-
```

The cumulative frequencies are:

```
In[60]:= CumFreqs[PbData]

Out[60]= {{500., 0.0769231}, {800., 0.153846}, {1000., 0.230769},
          {1200., 0.307692}, {1300., 0.384615}, {1300., 0.461538},
          {1700., 0.538462}, {2000., 0.615385}, {2000., 0.692308},
          {2400., 0.769231}, {3300., 0.846154}, {3500., 0.923077},
          {4200., 1.}}
```

The degree to which PbData is represented by a normal distribution can be evaluated graphically using **KSOneListPlot**

```
In[61]:= KSOneListPlot[PbData, 0, 5000,
           AxesLabel → {"Pb", "Cum\nFreq"}]

From In[61]:=
```



```
Out[61]= -Graphics-
```

or quantitatively using **KSOneList**

```
In[62]:= KSOneList[PbData]

Out[62]= 0.175706
```

The calculated K-S value has the probability

```
In[63]:= KSProb[KSOneList[PbData], Length[PbData]]

Out[63]= 0.776422
```

Therefore, we are justified in concluding that the lead data are **not** normally distributed only if we are willing to take a 78% chance of committing a Type I error.

Are the lead data better represented by a lognormal distribution? The easiest way to evaluate the possibility is to plot the logarithms of the data (noting that the minimum and maximum values must now be given as logarithms and that taking the logarithm of zero produces an error)

```
In[64]:= KSOneListPlot[Log[PbData],Log[300],
           Log[5000],AxesLabel → {"Pb", "Cum\nFreq"}]
```

From In[64]:=

Cum
Freq

1
0.8
0.6
0.4
0.2
6.5 7 7.5 8 8.5 Pb

Out[64]= -Graphics-

```
In[65]:= KSOneList[Log[PbData]]
```

Out[65]= 0.110146

```
In[66]:= KSProb[%, Length[PbData]]
```

Out[66]= 0.995512

The plot, K-S statistic, and K-S probability all suggest that the lead data are better represented by a lognormal than a normal distribution. The likelihood of committing a Type I error by rejecting the null hypothesis that there is no difference between the empirical distribution and the lognormal distribution is > 99%.

Similarly, the two lead data sets can be compared using K-S plots and statistics.

```
In[67]:= KSTwoListPlot[Log[PbData],Log[PbData2],
           Log[400],Log[5000],AxesLabel → {"Pb", "Cum\nFreq"}]
```

From In[67]:=

Cum
Freq

1
0.8
0.6
0.4
0.2
6.5 7 7.5 8 8.5 Pb

Out[67]= -Graphics-

KSTwoList calculates the K-S statistic by dividing the range between the minimum and maximum values in the data sets by a large user-specified number, which is 100 in the example above. This number should be larger than the number of data points in the two lists being compared.

```
In[68]:= KSTwoList[Log[PbData], Log[PbData2], 100]

Out[68]= 0.307692

In[69]:= KSProb[%, 13]

Out[69]= 0.138274
```

The greatest separation between the two cumulative frequency plots occurs between values of 1300 and 1700 ppm and is, as calculated by the K-S function, approximately 0.31. There is a 14% chance of committing a Type I error if we reject the null hypothesis that the two empirical distributions are the same.

4.8 Generating Random Numbers from Probability Distributions

One of the keys to probabilistic simulations of geologic processes is the ability to randomly sample probability distributions that are inferred to represent field data, laboratory data, or some empirically observed process. For example, if there is evidence to suggest that a variable is normally distributed, then it will be very useful to be able to select tens, hundreds, or perhaps even thousands of values at random from an underlying normal distribution with a mean and variance identical to the observed data. Think of the process as a sophisticated version of random number generator that you may have used in a spreadsheet program. Instead of selecting random numbers between 0 and 1, each with an equal likelihood of being generated, the likelihood of a given value being generated depends on the underlying probability distribution.

Say that we want to draw a samples at random from a standard normal distribution with zero mean and unit standard deviation. This is accomplished in *Mathematica* by typing:

```
In[70]:= Random[NormalDistribution[0, 1]]

Out[70]= -0.752627
```

It may not be terribly useful to generate just one value, but it might be very useful to be able to draw tens, hundreds, or thousands of values when simulating geologic processes. To generate a table of 10 random values from the same distribution, type

```
In[71]:= RandomArray[NormalDistribution[0, 1], {10}]

Out[71]= {-0.756926, 0.949171, 2.04928, -0.131767, 1.11098,
          -0.549414, 0.151664, -0.498526, -0.862757, 1.20414}
```

How about something more substantial? Let's generate a table of 100 values and give that table the variable name **RandomValues**.

```
In[72]:= RandomValues
           = RandomArray[NormalDistribution[0, 1], {100}];
```

As usual, the semi-colon can be used to supress output. The values are calculated and stored, but not displayed. Here's a histogram of the values:

```
In[73]:= Histogram[RandomValues,
           HistogramCategories → 20, HistogramRange → {-4, 4},
           ApproximateIntervals → No, BarStyle → GrayLevel[0.6]]
```

From In[73]:=

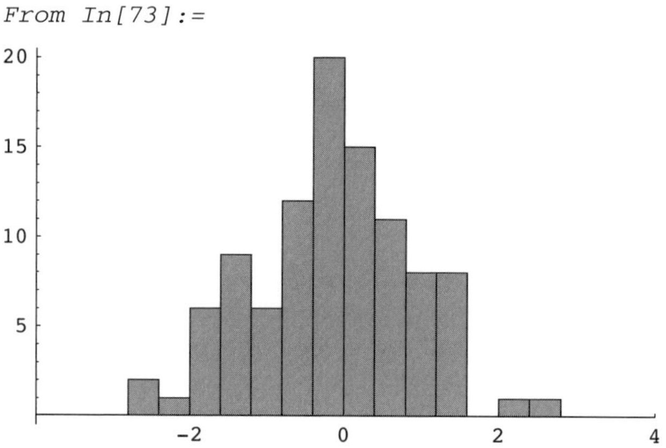

Out[73]= -Graphics-

Experiment with *Mathematica* and try generating and plotting several different sets of random numbers. The results will be overwritten each time you execute the **RandomValues** = ... line, so give each set a different name (e.g., RandomValues2, RandomValues3, etc.) if you don't want to loose your previous results.

To visually compare the generated distribution with its underlying theoretical distribution, first re-scale the histogram so that the total area of the bars is 1, just as it must be for any PDF. This is accomplished with the HistogramScale->1 option.

```
In[74]:= RandomValueHistogram =
           Histogram[RandomValues, HistogramScale → 1,
             HistogramRange → {-4, 4}, HistogramCategories → 20,
             ApproximateIntervals → No,
             BarStyle → GrayLevel[0.6]]
```

From In[74]:=

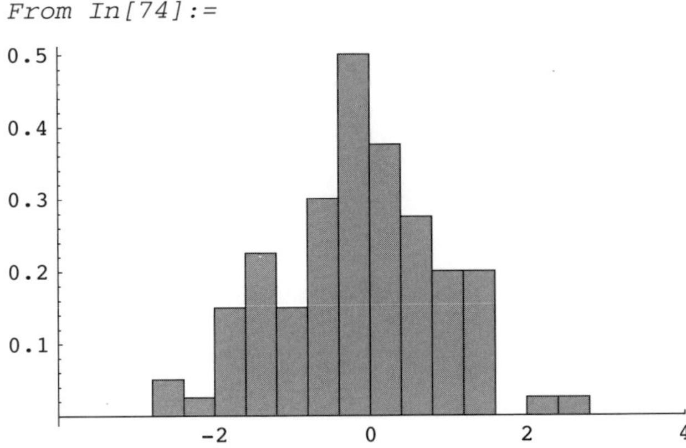

Out[74]= -Graphics-

Now, plot of the PDF

```
In[75]:= Plot[PDF[NormalDistribution[0,1],x],
           {x, -4, 4}, PlotStyle → Thickness[0.005]]
```

From In[75]:=

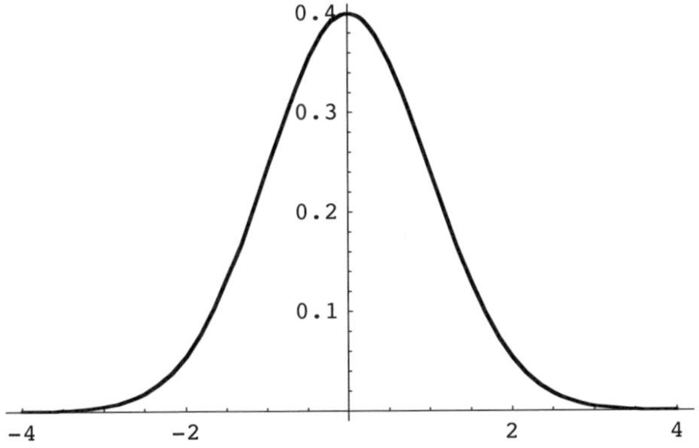

Out[75]= -Graphics-

and superimpose the two

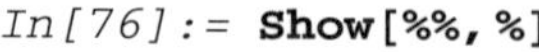

```
In[76]:= Show[%%, %]
```

```
From In[76]:=
```

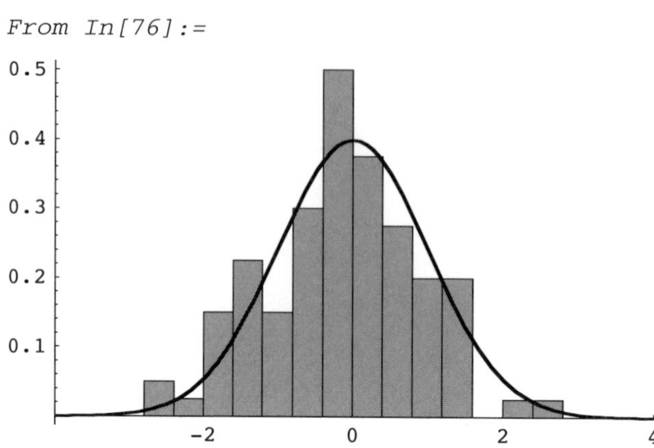


```
Out[76]= -Graphics-
```

> **Computer Note:** Is the random sample a good approximation of the PDF? Repeat the exercise, particularly with different sample sizes, to get a feel for the variability inherent in sets of randomly selected values.

The same procedures can be followed to generate random numbers from any of *Mathematica*'s standard probability distributions. Below is an example using 500 values drawn from a beta distribution.

```
In[77]:= RandomArray[BetaDistribution[2, 7], {500}];

In[78]:= Histogram[%, HistogramScale → 1,
           BarStyle → GrayLevel[0.6]]
```

```
From In[78]:=
```

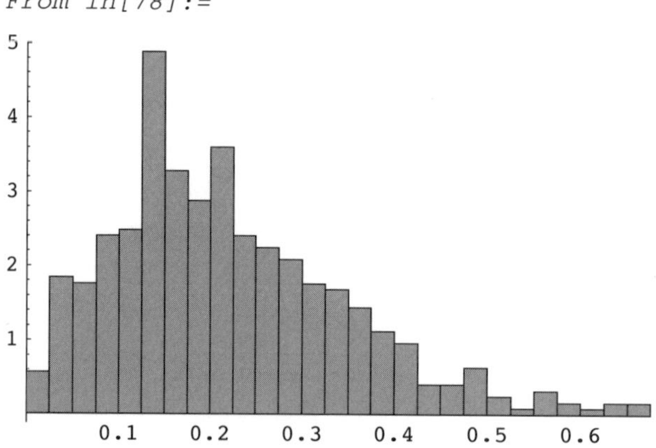


```
Out[78]= -Graphics-
```

```
In[79]:= Plot[PDF[BetaDistribution[2,7],x],
           {x, 0, 1}, PlotStyle → Thickness[0.005]]
```

From In[79]:=

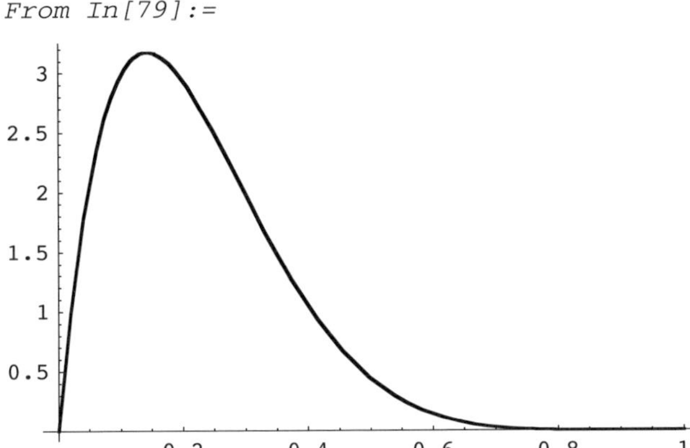


```
Out[79]= -Graphics-
```

```
In[80]:= Show[%%, %]
```

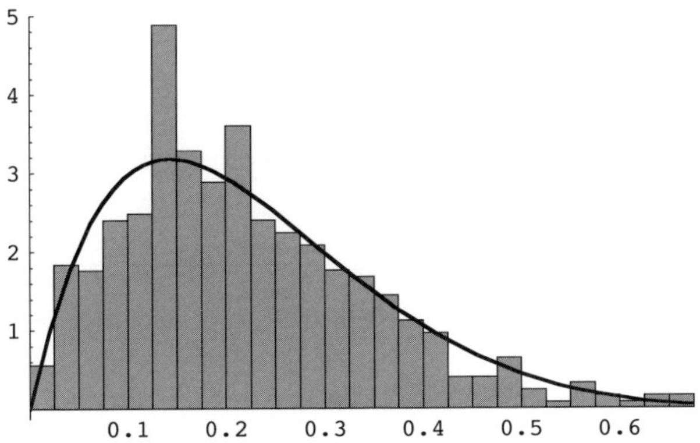


```
Out[80]= -Graphics-
```

It might sometimes be useful to generate binary {0,1} values, for example to denote the perfectly random occurrence of a "yes-no" process. For example, did a landslide occur in a GIS raster or not? Did an earthquake occur in a given time period or not? This can be done using

```
In[81]:= Random[Integer]
```

```
Out[81]= 0
```

which generates either 0 or 1, each with a 0.50 probability of occurrence.

4.9 Care and Feeding of the Random Number Generator

All random number generators are the results of mathematical algorithms based upon some initial input, or seed, meaning that they can never produce truly random numbers. Eventually they will begin to repeat themselves. The best that we, as users, can look for is a generator that takes a long time before it begins repeating. Each time that *Mathematica* is started up, it reseeds the random number generator with the time of day. If for some reason you want to generate the same set of random numbers repeatedly, simply reseed the generator with the same value with the SeedRandom[] function each time. For example, the following expression does two things: First, it seeds the random number generator with the integer 5. Second, it generates a list of random numbers using that seed.

```
In[82]:= SeedRandom[5];
         RandomArray[NormalDistribution[0, 1], 5]

Out[82]= {0.401391, -0.564765, -0.793385, 0.59151, -1.68444}
```

A second execution of the same command will produce a different list

```
In[83]:= RandomArray[NormalDistribution[0, 1], 5]

Out[83]= {0.579222, 1.58998, 0.471983, -0.0545941, -0.542878}
```

Reset the random seed to 5, though, and the first list is generated again

```
In[84]:= SeedRandom[5];
         RandomArray[NormalDistribution[0, 1], 5]

Out[84]= {0.401391, -0.564765, -0.793385, 0.59151, -1.68444}
```

One way to test for obvious problems with a random number generator is to generate a lot of random numbers, plot them, and look a small piece of the plot for any patterns or lattice-like structures in the plot. The following two commands generate 1,000,000 random points and then plots those within the range $0.001 \leq x \leq 0.002$. Are there any patterns or lattice structures evident?

```
In[85]:= Table[{Random[], Random[]}, {i, 1000000}];
```

```
In[86]:= ListPlot[%, PlotRange → {{0.001, 0.002}, {0, 1}}]

From In[86]:=
```

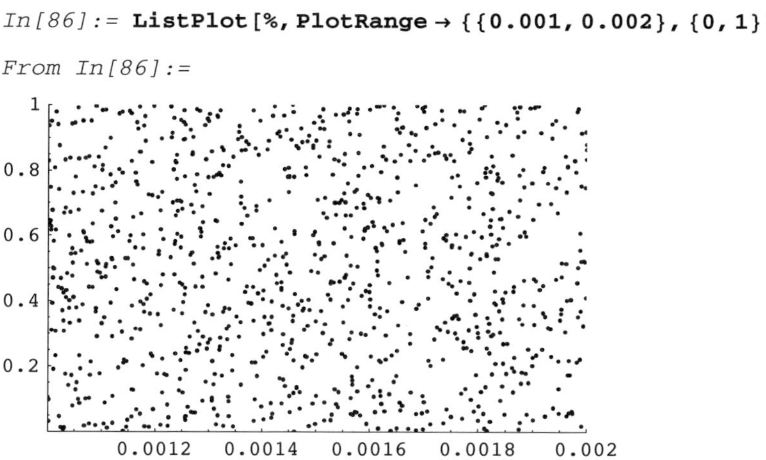


```
Out[86]= -Graphics-
```

4.10 Illustrating the Central Limit Theorem

The Central Limit Theorem, which is at the core of modern statistical theory, states that the sums or means of a large number of samples selected at random from almost any non-normal distribution will tend to become normally distributed as the sample size becomes large. Sums and means are interchanable in this context because one is simply the other scaled by the number of observations. Moreover, the mean of the means will approach the mean of the underlying distribution and the variance of the means will be that of the population divided by the number of samples. In most cases, the truth of the Central Limit Theorem starts becoming obvious when the sample size exceeds 25 or so. Cases in which the Central Limit Theorem does not apply are rare, and are extremely unlikely to be encountered in geologic applications. The most frequently cited violation of the Central Limit Theorem is the Cauchy distribution, which is symmetric but for which a mean and variance cannot be calculated because of its mathematical form.

The Central Limit Theorem can be illustrated by generating sets of random numbers from non-normal distributions, calculating their means, and plotting the results. First, create a table to hold some results. It has zero length and contains only Null values, but this step is necessary in order to create the table as a variable

```
In[87]:= CentralLimitResults = Table[Null, {0}]

Out[87]= {}
```

Now, select N numbers randomly from a non-normal distribution, calculate their mean, and repeat the process M times. The example below uses a uniform distribution ranging over $-5 \leq X \leq 5$, $N = 50$, and $M = 50$.

```
In[88]:= Do[
           AppendTo[CentralLimitResults,
             Mean[Table[Random[
               UniformDistribution[-5, 5]], {50}]]],
           {i, 50}
         ]
```

```
In[89]:= Histogram[CentralLimitResults,
           BarStyle → GrayLevel[0.6]]
```

```
From In[89]:=
```

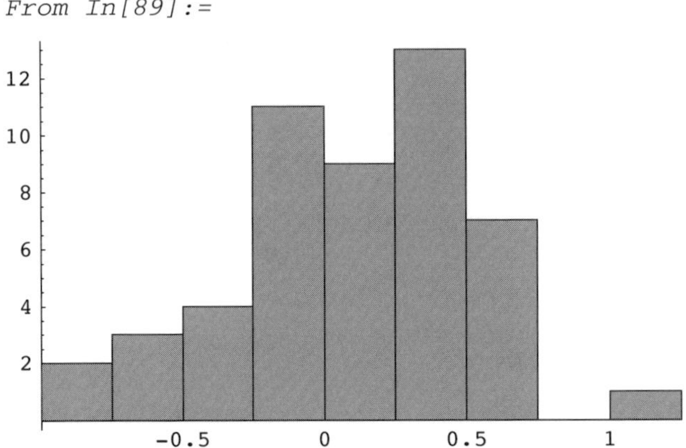


```
Out[89]= -Graphics-
```

As shown above, the results are beginning to look something like a normal distribution. They are definitely not uniformly distributed. How do the means and variances compare? The values for the underlying uniform distribution are:

```
In[90]:= Mean[UniformDistribution[-5., 5.]]
         Variance[UniformDistribution[-5., 5.]]

Out[90]= 0.

Out[90]= 8.33333
```

The mean and scaled variance of the mean values are:

```
In[91]:= Mean[CentralLimitResults]
         50 * Variance[CentralLimitResults]

Out[91]= 0.0722112

Out[91]= 8.97845
```

The results are close, but not in exact agreement. As an experiment, repeat the simulation several times to get a feel for the variability of results obtained for $N = 50$. Better results can be obtained by increasing the sample size to $N = 500$ and the number of samples to $M = 100$. Note that the first step is to clear previous values from the results table.

```
In[92]:= CentralLimitResults = Table[Null, {0}]
         Do[
           AppendTo[CentralLimitResults,
             Mean[Table[Random[
               UniformDistribution[-5, 5]], {100}]]],
           {i, 500}
         ]

Out[92]= {}

In[93]:= Histogram[CentralLimitResults,
           BarStyle → GrayLevel[0.6]]

From In[93]:=
```

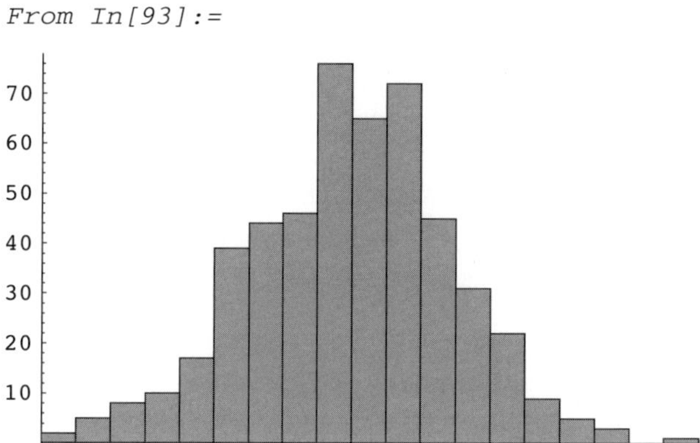


```
Out[93]= -Graphics-

In[94]:= Mean[CentralLimitResults]
           100 * Variance[CentralLimitResults]

Out[94]= -0.00364966

Out[94]= 8.55427
```

The agreement, although still not exact, is better than the first example and does serve to illustrate the very non-intuitive consequences of the Central Limit Theorem.

What are the implications of the Central Limit Theorem for geological applications? Field measurements or experimental results that are subjected to statistical analysis can be considered to be the sums of many independent factors and in large numbers **should** appoximate a normal distribution. In theory, therefore, one should be able to use functions such as **MeanTest** indiscriminantly because the Central Limit Theorem tells us that the normal distribution is just that: the one that random variables normally follow. They key word, though, is "should". Many variables of interest to geologists follow highly skewed distributions such as the lognormal distribution, so data sets should always be plotted to see if they at least come close to being normally distributed before using tests that apply to normal distributions.

Alternatively, nonparametric methods such as K-S statistics can be used to compare distributions.

Computer Note: Write a series of *Mathematica* statements to graphically illustrate the Central Limit Theorem for different sample sizes.

4.11 The Pitfalls of Undersampling

Distributions fitted to data are only as good as the data. Choosing too few data to adequately characterize the underlying distribution is known as **undersampling**. Its pitfalls can be graphically illustrated by repeatedly selecting a number of values from a specified distribution (in this case, a standard normal distribution with zero mean and unit standard deviation), calculating a sample mean and standard deviation, and then plotting the normal PDF with those values.

Here is an example consisting of 25 trials, in each of which only three values are selected at random from the underlying distribution:

```
In[95]:= GraphTable = Table[Null, {5}, {5}];

         BackgroundPlot =
           Plot[PDF[NormalDistribution[0, 1], x],
             {x, -5, 5}, Axes → None, DisplayFunction → Identity,
             PlotRange → {0, 1.5}, PlotStyle → {Thickness[0.02],
             GrayLevel[0.6]}];

         Do[
           Block[{pseudodata, ForegroundPlot}, pseudodata =
             Table[Random[NormalDistribution[0, 1]], {3}];
             ForegroundPlot =
               Plot[PDF[NormalDistribution[Mean[pseudodata],
                 StandardDeviation[pseudodata]], x], {x, -5, 5},
               Axes → None, PlotStyle → {Thickness[0.01]},
               DisplayFunction → Identity, PlotRange → {0, 1.5}];
             GraphTable[[i, j]] = Show[BackgroundPlot,
           ForegroundPlot]], {i, 5}, {j, 5}
         ]
         Show[GraphicsArray[GraphTable],
           DisplayFunction → $DisplayFunction]
```

From In[95]:=

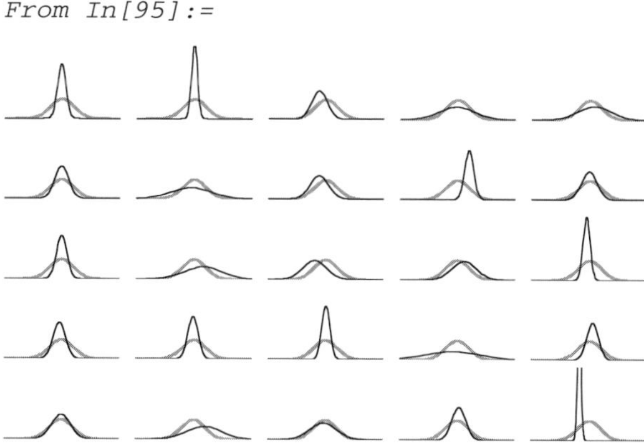

Out[95]= -GraphicsArray-

The gray curves are the underlying or "true" distributions and the black curves are the fitted distributions. In some cases, the agreement between the fitted and underlying distributions is good. In other cases, the fitted distribution is a very poor representation of the underlying distribution.

What is the result if we increase the number of samples to 10 per trial?

In[96]:=
```
BackgroundPlot =
  Plot[PDF[NormalDistribution[0, 1], x],
    {x, -5, 5}, Axes → None, DisplayFunction → Identity,
    PlotRange → {0, 1.5}, PlotStyle → {Thickness[0.02],
    GrayLevel[0.6]}];

Do[
  Block[{pseudodata, ForegroundPlot}, pseudodata =
    Table[Random[NormalDistribution[0, 1]], {10}];
    ForegroundPlot =
      Plot[PDF[NormalDistribution[Mean[pseudodata],
        StandardDeviation[pseudodata]], x], {x, -5, 5},
      Axes → None, PlotStyle → {Thickness[0.01]},
      DisplayFunction → Identity, PlotRange → {0, 1.5}];
    GraphTable[[i, j]] = Show[BackgroundPlot,
  ForegroundPlot]], {i, 5}, {j, 5}
]
Show[GraphicsArray[GraphTable],
  DisplayFunction → $DisplayFunction]
```

```
From In[96]:=
```

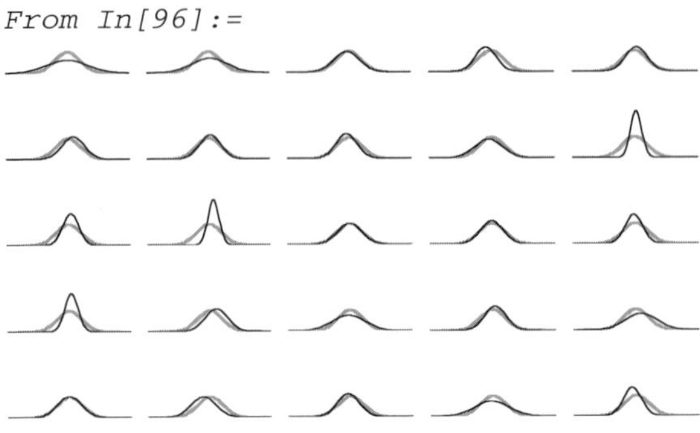

```
Out[96]= -GraphicsArray-
```

The agreement between fitted and underlying distributions is clearly better when 10 values are used. Choosing more, say 25 or 30, would produce even smaller differences. The exact number of samples required to adequately characterize an underlying distribution depends on the desired confidence level and the standard deviation of the underlying distribution. Statistics handbooks contain formulae for the estimation of the sample sizes required for specified confidence levels assuming that the data are normally distributed. For example, the number of samples necessary to determine the confidence interval ($\pm h$) around the mean of normally distributed data at the α level of significance first requires us to calculate the value for which the Student t distribution has only an $\alpha/2$ probability of being exceeded. The Student t distribution resembles the normal distribution, but its exact shape is controlled by the degrees of freedom ($dof = n - 1$ is used when n samples are used to estimate 1 parameter, in this case the mean). The two are virtually identical for large numbers of samples.

Below are four plots showing the Student t distribution for $n = 1$, 5, and 10 as well as a standard normal distribution ($\mu = 0$, $\sigma = 1$) for comparison.

```
In[97]:=
  Show[
    GraphicsArray[
      {{Plot[PDF[StudentTDistribution[1],x], {x, -10, 10},
        PlotRange → {0, 0.5}, DisplayFunction → Identity,
        Frame → True, Epilog → Text["t\nn = 1", {6, 0.35}]],
      Plot[PDF[StudentTDistribution[5],x], {x, -10, 10},
        PlotRange → {0, 0.5}, DisplayFunction → Identity,
        Frame → True, Epilog → Text["t\nn = 5", {6, 0.35}]],
      {Plot[PDF[StudentTDistribution[10],x], {x, -10, 10},
        PlotRange → {0, 0.5}, DisplayFunction → Identity,
        Frame → True, Epilog → Text["t\nn = 10", {6, 0.35}]],
      {{Plot[PDF[StudentTDistribution[0,1],x], {x, -10, 10},
        PlotRange → {0, 0.5}, DisplayFunction → Identity,
        Frame → True, Epilog → Text["t\nμ=0\nσ=1", {6, 0.35}]]}},
    ], DisplayFunction → $DisplayFunction
  ]
```

From In[97]:=

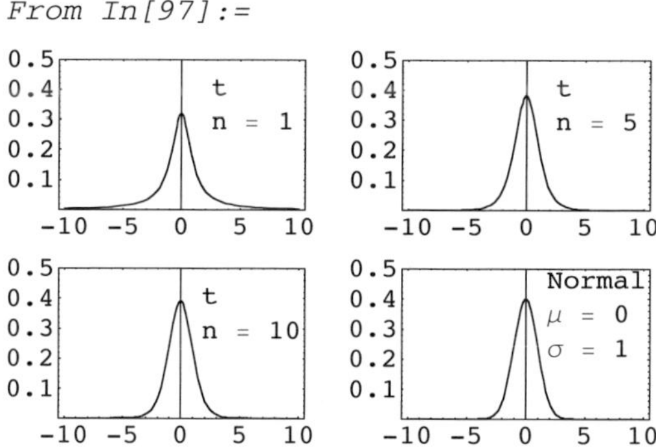

Out[97]= -GraphicsArray-

Here is a plot of **1 - CDF[StudentTDistribution[**dof**]]**, where $dof = n - 1$. In this case, *dof* ranges from 2 to 22 in steps of 3

In[98]:=

```
Show[
  Table[Plot[1 - CDF[StudentTDistribution[dof], x], {x, 0, 10},
    PlotRange→All, DisplayFunction→Identity], {dof, 2, 22, 3}]
  , DisplayFunction → $DisplayFunction,
  AxesLabel → {"x", "1 - CDF"}
]
```

From In[98]:=

1-CDF
0.5
0.4
0.3
0.2
0.1
2 4 6 8 10 x

Out[98]= -Graphics-

Notice that there is very little difference in the results for values with $dof \geq 5$. Reading from the graph (it isn't possible to solve the equations exactly because of the functions involved), the critical value for $\alpha = 0.05$ and $n - 1 = 2$ is approximately 4.3. To apply this method, we also need to have some estimate of the standard deviation, which we know to be 1. In practice, there would have to be some way of determining an *a priori* estimate, for example by calibration of an instrument used to collect the data. It could have also been estimated from the data, which would require the degrees of freedom to be reduced by 1.

The confidence level for our first numerical sampling experiment is (Crow et al., 1960, *Statistics Manual*)

```
In[99]:= h = Tcrit s / √n /. {Tcrit → 4.3, n → 3, s → 1.}

Out[99]= 2.4826
```

So, for three samples the confidence level surrounding the mean is 2.48261 ± 2.48. Not very encouraging! Repeating the exercise for 10 samples, the critical t value drops to about 2.1 and

```
In[100]:= h = Tcrit s / √n /. {Tcrit → 2.3, n → 10, s → 1.}

Out[100]= 0.727324
```

or $\bar{x} \pm 0.73$.

Conversely, what if we specify the confidence interval and wish to calculate the number of samples necessary to attain it? The equation above can be re-arranged to solve for n to estimate the sample size. If you've been reading carefully, you may have noticed that the value of Tcrit depends on n, so in theory this must be an iterative process in which we guess n, look up a value of Tcrit to calculate n, and repeat the process until n converges. In practice, the plot above shows that for 10 or more samples there is little change in the critical value so the first guess will often be good enough.

If we wish to determine the mean value with a confidence interval of ±0.1, then we should collect (Crow *et al.*, 1960)

```
In[101]:= Clear[h]

In[102]:= n = (Tcrit s / h)^2 /. {Tcrit → 2., h → 0.1, s → 1.}

Out[102]= 400.
```

That's right, 400 samples! Similar tests, with similar restrictions, exist for testing standard deviations.

In actual applications, the ability to collect data is commonly constrained by money, time, or both. How much are you, a client, or perhaps some regulators willing to pay in order to have more confidence in the result? And, how precise do you need to be? That depends in part on the sensitivity of a model to the random variable represented by the distribution as well as the risk (which is a function of the likelihood of an occurrence **and** its consequences) involved. There is no simple answer.

4.12 References and Recommended Reading

Carr, J.R., 2002, *Data Visualization in the Geosciences*: Prentice Hall.

Crovelli, R.A., 2000, *Probability Models for Estimation of Number and Costs of Landslides*: U.S. Geological Survey Open-File Report 00-249.

Crow, E.L., Davis, F.A., and Maxfield, M.W., 1960, *Statistics Manual*: Dover Publications.

Davis, J.C., 2002, *Statistics and Data Analysis in Geology (3d ed.)*: John Wiley & Sons.

Fisher, N.I., Lewis, T.L., and Embleton, B.J., 1987, *Statistical Analysis of Spherical Data:* Cambridge University Press.

Glynn, J. and Gray, T., 2000, *A Beginner's Guide to Mathematica Version 4*: Cambridge University Press.

Isaaks, E.H. and Srivastava, R.M., 1989, *Applied Geostatistics*: Oxford University Press.

Kock, G.S., Jr. and Link, R.F., 1980, *Statistical Analysis of Geological Data (Two Volumes Bound as One)*: Dover Publications.

Ross, S.M., 1985, *Introduction to Probability Models (4th ed.)*: Academic Press.

Wolfram, S., 1999, *The Mathematica Book (4th ed.)*: Cambridge University Press.

5 Probabilistic Simulation

5.1 *Mathematica* Packages You Will Need

Be sure to execute the following statement to ensure that you will have available all of the add-on and book-specific *Mathematica* functions used in this chapter.

```
In[1]:= Needs["Statistics`ContinuousDistributions`"]
        Needs["Statistics`DiscreteDistributions`"]
        Needs["Statistics`DescriptiveStatistics`"]
        Needs["Statistics`DataManipulation`"]
        Needs["Statistics`ConfidenceIntervals`"]
        Needs["Statistics`HypothesisTests`"]
        Needs["Statistics`NonlinearFit`"]
        Needs["Graphics`Graphics`"]
        Needs["Graphics`ImplicitPlot`"]
        Needs["CompGeosci`"]
```

Computer Note: The CompGeosci package will load correctly only if it is located in one of the directories in *Mathematica*'s standard file path. Execute the statement **`$Path`** to see a list of the default paths on your computer and place the file CompGeosci.m in one of those directories. The specific file paths may differ from one operating system to another. See Chapter 1 for more information about installing the CompGeosci package.

5.2 Flood Frequency Modeling

In this example, we'll use one empirical and two theoretical probability distributions to estimate the probability that a peak annual flood will be exceeded in any given year. Here are 101 years of annual peak flow data (in cfs, or cubic feet per second) for the Rio Grande at Embudo, New Mexico (the first USGS stream gauge, now marked by a roadside monument):

```
In[2]:=
      PeakFlow =
        {5890, 6270, 8790, 6860, 5300, 5160, 3110, 8980, 4700,
        1690, 5600, 7400, 2500, 16200, 14000, 2080, 7190, 7330,
        8560, 8600, 3580, 7280, 12700, 14400, 7500, 4820, 8780,
        1580, 5500, 9500, 5240, 5850, 2240, 1740, 7180, 4340, 832,
        5900, 3630, 6690, 5440, 2410, 1990, 12000, 10800, 2220,
        8770, 5380, 1950, 4080, 10200, 9990, 1470, 710, 8720,
        2000, 1860, 2200, 1020, 5000, 6840, 2760, 2320, 2340,
        3980, 966, 952, 5200, 1950, 3550, 3270, 3140, 2250,
        1860, 1090, 6620, 1050, 3700, 2290, 2480, 1490, 9000,
        5080, 2930, 5010, 5660, 6010, 8420, 7500, 9280, 1110,
        2540, 2480, 5730, 3330, 5580, 7410, 936, 4930, 2530,
        3310} //N;
```

5.2.1 Plotting the Data

What does the peak flow distribution look like? Here is a stem plot of the data made using the **ListStemPlot** function from the *Mathematica* package included with this book:

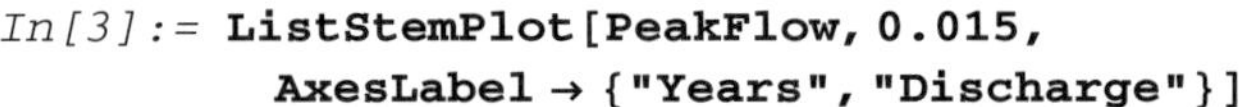

```
In[3]:= ListStemPlot[PeakFlow, 0.015,
          AxesLabel → {"Years", "Discharge"}]
```

From In[3]:=

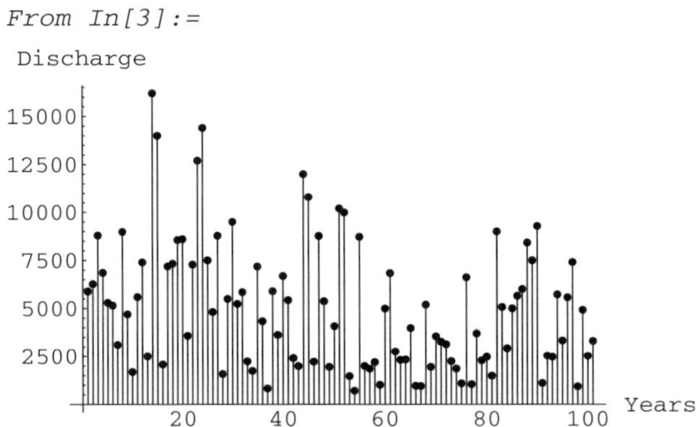

Out[3]= -Graphics-

An alternative might have been to use a vertical bar chart or **ListPlot** with **PlotJoined- > True**. Notice that the horizontal axis shows the number of each data point, not the year. The same data set can be shown in a histogram, in this case scaled so that it can be subsequently shown with a PDF.

```
In[4]:= PeakFlowHistogram = Histogram[PeakFlow,
          HistogramScale → 1, BarStyle → GrayLevel[0.6]]
```

From In[4]:=

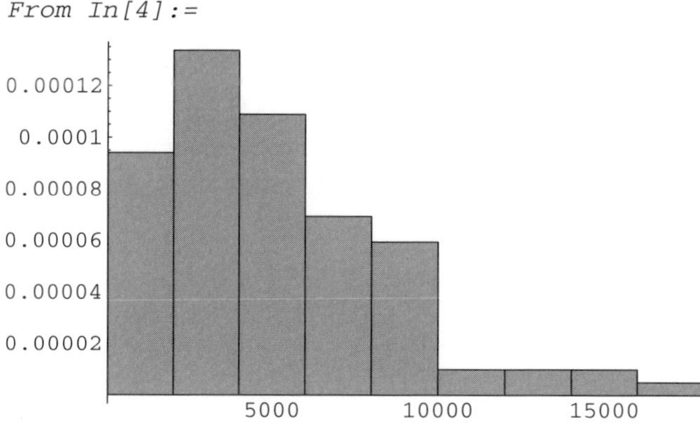

Out[4]= -Graphics-

5.2.2 Log-Normal and Extreme Value Distribution Fitting

The distribution shown in the previous histogram is obviously skewed, so how about trying to represent it with a log-normal distribution?

```
In[5]:= Histogram[Log[PeakFlow],HistogramScale → 1,
          BarStyle → GrayLevel[0.6]]
```

From In[5]:=

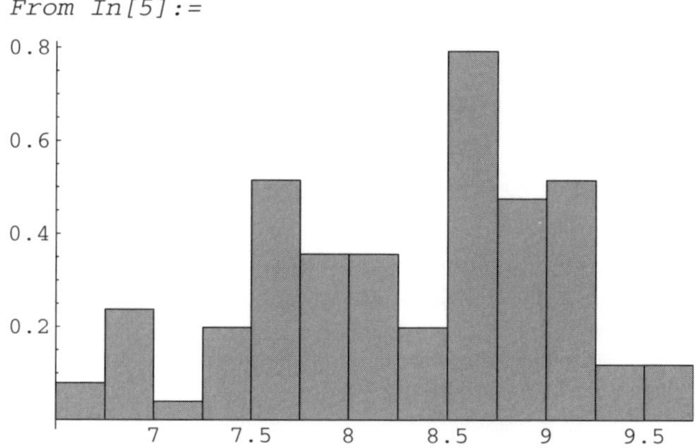

Out[5]= -Graphics-

It is difficult to say if this improved things, but at least the distribution might be a little more symmetric. Now, we can fit a log-normal distribution to the data using the method of moments.

```
In[6]:= meanval = Mean[Log[PeakFlow]]
```

Out[6]= 8.2935

```
In[7]:= dev = StandardDeviation[Log[PeakFlow]]

Out[7]= 0.736039
```

Recall that we've already used the variable names **meanval** and **dev**, so their previous values will be overwritten unless you have cleared them or restarted the kernel. Now, plot the PDF with this mean and standard deviation but suppress its output, then show it along with the first histogram.

```
In[8]:= Plot[PDF[LogNormalDistribution[meanval, dev], x],
          {x, 0, 17000}, PlotStyle → Thickness[0.008],
          DisplayFunction → Identity]

Out[8]= -Graphics-

In[9]:= Show[PeakFlowHistogram, %,
          DisplayFunction → $DisplayFunction]

From In[9]:=
```

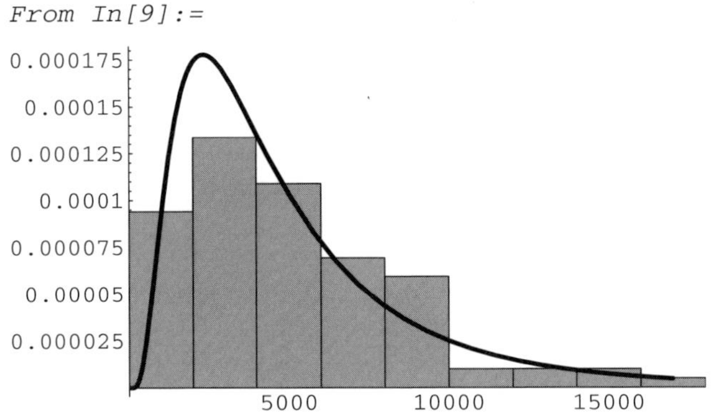


```
Out[9]= -Graphics-
```

The log-normal distribution appears to be a fair representation but, because the peak annual discharges are extreme values, perhaps we can do better with an extreme value distribution. The two extreme value distribution parameters α and β are related to the mean and standard deviation by (Chow et al., 1988)

$$In[10]:= \beta = \frac{\texttt{StandardDeviation[PeakFlow]}\sqrt{6.}}{\pi}$$

```
Out[10]= 2599.56

In[11]:= α = Mean[PeakFlow] - 0.5772 β

Out[11]= 3574.54
```

If you look this up in Chow et al. (1988), be aware that they use the parameters u and α, which correspond to our α and β. The two alphas are not equal! We'll use the *Mathematica* notation in this example. Alternatively, α and β could have been determined as described in Chapter 4. As above, we can plot the resulting PDF and superimpose it on the histogram

```
In[12]:= Plot[PDF[ExtremeValueDistribution[α, β], x],
           {x, 0, 17000}, PlotRange → All,
           PlotStyle → Thickness[0.008],
           DisplayFunction → Identity]

Out[12]= -Graphics-

In[13]:= Show[PeakFlowHistogram, %,
           DisplayFunction → $DisplayFunction]

From In[13]:=
```

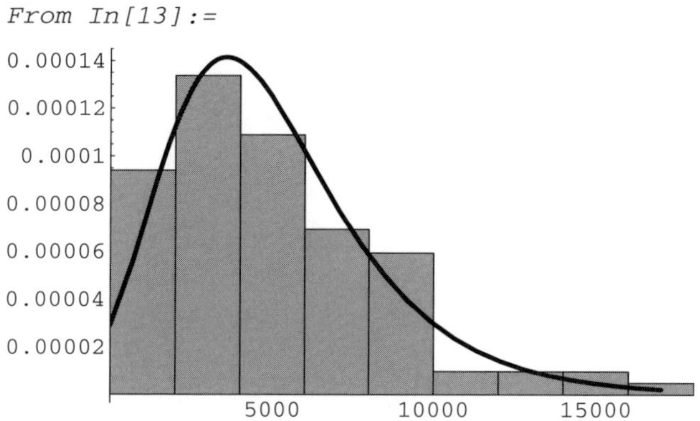


```
Out[13]= -Graphics-
```

The agreement between the observed peak flows and the theoretical distribution seems to have improved, particularly with regard to the height and location of the peak of the distribution.

5.2.3 Empirical Cumulative Distribution

A third method, which may be the most familiar, is to establish an empirical cumulative distribution using the measured discharges and without relying on any theoretical probability distribution. The Weibull formula, $P = m/(n+1)$, is often used for this in the United States (Chow *et al.*, 1988). The variable m is the rank of a given flood and n is the total number of data. This approach assumes that floods are ranked from largest to smallest and gives the probability that a given discharge will be exceeded. If the floods are ranked from smallest to largest, the same formula gives the probability that the discharge will not be exceeded. We'll do the latter, which will allow the results to be superimposed on the previous plot for comparison. First, sort or rank the discharges from smallest to largest and suppress the output (remove the semicolon if you would like to see the sorted list).

```
In[14]:= RankedFlow = Sort[PeakFlow];
```

Now, create a table containing each discharge and its corresponding probability of not being exceeded. Plot the results with discharge on the horizontal axis and the cumulative probability on the vertical axis. The first column in the table will be the peak discharge value of rank *m* and the second column will be the Weibull cumulative probability.

```
In[15]:= n = Length[PeakFlow];
         Table[{RankedFlow[[m]], N[m/(n + 1)]}, {m, n}];
         FrequencyPlot1 = ListPlot[%,
         PlotStyle → {GrayLevel[0.6], PointSize[0.02]}]
```

From In[15]:=

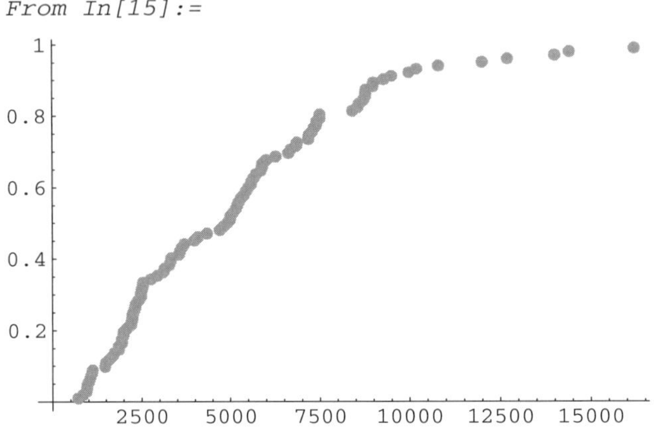

Out[15]= -Graphics-

5.2.4 Comparison of Results

One way to quickly compare the results produced by the empirical distribution, the log-normal distribution, and the extreme value distribution is to superimpose a series of cumulative frequency plots.

```
In[16]:= FrequencyPlot2 =
           Plot[CDF[ExtremeValueDistribution[α, β], x],
           {x, 0, 17000}, PlotRange → All,
           PlotStyle → {Thickness[0.008]},
           DisplayFunction → Identity]

Out[16]= -Graphics-

In[17]:= FrequencyPlot3 =
           Plot[CDF[LogNormalDistribution[meanval, dev], x],
           {x, 0, 17000}, PlotStyle → {Dashing[{0.02}],
           Thickness[0.008]}, DisplayFunction → Identity]

Out[17]= -Graphics-

In[18]:= Show[FrequencyPlot1, FrequencyPlot2, FrequencyPlot3,
           DisplayFunction → $DisplayFunction,
           FrameTicks → {{0, 4000, 8000, 12000, 16000},
           Automatic, {}, {}}]
```

From In[18]:=

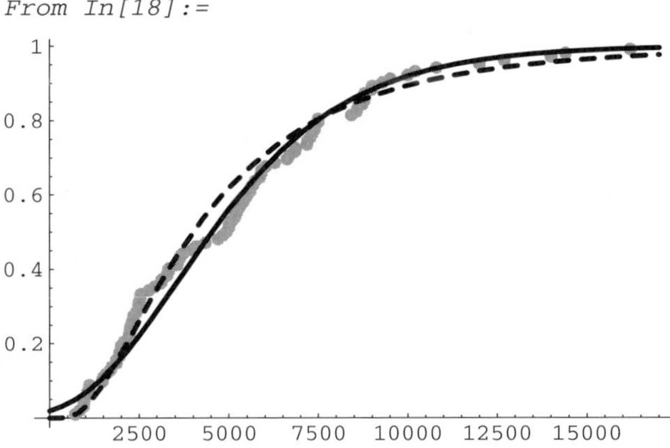

Out[18]= -Graphics-

A cumulative frequency plot made with the **CumFreqPlot** function introduced in Chapter 4 would have been virtually indistinguishable from the Weibull plot. In fact, it would have been a theoretically more correct plot because the Weibull plotting formula was a compromise made because it was difficult to plot a cumulative probability of 1 on the probability graph paper that was once used for this analysis (Chow *et al.*, 1988).

Computer Note: Look up information about Log-Pearson Type III distributions, which have been recommended as a standard for flood frequency analysis in the United States, and redo the calculations (the log-Pearson Type III distribution becomes the log-normal distribution if there is no skewness). Chow *et al.* (1998) and many other hydrology texts will have the necessary information.

Computer Note: Use the Kolmogorov-Smirnov functions introduced in Chapter 4 to compare the three distributions. You will have to conduct three comparisons, each with a different pair of lists. Do the K-S results support the visual inference that there is not much difference between the three distributions?

5.2.5 Exceedance Probability and Recurrence Intervals

What is the probability that a peak flow will exceed, say, 8,000 cfs in any given year? The answer is obtained by finding the complement of the CDFs evaluated at that discharge. The complement is necessary because the CDF will give the probability of a flow **less** than 8,000 cfs.

```
In[19]:= 1 - CDF[ExtremeValueDistribution[α, β], 8000.]

Out[19]= 0.166606
```

```
In[20]:= 1 - CDF[LogNormalDistribution[meanval, dev], 8000.]

Out[20]= 0.172975
```

The recurrence interval for a given peak annual discharge can be estimated by multiplying the cumulative probability of that discharge by the number of years of data. For example, the peak flow of 8,000 cfs would have estimated recurrence intervals of

```
In[21]:= Length[PeakFlow] *
           CDF[LogNormalDistribution[meanval, dev], 8000.]

Out[21]= 83.5296
```

and

```
In[22]:= Length[PeakFlow] *
           CDF[ExtremeValueDistribution[α, β], 8000.]

Out[22]= 84.1728
```

using the two theoretical CDFs. The actual peak flow data can also be used for the estimate, either by reading from the graph above or using the table of cumulative probabilities below:

```
In[23]:= CumFreqs[PeakFlow]

Out[23]=
{{710., 0.00990099}, {832., 0.019802}, {936., 0.029703}, {952., 0.039604},
{966., 0.049505}, {1020., 0.0594059}, {1050., 0.0693069},
{1090., 0.0792079}, {1110., 0.0891089}, {1470., 0.0990099},
{1490., 0.108911}, {1580., 0.118812}, {1690., 0.128713}, {1740., 0.138614},
{1860., 0.148515}, {1860., 0.158416}, {1950., 0.168317}, {1950., 0.178218},
{1990., 0.188119}, {2000., 0.19802}, {2080., 0.207921}, {2200., 0.217822},
{2220., 0.227723}, {2240., 0.237624}, {2250., 0.247525}, {2290., 0.257426},
{2320., 0.267327}, {2340., 0.277228}, {2410., 0.287129}, {2480., 0.29703},
{2480., 0.306931}, {2500., 0.316832}, {2530., 0.326733}, {2540., 0.336634},
{2760., 0.346535}, {2930., 0.356436}, {3110., 0.366337}, {3140., 0.376238},
{3270., 0.386139}, {3310., 0.39604}, {3330., 0.405941}, {3550., 0.415842},
{3580., 0.425743}, {3630., 0.435644}, {3700., 0.445545}, {3980., 0.455446},
{4080., 0.465347}, {4340., 0.475248}, {4700., 0.485149}, {4820., 0.49505},
{4930., 0.50495}, {5000., 0.514851}, {5010., 0.524752}, {5080., 0.534653},
{5160., 0.544554}, {5200., 0.554455}, {5240., 0.564356}, {5300., 0.574257},
{5380., 0.584158}, {5440., 0.594059}, {5500., 0.60396}, {5580., 0.613861},
{5600., 0.623762}, {5660., 0.633663}, {5730., 0.643564}, {5850., 0.653465},
{5890., 0.663366}, {5900., 0.673267}, {6010., 0.683168}, {6270., 0.693069},
{6620., 0.70297}, {6690., 0.712871}, {6840., 0.722772}, {6860., 0.732673},
{7180., 0.742574}, {7190., 0.752475}, {7280., 0.762376}, {7330., 0.772277},
{7400., 0.782178}, {7410., 0.792079}, {7500., 0.80198}, {7500., 0.811881},
{8420., 0.821782}, {8560., 0.831683}, {8600., 0.841584}, {8720., 0.851485},
{8770., 0.861386}, {8780., 0.871287}, {8790., 0.881188}, {8980., 0.891089},
{9000., 0.90099}, {9280., 0.910891}, {9500., 0.920792}, {9990., 0.930693},
{10200., 0.940594}, {10800., 0.950495}, {12000., 0.960396},
{12700., 0.970297}, {14000., 0.980198}, {14400., 0.990099}, {16200., 1.}}
```

The values bracketing 8000 cfs are 7500 and 4080 cfs, and a value for 8420 cfs can be easily interpolated

```
In[24]:= Interpolation[{{7500., 0.811881}, {8420., 0.821782}},
           InterpolationOrder → 1]

Out[24]= InterpolatingFunction[{{7500., 8420.}}, <>]
```

You may be wondering why it wouldn't have been easier to interpolate the entire cumulative frequencies list. The answer is that there are duplicate discharge values, which will return an error from **Interpolation**. The empirical recurrence interval is thus

```
In[25]:= %[8000.] * Length[PeakFlow]

Out[25]= 82.5435
```

In this case, there was little difference between the results returned by the three different methods.

5.3 Didn't We Just Have a 100 Year Flood?

Binomial and Poisson distributions can be used to predict the likelihood of an event such as an earthquake, flood, landslide, or debris flow occurring (or not occurring) within a given time frame. Costa and Baker (1981), Crovelli (2000), and Keaton (1994) used binomial models to simulate the temporal occurrence of landslides, floods, and debris flows. Hammond *et al.*, (1992) used a binomial model to simulate phreatic surface heights for input into a slope stability model. Crovelli also used a Poisson model to simulate landslide occurrence, and Poisson models are often used to simulate the occurrence of earthquakes through time (Reiter, 1990).

It is important to remember that binomial and Poisson models assume that each occurrence is completely independent of the others. That is to say, for example, what happened last year has no effect on what will happen this year. An example of a condition that violates the independence requirement would be a prolonged drought or series of unusually wet years, because the events of subsequent years would not be truly independent of each other. In seismic analyses, the condition of independence is violated when the data contains foreshocks or aftershocks. It is up to the geologist using the method to ensure that the restrictions are reasonably well satisified.

What is the probability of a stream having no peak annual discharges with a 100 year or greater recurrence interval in any given century? The 100 year flood has an annual probability of occurrence of 0.01, so the probability that no 100 year flood (0 events) will occur per century (100 years) is:

```
In[26]:= CDF[BinomialDistribution[100, 0.01], 0]

Out[26]= 0.366032
```

Similarly, the likelihood of having no 100 year floods in a century is estimated by a Poisson distribution to be:

```
In[27]:= CDF[PoissonDistribution[100 * 0.01], 0]

Out[27]= 0.367879
```

The total probability of having either zero or some other number of 100 year floods must be 1. Thus, the probability of **one or more** 100 year floods (*i.e.*, more than zero) per century is

```
In[28]:= 1 - %%

Out[28]= 0.633968
```

and according the the Poisson model

```
In[29]:= 1 - %%

Out[29]= 0.632121
```

The probability of exactly one 100 year flood per century is, according to the two distributions,

```
In[30]:= PDF[BinomialDistribution[100, 0.01], 1]

Out[30]= 0.36973
```

and

```
In[31]:= PDF[PoissonDistribution[100 * 0.01], 1]

Out[31]= 0.367879
```

The probability of having two or more 100 year floods in a century can be found from the information above, or directly from

```
In[32]:= 1 - CDF[BinomialDistribution[100, 0.01], 1]
         1 - CDF[PoissonDistribution[100 * 0.01], 1]

Out[32]= 0.264238

Out[32]= 0.264241
```

So far the two distributions have produced similar results. As discussed above, however, the binomial distribution overestimates the exceedance probability of events with short recurrence intervals over short periods of time. For example, consider the differences for a flood with a 2 year recurrence interval over a 3 year period:

```
In[33]:= 1 - CDF[BinomialDistribution[3, 1/2.], 0]

Out[33]= 0.875

In[34]:= 1 - CDF[PoissonDistribution[3 * 1/2.], 0]

Out[34]= 0.77687
```

5.4 Monte Carlo Simulation of a Wetting Front

Monte Carlo simulations involve the repeated calculation of results using values randomly selected from probability distributions. One selects a value for each of the random variables, calculates a result or realization for that particular set of random variables and stores it, then repeats the process hundreds or thousands of times (Harr, 1996; Cullen and Frey, 1999). The term Monte Carlo is generally believed to have been originated as a code word during the Manhattan Project at Los Alamos. The collection of realizations, known as an ensemble, constitutes a probability distribution of its own that can be used to calculate the probability that an event will occur. In the example given below, it is assumed that the variables are uncorrelated. That is to say, the value taken on by one variable has no relationship to the values taken on the others. It is also possible to formulate models involving correlated variables, although we will not do so.

The Monte Carlo simulation presented below is essentially the same as one used to estimate the time it would take for water seeping out of a newly installed leach field to reach the local water table located 10 m below, and which was used to support testimony in a water rights hearing. The original simulation was undertaken to help evaluate a claim that it would take decades for treated water from the drain field to recharge the water table, and that the water rights applicant should be denied credits for returning water to the system. The variables are: t = wetting front travel time, f = fillable porosity, K = saturated hydraulic conductivity upon rewetting, L = travel distance, Hw = depth of ponded water (assumed to be zero in this case), and Hcr is the critical pressure head that must be exceeded to wet the soil. The wetting front travel time is given by (Bouwer, 1978)

$$t = \frac{f}{K}\left[L - (Hw - Hcr)\mathrm{Log}\left[\frac{\mathrm{Hw} + L - \mathrm{Hcr}}{\mathrm{Hw} - \mathrm{Hcr}}\right]\right]$$

For lack of information to the contrary, it was assumed that f was uniformly distributed between 0.02 and 0.25, using values taken from published literature, and that Hcr was uniformly distributed between –0.1 and 1.0 m of water. The distance to the water table L = 10 m, was assumed to be known with certainty. In this case there was no ponding of water to help drive the wetting front, so Hw = 0. There were some hydraulic conductivity data available, and hydraulic conductivity is very often log-normally distributed, so a log-normal distribution was used for K. The following PDF is used to specify K in the simulation.

```
In[35]:= Plot[PDF[LogNormalDistribution[-17., 0.75], x],
           {x, 0, 3 10^-7}, PlotRange → All,
           AxesLabel → {"K", "PDF"}]
```

From In[35]:=

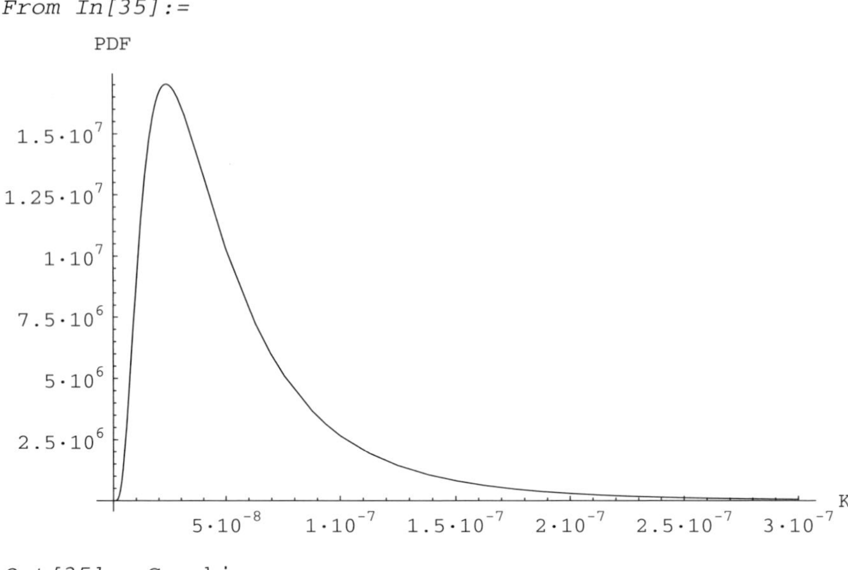

Out[35]= -Graphics-

Here is the Monte Carlo simulation of the travel time of a Green-Ampt wetting front. First, the number of trials, **ntrials**, is set to 1000. A blank array of zeroes, **t**, is then defined and filled. Values for **f**, **K**, and **Hcr** are selected at random and one realization calculated. The realization is converted from seconds to years and the process is repeated 999 more times.

```
In[36]:= ntrials = 1000;

         t = Table[0, {ntrials}];

         Do[
           Block[{f, K, L, Hw, Hcr},
             f = Random[UniformDistribution[0.02, 0.25]];
             K = Random[LogNormalDistribution[-17., 0.75]];
             L = 10.;
             Hw = 0.;
             Hcr = Random[UniformDistribution[-0.1, -1.]];
             t[[i]] = f/K (L - (Hw - Hcr) Log[(Hw + L - Hcr)/(Hw - Hcr)]);
             t[[i]] = t[[i]]/3600./24./365.25;
           ], {i, ntrials}
         ]
```

Below are the results, showing that travel time is likely to be on the order of months, not years.

```
In[37]:= Min[t]

Out[37]= 0.0312399
```

```
In[38]:= Max[t]

Out[38]= 9.51068

In[39]:= Histogram[t, HistogramCategories → 50,
           BarStyle → GrayLevel[0.6]]

From In[39]:=
```

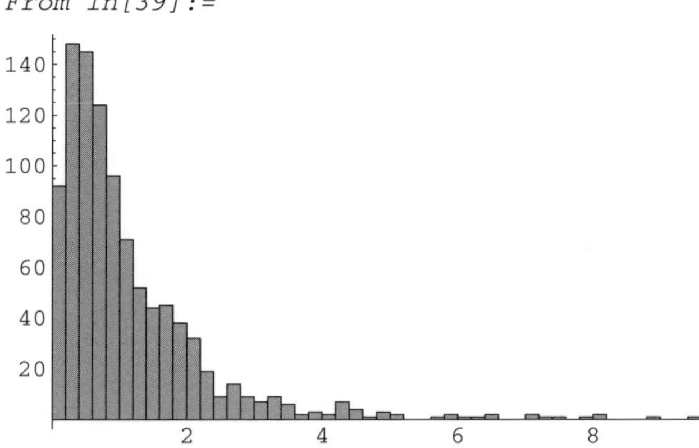


```
Out[39]= -Graphics-
```

Another way to illustrate the results is with cumulative plot, which shows a 60% probability that the travel time will be 1 year or less.

```
In[40]:= CumFreqPlot[t, 0, 7]

From In[40]:=
```

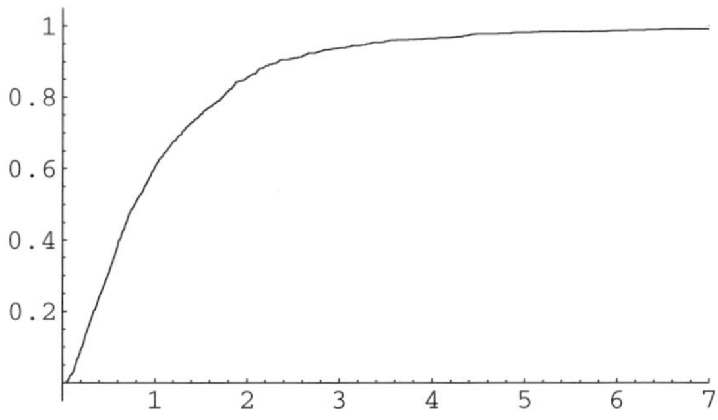


```
Out[40]= -Graphics-
```

> **Computer Note:** Experiment with different values of **`ntrials`** to see how it affects the results of the Monte Carlo simulation. Is there a wider range of results for 5 different trials with **`ntrials = 50`** than for 5 different trials with, say, **`ntrials = 1000`**?

5.5 Monte Carlo Analysis of Infinite Slope Stability

A common measure of performance in engineering and applied geology is the **factor of safety**, which is typically written as the ratio of resisting to driving forces or moments in a system. A slope, for example, will be stable if the resisting forces exceed the driving forces and the factor of safety is greater than 1. One of the most useful applications of probabilistic methods is to calculate the uncertainty or, conversely, the reliability associated with calculated factors of safety. Uncertainty has always been acknowledged by engineers and geoscientists, for example by using conservative or worst-case assumptions, conducting sensitivity analyses to evaluate results for different sets of input variables, and using acceptable values well above the theoretical threshold of safety. For example, a slope with a factor of safety of 1.01 would generally not be considered to be acceptably stable simply because there is too much room for error in the calculations. The minimum acceptable value is likely to depend on the risk involved, for example between a slope in a remote watershed and one in a heavily populated city.

We will use a simplified version of the infinite slope model to illustrate the use of Monte Carlo simulation to evaluate the uncertainty of a factor of safety. This model, which has been frequently used as a first approximation of slope stability in studies of watersheds and other large areas, assumes that the soil is of uniform thickness, uniform physical properties, and infinite laterial extent (Huang, 1983).

5.5.1 Static Factor of Safety

We will start by looking at the static factor of safety for slopes not subjected to seismic shaking. Let ϕ be the angle of internal friction (representing the frictional component of soil shear strength), β be the slope angle in degrees, and $0 \leq H \leq 1$ the dimensionless height of the prheatic surface above the base of the slide mass. Using these variables, the factor of safety against sliding for a cohesionless infinite slope with slope parallel groundwater flow and variable phreatic surface height is approximated by the following *Mathematica* function (assuming that the soil unit weight is twice the water unit weight):

```
In[41]:= FS[ϕ_, β_, H_] := (1. - H/2) * Tan[ϕ]/Tan[β]
```

For example,

```
In[42]:= FS[20.°, 20.°, 0.3]

Out[42]= 0.85
```

which is an unstable combination of variables. We can solve to find the value of **H** at which **FS** = 1, which is the critical water level above which sliding should occur, either analytically

```
In[43]:= Solve[FS[25.°, 20.°, H] == 1, H]

Out[43]= {{H → 0.438927}}
```

or by plotting it

```
In[44]:= Plot[FS[25. °, 20. °, H], {H, 0, 1},
           AxesLabel → {"H", "FS"}, AxesOrigin → {0, 0.65}]
```

From In[44]:=

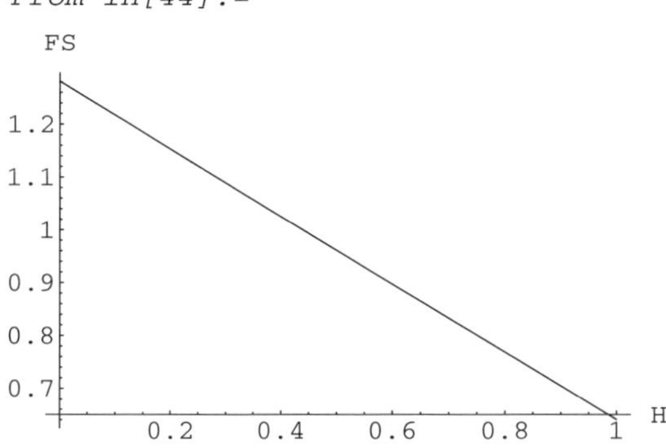

Out[44]= -Graphics-

Below is a Monte Carlo simulation similar to that used for the wetting front model just discussed. The logic is the same; only the variables have been changed.

```
In[45]:= ntrials = 1000;
         results = Table[0, {ntrials}];
         Do[
           Block[{ϕ, β, H},
             ϕ = Random[UniformDistribution[25. °, 35. °]];
             β = Random[UniformDistribution[20. °, 30. °]];
             H = Random[UniformDistribution[0.01, 1]];
             results[[i]] = FS[ϕ, β, H];
           ], {i, ntrials}
         ]
```

```
In[46]:= Histogram[results, BarStyle → GrayLevel[0.6]]
```

From In[46]:=

80
60
40
20
0.5 0.75 1 1.25 1.5 1.75

Out[46]= -Graphics-

The histogram is slightly skewed, suggesting that it might be reasonably represented by a log-normal distribution. The K-S statistic between the logarithm of the results and a log-normal distribution having the sample mean and standard deviation is

```
In[47]:= KSOneList[results]

Out[47]= 0.0451551
```

which can be verified by plotting **results** along with a log-normal distribution.

```
In[48]:= KSOneListPlot[Log[results], Floor[Min[Log[results]]],
           Ceiling[Max[Log[results]]],
           AxesLabel → {"Log\nFS", "Cum\nProb"}]

From In[48]:=
```

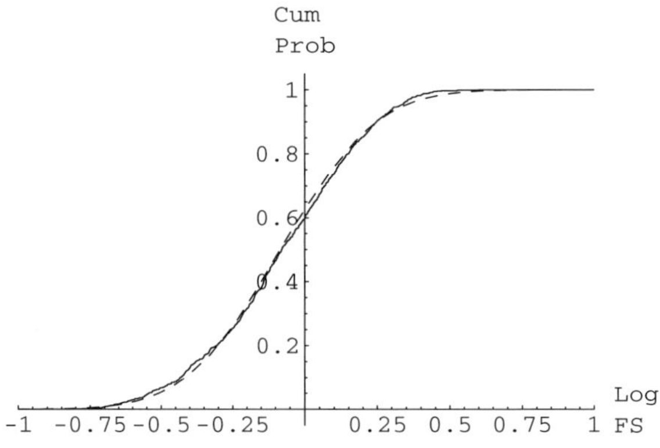


```
Out[48]= -Graphics-
```

According to the graph, the probability that $FS \le 1$ (or Log[FS] ≤ 0) is about 0.65. This is the probability of sliding. We can also calculate a probability by fitting a theoretical PDF, in this case a log-normal distribution, to the results.

```
In[49]:= meanval = Mean[Log[results]]
         dev = StandardDeviation[Log[results]]

Out[49]= -0.0834508

Out[49]= 0.258332

In[50]:= CDF[LogNormalDistribution[meanval, dev], 1]

Out[50]= 0.626667
```

Some practioners prefer not to use this approach because it requires one to assume that the results are adequately represented by some kind of theoretical probability distribution. In this case, it shouldn't present a problem.

A third alternative is to use something called a reliability index, which is defined as the difference between the calculated FS and some critical value (in this case, $FS = 1$) divided by the standard deviation of the results.

```
In[51]:= RI = (meanval - Log[1.])/dev

Out[51]= -0.323038
```

The reliability index tells how many standard deviations the calculated mean lies away from the critical value. Thus, a value of –0.36 says that the calculated mean lies 0.36 standard deviations below the critical value.

5.5.2 Effects of Changing Independent Variable Distributions

So far we have assumed that all of the variables contributing to the factor of safety are uniformly distributed. But, studies show that water levels in slopes susceptible to landsliding may be generally low most of the time and only occasionally high enough to trigger landsliding, for example during and immediately after heavy rainstorms (*e.g.*, Haneberg, 1991; Haneberg and Gökce, 1994). How do we account for the fact that most of the time pore water pressure will be too low to cause landsliding? One possibility is to simulate H as a log-normally distributed variable, although we could also simulate it as a Pareto or beta variable. We know that physically possible range of phreatic surface heights is (excluding the possibility of artesian pressures) $0 \leq H \leq 1$, but have no idea about its mean or standard deviation. It turns out that the standard deviation of a uniform distribution is

```
In[52]:= StandardDeviation[UniformDistribution[minval, maxval]]

Out[52]= (maxval - minval)/(2 Sqrt[3])
```

or, for values of 0.01 and 1,

```
In[53]:= % /. {maxval → Log[1.], minval → Log[0.01]}

Out[53]= 1.3294
```

The mean value is just the average of the minimum and maximum values

```
In[54]:= 0.5 (Log[1.] + Log[0.01])

Out[54]= -2.30259
```

And here is the PDF that now represents the water levels in the slope.

```
In[55]:= Plot[PDF[LogNormalDistribution[-2.31, 1.33], x],
           {x, 0, 1}, AxesLabel → {"H", "PDF"}]
```

From In[55]:=

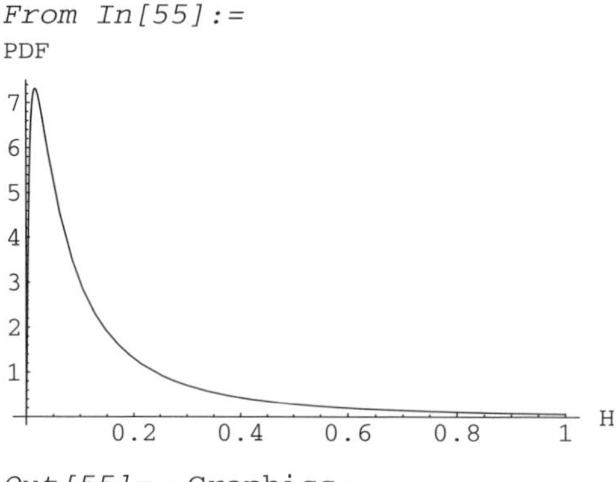

Out[55]= -Graphics-

Although the log-normal distribution is finite at its low end, it continues on to positive infinity. Thus, there will always be a small likelihood of selecting a value of *H* > 1, which will produce a negative factor of safety. This is the same Monte Carlo simulation as above, with only the *H* distribution changed

```
In[56]:= ntrials = 1000;
         results = Table[0, {ntrials}];
         Do[
           Block[{φ, β, H},
             φ = Random[UniformDistribution[25. °, 35. °]];
             β = Random[UniformDistribution[20. °, 30. °]];
             H = Random[LogNormalDistribution[-2.31, 1.33]];
             results[[i]] = FS[φ, β, H];
           ], {i, ntrials}
         ]
```

and its results in histogram form

```
In[57]:= Histogram[results, BarStyle → GrayLevel[0.6],
           HistogramRange → {-2, 2}]
```

From In[57]:=

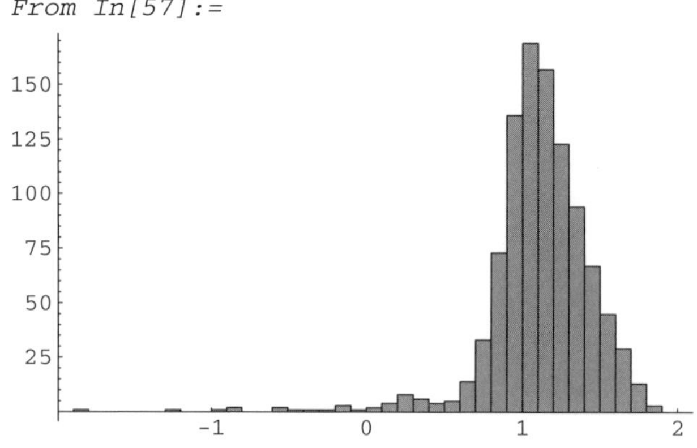

Out[57]= -Graphics-

There are a few physically unrealistic.negative values produced by values of $H > 1$. What can be done with them? There are at least three options: First, if the number of negative values is small enough, they can just be ignored on the assumption that they won't have much influence in an ensemble of hundreds or thousands of realizations. Second, the Monte Carlo routine can be changed to check for values of $H > 1$ (which is what causes a negative *FS* value) and replace those values with 1. This can produce an unsightly bulge in the *FS* histogram at the low end of the range. Third, H can be specified by a distribution that has a finite range, for example a beta distribution. We'll choose the first option and simply remove the offensive values.

To remove the negative values, first create a table of zero length and use **Null** as a placeholder. This needs to be done because *Mathematica* won't perform operations using a variable that doesn't yet exist, so we create the table and but give it no length or contents. Then, write a routine to loop through **Results**, check to see if each value is positive, and, if so, use **AppendTo** to add it to **NewResults**.

```
In[58]:= newresults = Table[Null, {0}]
         len = Length[results];
         Do[
           If[results[[i]] ≥ 0,
             AppendTo[newresults, results[[i]]]], {i, len}
         ]

Out[58]= {}
```

The number of values removed will differ each time the simulation is run, and there is always a chance that no negative values will have to be removed. Removing just a few values out of a thousand is unlikely to have an effect on any inferences made using the results. The number of non-negative values in this simulation is:

```
In[59]:= Length[newresults]

Out[59]= 985
```

Here is a K-S plot of the results, showing that they are in this case closely represented by a normal distribution:

```
In[60]:= KSOneListPlot[newresults, Floor[Min[newresults]],
           Ceiling[Max[newresults]],
           AxesLabel → {"Log\nFS", "Cum\nProb"}]
```

From In[60]:=

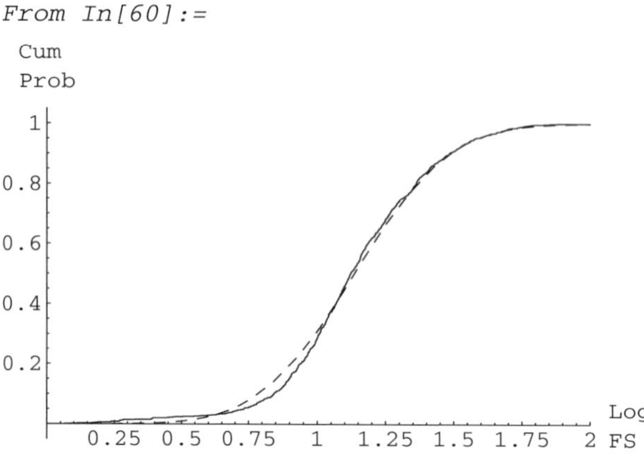


```
Out[60]= -Graphics-
```

The K-S statistic between **newresults** and a normal distribution having the sample mean and standard deviation is

```
In[61]:= KSOneList[newresults]
```

```
Out[61]= 0.0473539
```

which is an acceptable value. The assumption of log-normally distribution phreatic surface heights decreases the probability of sliding to about 0.30 (this value may differ each time the simulation is run), or about half that of the previous example. Similar results are obtained if we change other variables to distributions with central tendencies. Was our assumption of a log-normal distribution valid? This is a question that can only be answered with data from the field for a specific slope or, if we are willing to accept them, the results of a groundwater flow model. Something that must be avoided at all costs is the creative selection of distributions in order to produce desired results without regard to the underlying geologic reality.

> **Computer Note:** Does the K-S statistic decrease if the logarithms of the results are used? Perform a log transform on **newresults**, calculate a K-S statistic, and make a new K-S plot.

5.5.3 Conditional Probability: Earthquakes and Slope Stability

A similar approach can be taken if the probability of a given ground acceleration value occuring over a specified time period is known, as in the probabilistic earthquake hazard maps published by the US Geological Survey. The pseudostatic factor of safety of an infinite slope subjected to seismic acceleration is (Huang, 1983)

```
In[62]:= Clear[β, ϕ, H]

In[63]:= SeismicFS[ϕ_, β_, H_, Cs_] :=
           ((1 - H/2) Cos[β] - Cs Sin[β]) Tan[ϕ]
           -------------------------------------
                  Sin[β] + Cs Cos[β]
```

in which **Cs** is a coefficient of seismic acceleration given in terms of *g*, the gravitational acceleration. It is easy to show that this equation reduces to the static *FS* equation used above if *Cs* is zero. To wit,

```
In[64]:= SeismicFS[ϕ, β, H, 0]
```

```
Out[64]= (1 - H/2) Cot[β] Tan[ϕ]
```

Now, back to earthquakes. According to the USGS national earthquake hazard maps (http://eqint.cr.usgs.gov/eq/html/zipcode.shtml), a peak ground acceleration of 0.12 *g* has an 0.10 probability of being exceeded in 50 years in Socorro, New Mexico. We'll use that for an example. The Monte Carlo simulation is similar to the previous two except that a fixed **Cs** value is specified. Thus, the results will show the probability of a landslide given the specified value of **Cs**.

```
In[65]:= ntrials = 1000;
         results = Table[0, {ntrials}];
         Cs = 0.12;
         Do[
           Block[{ϕ, β, temp},
             ϕ = Random[UniformDistribution[30. °, 35. °]];
             β = Random[UniformDistribution[20. °, 25. °]];
             H = Random[LogNormalDistribution[-2.31, 1.33]];
             results[[i]] = SeismicFS[ϕ, β, H, Cs];
           ], {i, ntrials}
         ]
```

Here is a histogram showing the Monte Carlo simulation results:

```
In[66]:= SeismicHistogram=Histogram[results,HistogramScale→1,
           BarStyle → GrayLevel[0.6], HistogramRange → {-2, 2}]
```

```
From In[66]:=
```

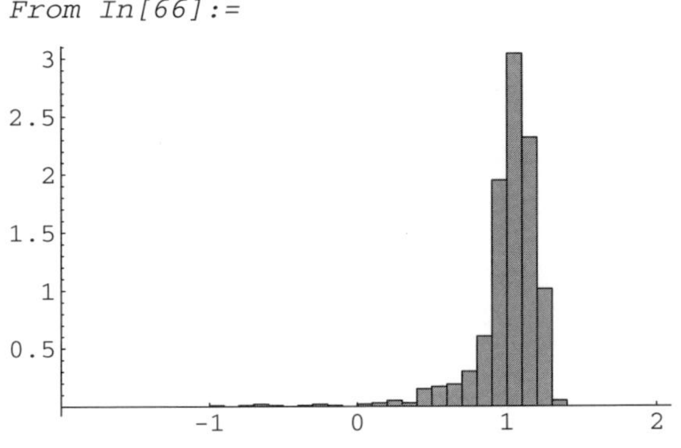


```
Out[66]= -Graphics-
```

As before, we'll censor the offensive negative values by simply removing them

```
In[67]:= newresults = Table[Null, {0}]
         len = Length[results];
         Do[
           If[results[[i]] ≥ 0,
           AppendTo[newresults, results[[i]]]], {i, len}
         ]

Out[67]= {}
```

and then take another look at the histogram of newly censored results.

```
In[68]:= SeismicHistogram = Histogram[newresults,
           HistogramScale → 1, BarStyle → GrayLevel[0.6],
           HistogramRange → {0, 1.5}]

From In[68]:=
```

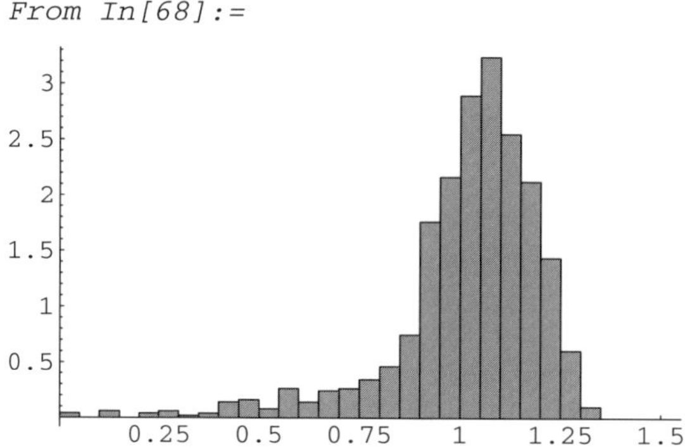


```
Out[68]= -Graphics-
```

Much better. Now on with the analysis.

```
In[69]:= meanval = Mean[newresults]

Out[69]= 1.01772

In[70]:= dev = StandardDeviation[newresults]

Out[70]= 0.186082

In[71]:= Plot[PDF[NormalDistribution[meanval, dev], x],
           {x, 0, 2}, PlotStyle → Thickness[0.008],
           PlotRange → All, DisplayFunction → Identity]

Out[71]= -Graphics-

In[72]:= Show[%, SeismicHistogram, %,
           DisplayFunction → $DisplayFunction]
```

From In[72]:=

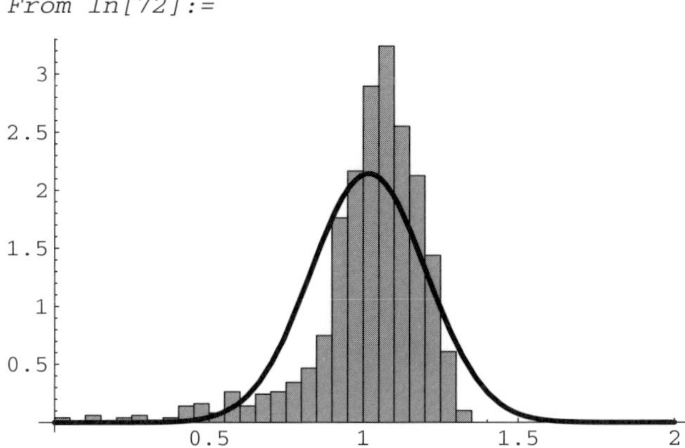

Out[72]= -Graphics-

Perhaps a log-normal distribution can provide a closer match. A quick way to tell is to calculate K-S stastics for both possibilities, which are

In[73]:= **KSOneList[newresults]**

Out[73]= 0.122114

In[74]:= **KSOneList[Log[newresults]]**

Out[74]= 0.244476

Therefore, the normal distribution appears to be the better choice of the two although it is not the only possibility. We can either use the empirical distribution just as produced by the Monte Carlo simulation or try to fit a different distribution.

> **Computer Note:** Fit a beta distribution to the results to see if it agrees more closely. Using the beta distribution as implemented by *Mathematica*, you will have to rescale the Monte Carlo output so that it ranges between 0 and 1. Alternatively, you can write your own implementation of the beta function PDF.

The probability of landsliding assuming a normal distribution is:

In[75]:= **CDF[NormalDistribution[meanval, dev], 1.]**

Out[75]= 0.462072

whereas a cumulative probability plot of the Monte Carlo results suggests a lower value of aproximately 0.35.

In[76]:= **CumFreqPlot[newresults, 0, 1.5]**

```
From In[76]:=
```

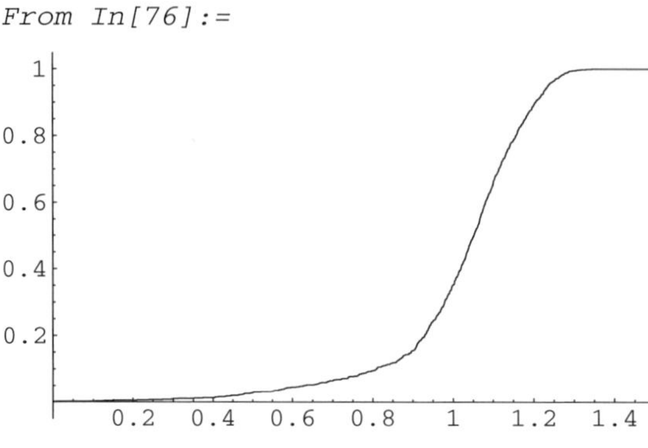


```
Out[76]= -Graphics-
```

Because the acceleration that we used (0.12 *g*) is a value inferred to have a 0.10 probability of being exceeded in 50 years, the conditional probability of a landslide due to an earthquake during a 50 year interval is (using the cumulative plot result)

```
In[77]:= 0.35 * 0.1

Out[77]= 0.035
```

5.6 Apparent Clast Size Distributions: The Outcrop Effect

Geologists must often use 1-D or 2-D information to make inferences about 3-D populations. For example, borehole logs or outcrop maps might be used in an attempt to predict the distribution of large clasts that will be encountered during excavation in melanges or glacial till (Medley, 2002 and references therein). Or, fabric data from thin sections might be used to infer properties such as the permeability or thermal diffusivity anisotropy of rocks (Haneberg *et al.*, 1994; Goodwin and Haneberg, 1996). The science of studying the 3-D attributes of objects usually observed in 2-D is known as **stereology**.

One way to understand the bias introduced by sampling 3-D objects from 2-D surfaces, which we will refer to as the **outcrop effect**, is to use Monte Carlo simulation to generate apparent grain size distributions from known distributions. This is relatively easy to do and can form the basis of numerical experiments that provide some insight into the bias introduced by the outcrop effect. The inverse problem of estimating the underlying clast size distributions from 2-D planar views or, even worse, 1-D transects is more formidable. For details, see Sahagian and Proussevitch (1998) or Heilbronner (2002). The methods described by both of these authors make use of Monte Carlo simulations to generate apparent grain size populations that allow one to work backwards from the apparent populations to derive the true underlying populations.

Flat-Lying Ellipsoids

Herbison-Evans (2002) gives a good description of the mathematics behind 3-D ellipsoid geometry. First, we'll need to define a vector containing the three coordinate axes.

```
In[78]:= X = {x, y, z}

Out[78]= {x, y, z}
```

Just to be flexible, we'll include a vector containing the distances that the center of the ellipsoid is removed from the coordinate system origin {0, 0, 0}. This isn't necessary if we only want to generate ellipses centered at the origin.

```
In[79]:= U = {Δx, Δy, Δz}

Out[79]= {Δx, Δy, Δz}
```

Next comes a shape matrix containing the semi-axes a, b, and c

```
In[80]:= V = {{1/a^2, 0, 0}, {0, 1/b^2, 0}, {0, 0, 1/c^2}}
```

Out[80]= $\left\{\left\{\frac{1}{a^2}, 0, 0\right\}, \left\{0, \frac{1}{b^2}, 0\right\}, \left\{0, 0, \frac{1}{c^2}\right\}\right\}$

In matrix form, **V** looks like

```
In[81]:= MatrixForm[V]
```

Out[81]= $\begin{pmatrix} \frac{1}{a^2} & 0 & 0 \\ 0 & \frac{1}{b^2} & 0 \\ 0 & 0 & \frac{1}{c^2} \end{pmatrix}$

We can now assemble **X**, **U**, and **V** into the equation defining a flat lying 3-D ellipsoid. We'll consider the more general (and more complicated) problem of a rotated ellipsoid further on in this chapter.

```
In[82]:= FlatEllipsoid = (X - U).V.(X - U)
```

Out[82]= $\frac{(x - \Delta x)^2}{a^2} + \frac{(y - \Delta y)^2}{b^2} + \frac{(z - \Delta z)^2}{c^2}$

or, if there are no offsets,

```
In[83]:= (X - U).V.(X - U) /.{Δx → 0, Δy → 0, Δz → 0}
```

Out[83]= $\frac{x^2}{a^2} + \frac{y^2}{b^2} + \frac{z^2}{c^2}$

Unfortunately, **`Plot`** can't be used to draw a picture of a two-dimensional cross-section of the an ellipsoid because the equation isn't a single-valued functional relationship. *Mathematica*, however, contains an add-on function called **`ImplicitPlot`** that is designed specifically for this kind of problem. Because we're plotting a 3-D object on a 2-D plane, we'll need to specify the value of the third coordinate (in this case, $z = 0$ to produce a slice through the center of the ellipsoid). Here is an example of the elliptical cross-section normal to the z axis for an ellipsoid with $a = 5$, $b = 3$, and $c = 1$:

```
In[84]:= ImplicitPlot[
           FlatEllipsoid == 1 /. {a → 5., b → 3., c → 1., Δx → 0,
           Δy → 0, Δz → 0., z → 0},
           {x, -5, 5}, AspectRatio → 3/5]
```

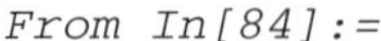

From In[84]:=

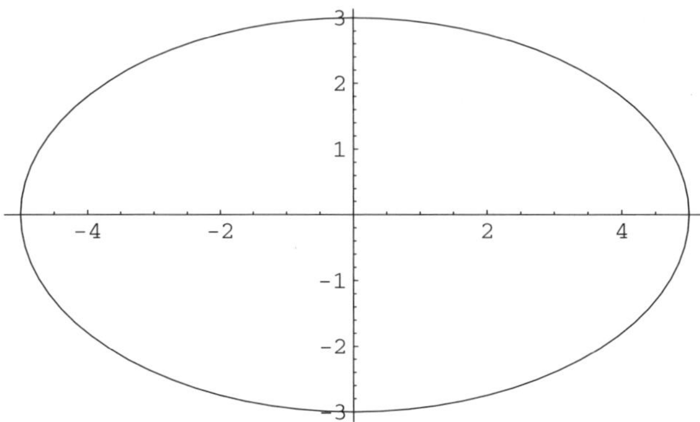


```
Out[84]= -Graphics-
```

The technique can be extended into the realm of probabilistic simulation by letting Δz (the out-of-plane coordinate offset) become a random variable. In essence, the outcrop face or thin section becomes the plane defined by the x and y axes (*i.e.*, $z = 0$), and an apparent clast size population is generated by moving the ellipsoid back and forth along the z axis by varying Δz. Below is a routine to generate 2-D pictures of 100 ellipsoidal clasts whose apparent grain sizes are due to their offset from the x-y plane.

```
In[85]:= RandomOffsets = Table[Random[
           UniformDistribution[-1, 1]], {i, 100}];

         Show[
           GraphicsArray[
             Table[ImplicitPlot[FlatEllipsoid == 1 /.
               {a → 5., b → 3., c → 1., Δx → 0., Δy → 0.,
               Δz → RandomOffsets[[i * j]], z → 0.},
               {x, -5., 5.}, Axes → None, PlotRange → {{-5.5, 5.5},
               {-3.5, 3.5}}, AspectRatio → 0.6,
               DisplayFunction → Identity], {i, 10}, {j, 10}
             ]
           ], DisplayFunction → $DisplayFunction
         ]
```

```
From In[85]:=
```

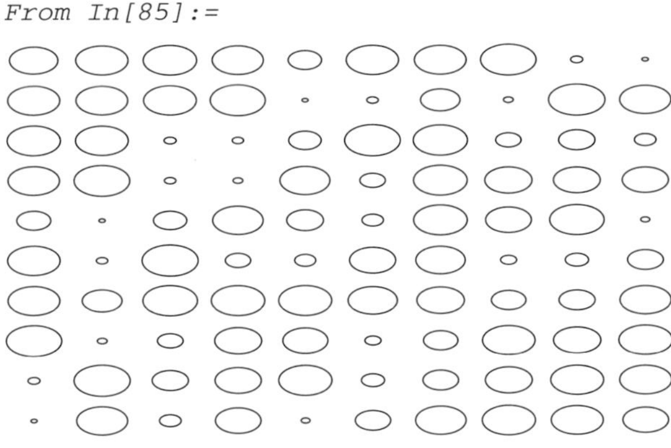

```
Out[85]= -GraphicsArray-
```

Computer Note: The ellipse-plotting routine above is fairly slow, taking about 21 seconds to execute on my computer. That's long enough to make some people impatient, but not quite long enough to step out for a cup of coffee. If there were a simple way of determining the apparent axis lengths of the ellipse formed when the ellipsoid intersects a plane, the ellipses could have been drawn more quickly using **Circle**[*x*, *y*, {*a*,*b*}]. Although, as shown below, it is possible to derive simple expressions for *a* and *b* in special cases such as flat-lying ellpsoids, in general the problem is much more difficult.

How would you describe this apparent clast size distribution if you saw it in an outcrop face or thin section? Would you have inferred that it represented a population of identically sized clasts? What kind of implications does this have for day-to-day fieldwork?

With a little more work, we can also generate an apparent clast size distribution curve. This is done by solving **FlatEllipsoid == 1** for **x** with **y = 0** (which will yield the maximum *x* dimension, or apparent *a*) and then for **y** with **x = 0** (which will yield the maximum *y* dimension, or apparent *b*). Here's how:

```
In[86]:= NSolve[1 == FlatEllipsoid /.
            {Δx → 0, Δy → 0, y → 0, z → 0}, x]
```

$$\text{Out[86]= } \left\{\left\{x \to -1.\sqrt{a^2 - \frac{1.\ a^2\,\Delta z^2}{c^2}}\right\}, \left\{x \to \sqrt{a^2 - \frac{1.\ a^2\,\Delta z^2}{c^2}}\right\}\right\}$$

```
In[87]:= NSolve[1 == FlatEllipsoid /.
            {Δx → 0, Δy → 0, x → 0, z → 0}, y]
```

$$\text{Out[87]= } \left\{\left\{y \to -1.\sqrt{b^2 - \frac{1.\ b^2\,\Delta z^2}{c^2}}\right\}, \left\{y \to \sqrt{b^2 - \frac{1.\ b^2\,\Delta z^2}{c^2}}\right\}\right\}$$

Thus, the a and b values of the apparent ellipse are scaled by a uniform factor of $\sqrt{1-(\Delta z/c)^2}$ as long as the clasts are all aligned with their x and y axes parallel to the outcrop plane. **This pleasantly simple result is, unfortunately, not correct if the clasts have random orientations.**

The next problem is to define what we mean by clast size, which is a non-trivial problem. Is it the longest axis? The intermediate axis, as many people argue when interpreting the results of a sieve analysis? The shortest axis? One way to distill the sizes of the apparent ellipses into a 1-D measurement that is somewhat akin to that used when sediments are sieved is to take the radii of circles having the same area as the ellipses generated by the Monte Carlo simulation. The following user-defined fuction takes a, b, c, and Δz as input and returns an apparent grain size

```
In[88]:= ClastSize[a_, b_, c_, Δz_] := Sqrt[a Sqrt[1. - (Δz/c)^2] b Sqrt[1. - (Δz/c)^2]]
```

Now, generate a set of 100 apparent clast sizes using the same list of random offsets as we used in the graph.

```
In[89]:= ClastSizeResults = Table[
            ClastSize[5., 3., 1., RandomOffsets[[i]]], {i, 100}
          ];
```

How does the mean of the apparent grain sizes compare to their true size?

```
In[90]:= Mean[ClastSizeResults]

Out[90]= 2.85976
```

The true clast size is taken to be the radius of a sphere having the same volume as the ellipsoid, 4/3 $\pi\, a\, b\, c$, or

```
In[91]:= TrueClastSize = (a b c)^(1/3) /. {a → 5., b → 3., c → 1.}

Out[91]= 2.46621
```

The error in estimated mean grain size introduced by the outcrop effect is thus

```
In[92]:= 100 Percent (Mean[ClastSizeResults] - TrueClastSize)/TrueClastSize

Out[92]= 15.9578 Percent
```

The error that occurs if the equivalent radii calculated from the outcrop dimensions of the ellipsoids are used to estimate the true total volume of the clasts (remember, the clasts are all the same size; they just have different offsets from the x-y plane).

```
In[93]:= len = Length[ClastSizeResults]

Out[93]= 100
```

In[94]:= $\frac{4}{3}\pi\sum_{i=1}^{len}$ **ClastSizeResults[[i]]**^{3}

Out[94]= 13022.8

The true volume of the 100 clasts is

In[95]:= **100.** $\frac{4}{3}\pi$ **TrueClastSize**^{3}

Out[95]= 6283.19

and the error in the estimated volume is therefore

In[96]:= **100 Percent** $\frac{\%\% - \%}{\%}$

Out[96]= 107.265 Percent

The overestimate is in part an artifact of the clast aspect ratios that we chose, with the largest possible cross-sectional area parallel to the outcrop plane. Although this value gives a good indication of the magnitude of error that can be introduced by the outcrop effect, it is specific to one simulation based on one grain size with an orientation that maximizes one component of the error. The error introduced for other grain shapes, particularly if they are not drawn from a uniform distribution or if they are randomly oriented, may be significantly different.

Why does the outcrop **over** predict the true clast size when the outcrop effect causes the clasts to appear smaller? Because there are two factors at play. First, the outcrop effect does make the clasts appear smaller. Second, we chose the clast orientation such that the short dimension is perpendicular to the outcrop plane. The result is that the two longest semi-axes are used to calculate an equivalent radius from the elliptical area, whereas all three semi-axes are used to calculate an equivalent radius from the ellipsoidal volume. In this case, the error introduced by ignoring the third dimension when calculating the equivalent radius outweighs that introduced by the oucrop effect.

Here is a histogram of the apparent grain size distribution:

In[97]:= **Histogram[ClastSizeResults, BarStyle → GrayLevel[0.6]]**

From In[97]:=

Out[97]= -Graphics-

and a cumulative frequency plot

```
In[98]:= CumFreqPlot[ClastSizeResults, Min[ClastSizeResults],
            Max[ClastSizeResults],
            AxesLabel → {"Equiv.\nSize", "Cum.\nProb"},
            AxesOrigin → {0, 0}, PlotRange → {{0, 4}, {0, 1}}]
```

From In[98]:=

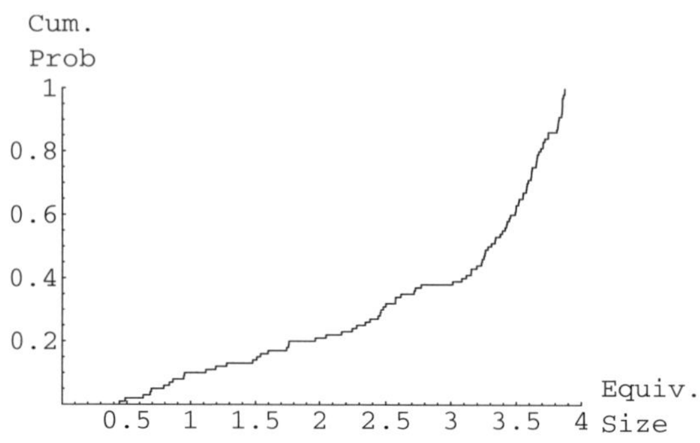

Out[98]= -Graphics-

5.6.1 Randomly Rotated Ellipsoids

The problem becomes more complicated if the clasts have random orientation, in which case the simulation must also include random amounts of rotation around each of the three axes using a 3-D rotation matrix. The rotation matrix can be specified using several different conventions (http://mathworld.wolfram.com/EulerAngles.html), but the convention built into *Mathematica* is not easy to grasp intuitively. So, we'll use a rotation matrix in which the three components of rotation correspond to the roll, pitch, and yaw that I experience in my kayak. These should also be familiar to anyone who has sailed or flown an airplane (or has a flight simulator video game). The roll angle (rotation around the x axis) is given by ψ, the pitch angle (rotation around the z axis) is given by by ϕ, and the yaw angle by θ. The total rotation is then found as the dot product of all three individual components

```
In[99]:= Clear[R];
         Rx = {{1, 0, 0}, {0, Cos[ψ], Sin[ψ]}, {0, - Sin[ψ], Cos[ψ]}};
         Ry = {{Cos[θ], 0, Sin[θ]}, {0, 1, 0}, {- Sin[θ], 0, Cos[θ]}};
         Rz = {{Cos[ϕ], Sin[ϕ], 0}, {- Sin[ϕ], Cos[ϕ], 0}, {0, 0, 1}};
         R = Rx.Ry.Rz;
```

As formulated above, the rotation matrix **R** performs rotations around the fixed coordinate axes. Reversing the order would perform the rotations around the axes of the ellipsoid, which change after each of the three incremental rotations.

The equation describing an ellipsoid with arbitrary orientation is (Herbison-Evans, 2002)

```
In[100]:= RotatedEllipsoid = (X - U).Transpose[R].V.R.(X - U)

Out[100]=
(z - Δz)
 (1/a^2 (Sin[θ] ((x - Δx) Cos[θ] Cos[ϕ] + (z - Δz) Sin[θ] + (y - Δy) Cos[θ] Sin[ϕ])) +
  1/c^2 (Cos[θ] Cos[ψ] ((z - Δz) Cos[θ] Cos[ψ] +
        (y - Δy) (-Cos[ψ] Sin[θ] Sin[ϕ] - Cos[ϕ] Sin[ψ]) +
        (x - Δx) (-Cos[ϕ] Cos[ψ] Sin[θ] + Sin[ϕ] Sin[ψ]))) + 1/b^2 (Cos[θ] Sin[ψ]
      ((z - Δz) Cos[θ] Sin[ψ] + (x - Δx) (-Cos[ψ] Sin[ϕ] - Cos[ϕ] Sin[θ] Sin[ψ]) +
        (y - Δy) (Cos[ϕ] Cos[ψ] - Sin[θ] Sin[ϕ] Sin[ψ])))) +
(x - Δx) (1/a^2 (Cos[θ] Cos[ϕ] ((x - Δx) Cos[θ] Cos[ϕ] +
        (z - Δz) Sin[θ] + (y - Δy) Cos[θ] Sin[ϕ])) +
  1/c^2 ((-Cos[ϕ] Cos[ψ] Sin[θ] + Sin[ϕ] Sin[ψ])
      ((z - Δz) Cos[θ] Cos[ψ] + (y - Δy) (-Cos[ψ] Sin[θ] Sin[ϕ] - Cos[ϕ] Sin[ψ]) +
        (x - Δx) (-Cos[ϕ] Cos[ψ] Sin[θ] + Sin[ϕ] Sin[ψ]))) +
  1/b^2 ((-Cos[ψ] Sin[ϕ] - Cos[ϕ] Sin[θ] Sin[ψ])
      ((z - Δz) Cos[θ] Sin[ψ] + (x - Δx) (-Cos[ψ] Sin[ϕ] - Cos[ϕ] Sin[θ] Sin[ψ]) +
        (y - Δy) (Cos[ϕ] Cos[ψ] - Sin[θ] Sin[ϕ] Sin[ψ])))) +
(y - Δy) (1/a^2 (Cos[θ] Sin[ϕ] ((x - Δx) Cos[θ] Cos[ϕ] +
        (z - Δz) Sin[θ] + (y - Δy) Cos[θ] Sin[ϕ])) +
  1/c^2 ((-Cos[ψ] Sin[θ] Sin[ϕ] - Cos[ϕ] Sin[ψ])
      ((z - Δz) Cos[θ] Cos[ψ] + (y - Δy) (-Cos[ψ] Sin[θ] Sin[ϕ] - Cos[ϕ] Sin[ψ]) +
        (x - Δx) (-Cos[ϕ] Cos[ψ] Sin[θ] + Sin[ϕ] Sin[ψ]))) +
  1/b^2 ((Cos[ϕ] Cos[ψ] - Sin[θ] Sin[ϕ] Sin[ψ])
      ((z - Δz) Cos[θ] Sin[ψ] + (x - Δx) (-Cos[ψ] Sin[ϕ] - Cos[ϕ] Sin[θ] Sin[ψ]) +
        (y - Δy) (Cos[ϕ] Cos[ψ] - Sin[θ] Sin[ϕ] Sin[ψ]))))
```

That is quite a mess, but fear not. There is sense to be made of it.

As above, the rotated ellipsoid equation can be used to generate an array of apparent clast shapes. The true clast shape is the same as above, but this time the roll, pitch, and yaw are allowed to range over intervals of ±10°. The angles used for the simulation of a real geologic material would almost certainly depend on its genesis and be constrained by the results of field or petrographic fabric analysis. One might expect a melange formed by shearing to have a different degree of angular dispersion than, say, an ablation till consisting of material dropped into place as a glacier recedes.

```
In[101]:= Randomθ = Table[Random[Real, {-10.°, 10.°}], {100}];
          Randomϕ = Table[Random[Real, {-10.°, 10.°}], {100}];
          Randomψ = Table[Random[Real, {-10.°, 10.°}], {100}];

          Show[
            GraphicsArray[
              Table[
                ImplicitPlot[
                  RotatedEllipsoid ==
                        1/.{a → 5., b → 3., c → 1., Δx → 0., Δy → 0.,
                      z → 0.,
                      θ → Randomθ[[i * j]],
                      ψ → Randomψ[[i * j]],
                      ϕ → Randomϕ[[i * j]],
                      Δz → RandomOffsets[[i * j]]}, {x, -6., 6.},
                  Axes → None, PlotRange → {{-6., 6.}, {-4., 4.}},
                  AspectRatio → 4/6.,
                  DisplayFunction → Identity
                ], {i, 10}, {j, 10}
              ]
            ], DisplayFunction → $DisplayFunction
          ]
```

```
From In[101]:=
```

```
Out[101]= -GraphicsArray-
```

Although the procedure is a little more complicated than for the flat-lying ellipsoids, it isn't too difficult to generate an apparent clast size distribution curve for the randomly rotated ellipses. See the optional interlude at the end of the chapter if you're wondering how eigenvalues are related to the ellipse areas.

```
In[102]:= RotatedClastResults = Table[Null, {0}];

          Do[
            Block[{temp1, temp2},
              temp1 = Expand[
                  RotatedEllipsoid /.
                       {Δx → 0, Δy → 0, z → 0, a → 5., b → 3., c → 1.,
                     Δz → RandomOffsets[[i]],
                     θ → Randomθ[[i]],
                     ψ → Randomψ[[i]],
                     ϕ → Randomϕ[[i]]}
                ];
              temp2 = temp1 /. {x → 0, y → 0};
              EllipseEqn = Expand[temp1/temp2];

              AandB = 1/Sqrt[Eigenvalues[
                     {{Coefficient[EllipseEqn, x^2],
                       Coefficient[EllipseEqn, x y]/2},
                       {Coefficient[EllipseEqn, x y]/2,
                       Coefficient[EllipseEqn, y^2]}}]];
              AppendTo[RotatedClastResults,
                       √(AandB[[1]] AandB[[2]])]
            ], {i, 100}
          ]
```

```
In[103]:= Histogram[RotatedClastResults,
            BarStyle → GrayLevel[0.6]]
```

From In[103]:=

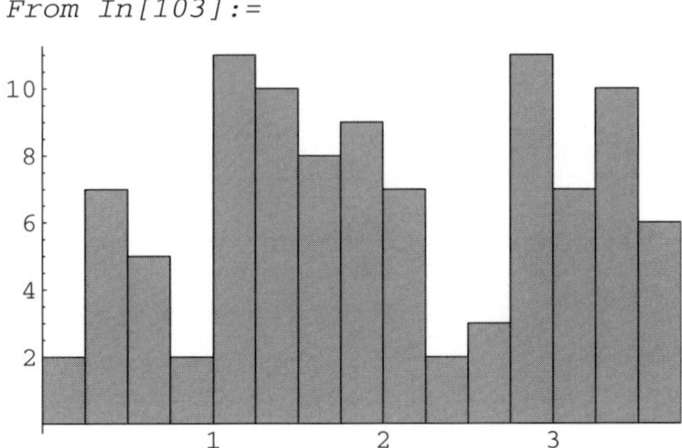

Out[103]= -Graphics-

```
In[104]:= CumFreqPlot[RotatedClastResults, 0, 4,
            AxesLabel → {"Equiv.\nSize", "Cum.\nProb"}]

From In[104]:=
```



```
Out[104]= -Graphics-
```

The apparent mean and its error are:

```
In[105]:= Mean[RotatedClastResults]

Out[105]= 2.00471
```

and

$$\text{In[106]:= } \mathbf{100\ Percent} \frac{\mathbf{Mean[RotatedClastResults] - TrueClastSize}}{\mathbf{TrueClastSize}}$$

```
Out[106]= -18.7132 Percent
```

which means that, in this case, adding random orientations changed the error in the apparent mean from about +20% to about –21% (the exact values will differ if you perform the simulation on your own computer with a different random seed). In terms of total clast volume,

$$\text{In[107]:= } \frac{4}{3}\pi \sum_{i=1}^{100} \mathbf{RotatedClastResults[[i]]}^3$$

```
Out[107]= 6028.1
```

$$\text{In[108]:= } \mathbf{100.}\ \frac{4}{3}\pi\, \mathbf{TrueClastSize}^3$$

```
Out[108]= 6283.19
```

$$\text{In[109]:= } \mathbf{100\ Percent} \frac{\%\% - \%}{\%}$$

```
Out[109]= -4.05987 Percent
```

Thus, when the possibility of random orientation is considered the outcrop effect causes the total clast volume to be underestimated instead of overestimated (as for the flat-lying ellipsoids).

The results will be different for larger degrees of angular variation. For example, here is an extreme case in which each of the three angles is allowed to vary over a range of ±89°:

```
In[110]:= Randomθ = Table[Random[Real, {-89.°, 89.°}], {100}];
          Randomϕ = Table[Random[Real, {-89.°, 89.°}], {100}];
          Randomψ = Table[Random[Real, {-89.°, 89.°}], {100}];

          Show[
            GraphicsArray[
              Table[
                ImplicitPlot[
                  RotatedEllipsoid == 1/.
                        {a → 5., b → 3., c → 1., Δx → 0., Δy → 0.,
                      z → 0.,
                      θ → Randomθ[[i * j]],
                      ψ → Randomψ[[i * j]],
                      ϕ → Randomϕ[[i * j]],
                      Δz → RandomOffsets[[i * j]]}, {x, -6., 6.},
                  Axes → None, PlotRange → {{-6., 6.}, {-6., 6.}},
                  AspectRatio → 1.,
                  DisplayFunction → Identity
                ], {i, 10}, {j, 10}
              ]
            ], DisplayFunction → $DisplayFunction
          ]

From In[110]:=
```

```
Out[110]= -GraphicsArray-
```

The apparent clast size distribution is:

```
In[111]:= RotatedClastResults2 = Table[Null, {0}];
          Do[
            Block[{temp1, temp2},
              temp1 = Expand[
                  RotatedEllipsoid /.
                        {Δx → 0, Δy → 0, z → 0, a → 5., b → 3., c → 1.,
                      Δz → RandomOffsets[[i]],
                      θ → Randomθ[[i]],
                      ψ → Randomψ[[i]],
                      ϕ → Randomϕ[[i]]}
                ];
              temp2 = temp1 /. {x → 0, y → 0};
              EllipseEqn = Expand[temp1/temp2];
              AandB = 1/Sqrt[Eigenvalues[
                      {{Coefficient[EllipseEqn, x^2],
                        Coefficient[EllipseEqn, x y]/2},
                        {Coefficient[EllipseEqn, x y]/2,
                        Coefficient[EllipseEqn, y^2]}}]];
              AppendTo[RotatedClastResults2,
                        √(AandB[[1]] AandB[[2]])]
            ], {i, 100}
          ]

In[112]:= InclinedClastSizeResults2 = Table[
              √(π inclinednewa inclinednewb) /.
                {Δx → 0., Δy → 0., Δz → RandomOffsets[[i]], a → 5.,
                  b → 3., c → 1., ψ → Randomψ[[i]],
                  θ → Randomθ[[i]], ϕ → Randomϕ[[i]]}, {i, 100}
            ];

In[113]:= Histogram[RotatedClastResults2,
            BarStyle → GrayLevel[0.6]]
```

From In[113]:=

Out[113]= -Graphics-

```
In[114]:= CumFreqPlot[RotatedClastResults2,
            Min[RotatedClastResults2],
            Max[RotatedClastResults2],
            AxesLabel → {"Equiv.\nSize", "Cum.\nProb"}]
```

From In[114]:=


```
Out[114]= -Graphics-
```

What about the mean clast size and error due to the outcrop effect?

```
In[115]:= Mean[RotatedClastResults2]

Out[115]= 0.535597
```

$$In[116]:= 100\ \mathbf{Percent}\frac{\mathbf{Mean[RotatedClastResults2] - TrueClastSize}}{\mathbf{TrueClastSize}}$$

```
Out[116]= -78.2826 Percent
```

The error in total clast volume is, in this simulation,

$$In[117]:= \frac{4}{3}\pi\sum_{i=1}^{100}\mathbf{RotatedClastResults2[[i]]}^3$$

```
Out[117]= 265.884
```

$$In[118]:= 100.\ \frac{4}{3}\pi\ \mathbf{TrueClastSize}$$

```
Out[118]= 1033.04
```

$$In[119]:= 100\ \mathbf{Percent}\frac{\%\% - \%}{\%}$$

```
Out[119]= -74.2621 Percent
```

We can safely conclude that variable clast orientation can combine with the outcrop effect to produce significant errors in the mean clast size and clast volume estimates. In the three simulations above, the errors in estimated mean values range from –75% to +25%. The errors in estimated total clast volume range over an even broader range, from about –56% to +113%. That's worth thinking about next time you're looking at an outcrop.

Optional Epilog: Calculating the Area of a Rotated Ellipse

It is easy to compute the area of an ellipse given the two semi-axes a and b. In the outcrop effect problem, however, the intersection of each random 3-D ellipsoid with the outcrop face produces an 2-D ellipse. For example, an ellipsoid with semi-axes $a = 5$, $b = 3$, and $c = 1$ that is rotated 20° about the z axis and intersects the x-y plane at $z = 0$ produces the ellipsoid

```
In[120]:= EllipseEqn = 0.0483 x^2 - 0.0457 x y + 0.103 y^2

Out[120]= 0.0483 x^2 - 0.0457 x y + 0.103 y^2
```

Here is a plot of the ellipse formed by the intersection:

```
In[121]:= ImplicitPlot[EllipseEqn == 1, {x, -5, 5}]

From In[121]:=
```

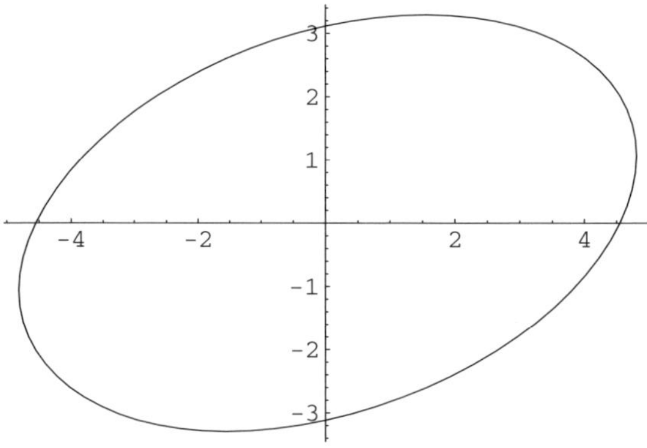


```
Out[121]= -Graphics-
```

The semi-axes correspond the the square roots of the reciprocals of the two eigenvalues shown below.

```
In[122]:= 1./
           Sqrt[
             Eigenvalues[
               {{Coefficient[EllipseEqn, x^2],
                 Coefficient[EllipseEqn, x y]/2.},
               {Coefficient[EllipseEqn, x y]/2.,
                 Coefficient[EllipseEqn, y^2]}}]
           ]

Out[122]= {2.9976, 4.99932}
```

From which the area can be calculated using the usual $A = \pi\, a\, b$.

5.7 References and Recommended Reading

Chow, V.T., Maidment, D.R., and Mays, L.W., 1988, *Applied Hydrology*: McGraw-Hill.

Costa, J.E., and Baker, V.R., 1981, *Surficial Geology*: John Wiley & Sons.

Crovelli, R.A., 2000, *Probability Models for Estimation of Number and Costs of Landslides*: U.S. Geological Survey Open-File Report 00-249, online version. http://pubs.usgs.gov/of/2000/ofr-00-0249/ProbModels.html.

Bouwer, H., 1978, *Groundwater Hydrology*: John Wiley & Sons.

Cullen, A.C. and Frey, H.C., 1999, *Probabilistic Techniques in Exposure Assessment*: Plenum Press.

Goodwin, L.B. and Haneberg, W.C., 1996, Deformational fabrics and inferred permeability of faulted sands from the Rio Grande rift, New Mexico: *Geological Society of America Abstracts with Programs, 1996 Annual Meeting*, v. 28, p. 255

Hammond, C., Hall, D., Miller, S., and Swetik, P., 1992, *Level I Stability Analysis (LISA) Documentation for Version 2.0*: U.S. Forest Service, Intermountain Research Station, General Technical Report INT-285.

Haneberg, W.C., 1991, Observation and analysis of short-term pore pressure fluctuations in a thin colluvium landslide complex near Cincinnati, Ohio: *Engineering Geology*, v. 31, p. 159–184.

Haneberg, W.C. and Gökce, A.Ö., 1994, *Rapid water-level fluctuations in a thin colluvium landslide west of Cincinnati, Ohio*: U.S. Geological Survey Bulletin 2059-C.

Haneberg, W.C., Goodwin, L. B., and Ferranti, C. J., 1994, Pseudotachylyte in a metamorphic core complex— analytical modeling of the effect of compositional variation on frictional melting: Geological Society of America, *1994 Annual Meeting Abstracts with Programs*, v. 26, n. 7, p. 269.

Harr, M.E., 1996, *Reliability-Based Design in Civil Engineering*: Dover Publications.

Heilbronner, R., 2002, *How to Derive Size Distributions of Particles from Size Distributions of Sectional Areas*: www.unibas.ch/earth/micro/manuals/ GrainSize_manual/GrainSize.pdf

Herbison-Evans, D., 2002, *Animated Cartoons by Computer Using Ellipsoids*: University of Sydney, Basser Department of Computer Science, Technical Report 94. http://linus.it.uts.edu.au/~don/pubs/cartoon.html

Huang, Y.H., 1983, *Stability Analysis of Earth Slopes*: Van Nostrand Reinhold.

Keaton, J.R., 1994, Risk-based probabilistic approach to site selection: *Bulletin of the Association of Engineering Geologists*, v. 31, p. 217-229.

Medley, E., 2002, Estimating block size distributions of melanges and similar block-in-matrix rocks (bimrocks), *in* Hammah, R., Bawden, W., Curran, J. and Telesnicki, M., editors, *Proceedings of 5th North American Rock Mechanics Symposium (NARMS)*, Toronto, Canada, July 2002: University of Toronto Press, p. 509-606E.

Reiter, L. 1990, *Earthquake Hazard Analysis*: Columbia University Press.

Sahagian, D. and Proussevitch, A.A., 1998, 3D particle size distributions fom 2D observations: stereology for natural applications: *Journal of Volcanology and Geothermal Research*, v. 84, p. 173–196

Wolfram, S., 1999, *The Mathematica Book (4th ed.)*: Cambridge University Press.

6 Interpolation and Regression

6.1 *Mathematica* Packages You Will Need

```
In[1]:= Needs["Graphics`"]
        Needs["Statistics`DescriptiveStatistics`"]
        Needs["Statistics`MultiDescriptiveStatistics`"]
        Needs["Statistics`LinearRegression`"]
        Needs["Statistics`NonlinearFit`"]
        Needs["Statistics`HypothesisTests`"]
        Needs["CompGeosci`"]
```

> **Computer Note:** The CompGeosci package will load correctly only if it is located in one of the directories in *Mathematica*'s standard file path. Execute the statement **`$Path`** to see a list of the default paths on your computer and place the file CompGeosci.m in one of those directories. The specific file paths may differ from one operating system to another. See Chapter 1 for more information about installing the CompGeosci package.

6.2 Interpolation or Regression: Which is Appropriate?

Interpolation and regression both involve the fitting of curves or surfaces to data, and both are equally easy to use in *Mathematica*. The choice of one or the other will depend on the nature of the data and the reason for fitting a curve or surface. Interpolation yields polynomial curves that pass exactly through each data point, and is useful when there is reason to infer that errors in the data are small or non-existent. It is, in essence, an exercise in connecting the dots. Regression, in contrast, yields a line or surface that minimizes the errors between the data and a modeled curve or surface according to some predefined criterion. It is useful when there is reason to infer that the data contain a substantial component of error or when the objective is to fit a function of a specific form to the data. For example, one might use interpolation to fill in the blanks within a grid of surveyed elevations and regression might be used to fit a plane representing the regional slope of the elevation data. Each of these tasks demands a different tool even though the data are the same.

6.3 Interpolation

6.3.1 Finding a Single Interpolating Polynomial

The *Mathematica* function **InterpolatingPolynomial** returns an $n-1$ order equation representing the polynomial passing through n data. For example, consider the set of 5 equally spaced elevation measurements below:

```
In[2]:= data = {177.5, 178., 178.8, 180.6, 182.6}

Out[2]= {177.5, 178., 178.8, 180.6, 182.6}
```

The 4th order polynomial that passes exactly through each data point is, specifying x as the independent variable,

```
In[3]:= f = Simplify[InterpolatingPolynomial[data, x]]

Out[3]= -0.0625 (-10.6878 + x) (4.56503 + x) (57.4215 - 5.7439 x + x^2)
```

Mathematica will accept **InterpolatingPolynomial** arguments either as a list of dependent variables, in which case it is assumed that the independent variables are 1, 2, 3..., or as a list of $\{x, f(x)\}$ pairs. The data and the interpolated curve can be compared by first plotting the data

```
In[4]:= dataplot = ListPlot[data, PlotStyle → PointSize[0.015],
          AxesLabel → {"x", "f(x)"}]

From In[4]:=
```

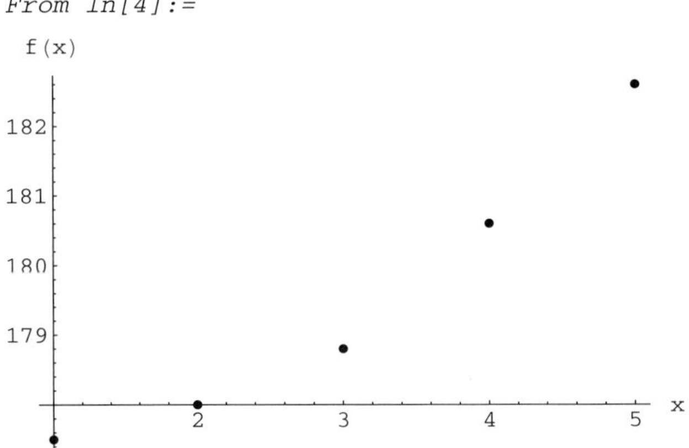


```
Out[4]= -Graphics-
```

and then the interpolated curve. The plot range has been deliberately chosen to exceed the range of the independent variable used to obtain the interpolation polynomial.

```
In[5]:= polyplot = Plot[f, {x, 0, 6}, AxesLabel → {"x", "f(x)"}]
```

```
From In[5]:=
```

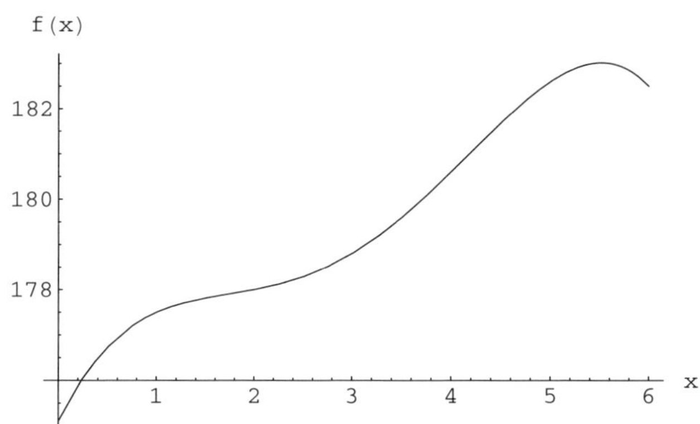


```
Out[5]= -Graphics-
```

Combining the two using **Show** illustrates that the interpolating polynomial does indeed pass exactly through each point and produces a reasonable result for values of $1 \leq x \leq 5$. Outside of the range of the data, however, the interpolated curve contains twists and turns that are not supported by the data and almost certainly not desirable. It is almost never a good idea to use an interpolated function outside of the range of the data from which it was derived!

```
In[6]:= Show[dataplot, polyplot]
```

```
From In[6]:=
```

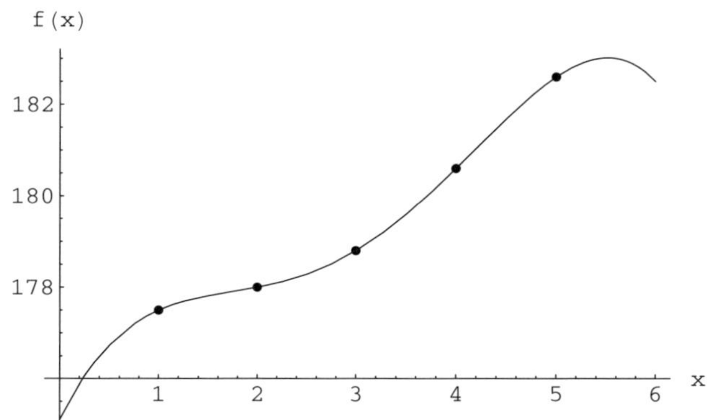


```
Out[6]= -Graphics-
```

InterpolatingPolynomial can also work with data points that are not uniformly spaced as long as their x values are known. To illustrate this, explicitly assign an x coordinate to each value in **data**.

```
In[7]:= data = Table[{i, data[[i]]}, {i, Length[data]}]

Out[7]= {{1, 177.5}, {2, 178.}, {3, 178.8}, {4, 180.6}, {5, 182.6}}
```

Now, remove one of the values using **Drop**

```
In[8]:= data = Drop[data, {3}]

Out[8]= {{1, 177.5}, {2, 178.}, {4, 180.6}, {5, 182.6}}
```

and perform the interpolation.

```
In[9]:= Simplify[InterpolatingPolynomial[data, x]]

Out[9]= -0.00833333 (-47.4224 + x) (449.408 + 8.42236 x + x^2)
```

Note that, because the number of points has been reduced, the result is a 3rd order polynomial. The line interpolated from the reduced data set can be compared to the original data set by superimposing plots.

```
In[10]:= Plot[%, {x, 1, 5}, AxesLabel → {"x", "f(x)"},
           DisplayFunction → Identity]

Out[10]= -Graphics-

In[11]:= Show[%, dataplot, DisplayFunction → $DisplayFunction]

From In[11]:=
```

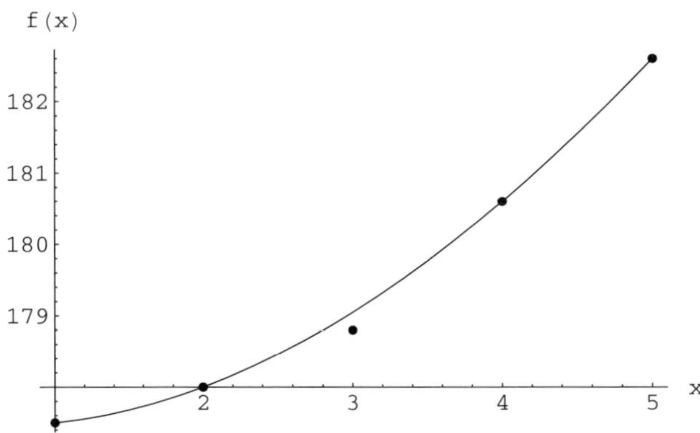


```
Out[11]= -Graphics-
```

The new interpolated line passes exactly through points 1, 2, 4, and 5 but does not pass through point 3 (which was dropped from the data set).

Problems with High-Order Polynomials

One of the drawbacks to using a single polynomial that passes through each data point is that the order of the polynomial will increase with the number of data, which can lead to unreasonably large fluctuations in the interpolated curve between the data points. To illustrate, we can use a data set consisting of 20 elevation measurements (the first 5 of which are the same as before).

```
In[12]:= data = {177.5, 178., 178.8, 180.6, 182.6, 184.8, 187.3,
           190.2, 194.3, 198.8, 201.6, 202.6, 202.9, 203.8, 205.4,
           207., 211.9, 217.1, 221.1, 222.6}

Out[12]= {177.5, 178., 178.8, 180.6, 182.6, 184.8, 187.3, 190.2,
           194.3, 198.8, 201.6, 202.6, 202.9, 203.8,
           205.4, 207., 211.9, 217.1, 221.1, 222.6}
```

InterpolatingPolynomial will produce a 19th order polynomial.

```
In[13]:= f = Simplify[InterpolatingPolynomial[data, x]]
```

Out[13]= $-13268.4 + 47251.7\,x - 72328.3\,x^2 + 65041.4\,x^3 - 38910.5\,x^4 + 16574.7\,x^5 - 5242.59\,x^6 + 1265.62\,x^7 - 237.501\,x^8 + 35.057\,x^9 - 4.09788\,x^{10} + 0.380203\,x^{11} - 0.0279409\,x^{12} + 0.00161515\,x^{13} - 0.0000724897\,x^{14} + 2.47209\times10^{-6}\,x^{15} - 6.18551\times10^{-8}\,x^{16} + 1.0701\times10^{-9}\,x^{17} - 1.14302\times10^{-11}\,x^{18} + 5.67717\times10^{-14}\,x^{19}$

As above, we can compare the data to the interpolated curve by superimposing plots

```
In[14]:= dataplot = ListPlot[data, PlotStyle → PointSize[0.015],
            AxesLabel → {"x", "f(x)"}, DisplayFunction → Identity];
          polyplot = Plot[f, {x, 1, 20}, AxesLabel → {"x", "f(x)"},
            DisplayFunction → Identity];
          Show[dataplot, polyplot,
            DisplayFunction → $DisplayFunction]
```

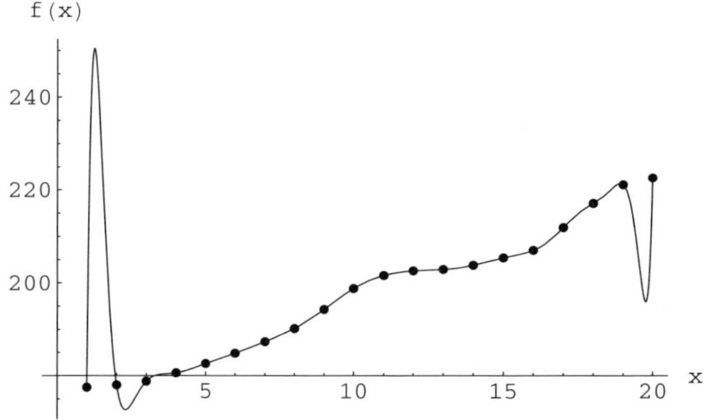


```
Out[14]= -Graphics-
```

The polynomial constains oscillations that are not supported by the data, but it does pass exactly through each point. Although the result is correct in the sense that it fulfills its mathematical obligation to pass through each point in **data**, it is poor because it adds oscillations that are not consistent with the general behavior of the data. Therefore, the use of **InterpolatingPolynomial** will almost always be limited to situations where the objective is to obtain an algebraic expression for a curve passing through just a few points.

6.3.2 Piecewise Polynomial Interpolation

When dealing with more than a handful of data, the objective of interpolation is generally to generate a well-behaved curve rather than to generate a single polynomial. This can be accomplished by numerically interpolating a series of low-order polynomial curves through successive points (*i.e.*, piecewise). Again using **data** to illustrate, the interpolation is performed using **Interpolation**.

```
In[15]:= f2 = Interpolation[data]

Out[15]= InterpolatingFunction[{{1., 20.}}, <>]
```

Interpolation will accept lists of data with or without independent variable values. If the values are not specified, they are assumed to be 1, 2, 3... The result of **Interpolation** is returned is a *Mathematica* function known as an interpolating function. It can be used just like any other function, for example

```
In[16]:= Plot[f2[x], {x, 1, 20}, AxesLabel → {"x", "f(x)"},
           DisplayFunction → Identity];
           Show[%, dataplot, DisplayFunction → $DisplayFunction]

From In[16]:=
```

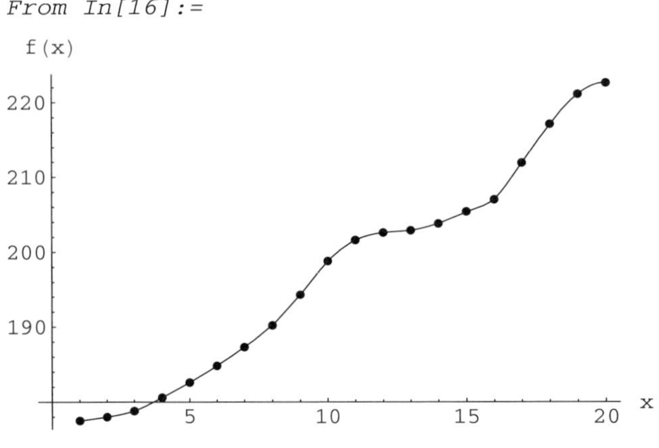


```
Out[16]= -Graphics-
```

The result represents the data much better than did the 19th order polynomial because **Interpolation** defaults to a succession of 3rd order polynomials passing through adjacent points. Why 3rd order? Because it is the lowest order polynomial for which curvature will be continuous (*i.e.*, its second derivative is not zero). Lower order polynomials can be used, but the result may be a jagged curve if there are many changes in slope. To specify a different order polynomial, use the option**InterpolationOrder** → n, where n is the desired order.

Interpolation will also work with irregularly spaced data sets as long as values for the independent variable are supplied. Refer to the *Mathematica* documentation for details.

> **Computer Note:** Following the example used for **Interpolating-Polynomial**, add x coordinates to **data**, drop at least one of its points, and then use **Interpolation** on the irregularly spaced result.

Mathematica can also interpolate multidimensional sets of equally spaced or gridded data. Consider the following table of gridded taken from a digital elevation model (Chapter 7 contains more information about using *Mathematica* to plot and analyze digital elevation data).

```
In[17]:= data2 = {{205.8, 208.3, 213.7, 218.5, 221.3, 222.4},
           {206.5, 210., 215.5, 220.3, 222.6, 223.},
           {207., 211.9, 217.1, 221.1, 222.6, 221.9},
           {207.6, 212.9, 217.8, 220.6, 220.3, 219.2},
           {207.5, 212.6, 215.8, 217., 216.4, 215.1},
           {205.8, 209.8, 212.5, 212.2, 211.2, 209.4}};
```

Mathematica uses linear interpolation to connect the points of known elevation.

```
In[18]:= ListPlot3D[data2, ColorOutput → GrayLevel]
```

```
From In[18]:=
```

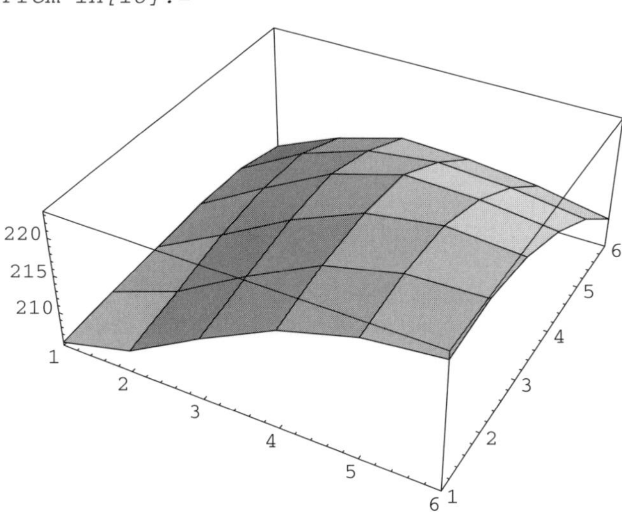


```
Out[18]= -SurfaceGraphics-
```

Suppose that the elevation data are located on 30 m centers, but that there is a need to estimate elevation values every 10 m. This can be most easily accomplished using **ListInterpolation**

```
In[19]:= ListInterpolation[data2]

Out[19]= InterpolatingFunction[{{1., 6.}, {1., 6.}}, <>]
```

The result is an interpolating function similar to that obtained from 1D interpolation. When the interpolation function is plotted, however, the order of the two spatial coordinates must be reversed so that the orientation corresponds to the surface in produced above by **ListPlot3D**. The discrepancy arises because **ListPlot3D** assumes coordinates are given by row and then column, whereas **Plot3D** assumes that they are given first by the x coordinate (which corresponds to the column number) and then the y coordinate (which corresponds to the row number).

```
In[20]:= Plot3D[%[x, y], {y, 1, 6}, {x, 1, 6},
          ColorOutput → GrayLevel, PlotPoints → 31]

From In[20]:=
```

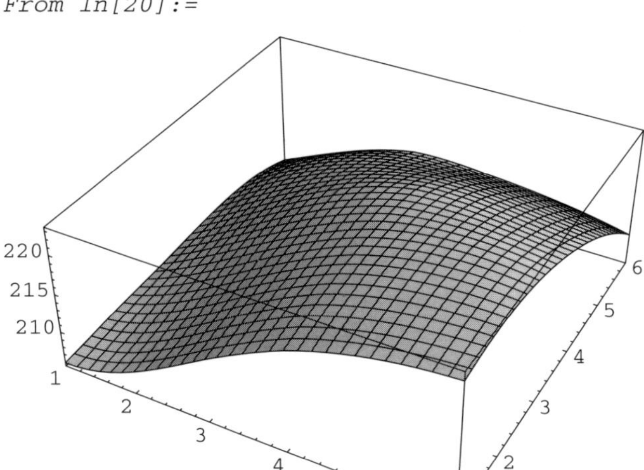


```
Out[20]= -SurfaceGraphics-
```

The ranges of the dependent variables can also be given to **ListInterpolation**. For example, if the elevation values in **data2** are gridded on 30 m centers beginning with the point $x = 100$ m and $y = 200$ m, then $100 \le x \le 250$ m and $200 \le y \le 350$ m.

```
In[21]:= ListInterpolation[data2, {{100, 250}, {200, 350}}]

Out[21]= InterpolatingFunction[{{100., 250.}, {200., 350.}}, <>]
```

The result is an interpolating function in terms of the specified coordinate ranges.

```
In[22]:= Plot3D[%[x, y], {y, 200, 350}, {x, 100, 250},
          ColorOutput → GrayLevel, PlotPoints → 31]
```

From In[22]:=

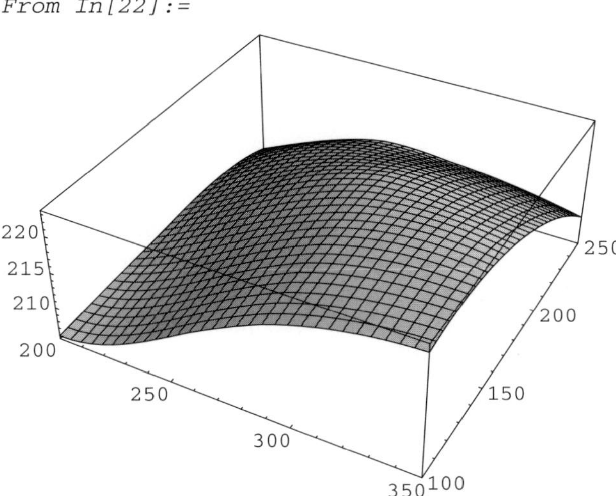

Out[22]= -SurfaceGraphics-

ListInterpolation works with arrays of any dimension, but the data must be regularly spaced or gridded. Although **Interpolation** will accept irregularly spaced values in one dimension, it will not do so in two or more dimensions. Chapter 7 discusses different gridding methods that can be used to interpolate irregularly spaced values in two or more dimensions.

6.4 Linear Regression

The objective of linear regression is to determine the coefficients of a linear polynomial curve or surface that minimize the difference between the data and the curve or surface. In that regard, linear regression is significantly different than interpolation because a regression curve does not necessarily pass exactly through all of the points. In fact, it would be very unsual for the curve to do so. The classic application of linear regression is to fit curves to experimental or field data that are inferred to obey some particular function but are known to contain errors only in the dependent variable. Another application might be to identify trends in noisy data, for example by calculating an average slope from the elevation data set used to illustrate interpolation methods. In that case, deviations from the regression line, known as residuals, would not represent errors in the sense that they are incorrect values; instead, they would represent small scale variability (noise) superimposed on a larger scale trend. Carr (2002) gives an interesting explanation of the historical evolution of the term regression.

Linear regression is restricted to functions of the independent variable (or variables) in which the coefficients are linear, for example

$$y = c_1 + c_2 x$$

$$y = c_1 + c_2 x c_3 x^2$$

$$y = c_1 + c_2 x + c_3 z$$

$$y = c_1 + c_2 \sin(\pi x/L) + c_3 \sin(2\pi x/L)$$

When the function being fitted is nonlinear in the independent variable x, as in the second example above, the procedure is known as **polynomial regression** (although it may sometimes be incorrectly called nonlinear regression). Despite the nonlinearity in x, polynomial regression is a form of linear regression because the coefficients remain linear. Middleton (2000) gives a good description of the differences between nonlinear functions with linear coefficients and functions with nonlinear coefficients. When the function contains two or more independent variables, as in the third example above, the procedure is known as **multiple regression**. It is also possible to perform polynomial multiple linear regressions. An example of a function that cannot be fitted using linear regression is $y = c_1 \exp(-c_2 x)$. Exponential relationships of this form have been used to characterize depth *vs.* porosity relationships in sediments and sedimentary rocks, so they are of interest to a variety of geoscientists. Nonlinear functions can be fitted using the nonlinear regression methods discussed further on in this chapter.

6.4.1 Derivation of Linear Least Squares Equations

Mathematica includes several functions that fit curves and surfaces to data, but it is instructive to work through the linear regression calculations step-by-step before introducing functions such as **Fit** and **Regress**. This will provide a basic understanding of the calculations while at the same time illustrating how *Mathematica*'s symbolic manipulation capabilities can be used to derive equations of interest to geoscientists.

To illustrate how simple linear regression lines are determined, we will use some rainfall and groundwater level data from a landslide along the Ohio River valley near Cincinnati, Ohio (Haneberg and Gökce, 1994). The data consist of rainfall (in mm) and resulting water level changes (in cm) from 14 separate precipitation events during March, April, and May 1980.

```
In[23]:= data = {{1.94, 2.5}, {3.33, 1.89}, {3.22, 1.67},
           {5.67, 1.31}, {4.72, 1.02}, {3.89, 0.96}, {2.78, 1.1},
           {10.56, 0.15}, {9.44, 3.92}, {12.78, 5.23},
           {14.72, 4.22}, {13.61, 3.63}, {20.39, 4.32},
           {38.89, 5.89}};

In[24]:= len = Length[data]

Out[24]= 14
```

An exploratory plot shows that the data follow a trend but also contain a fair amount of scatter.

```
In[25]:= dataplot = ListPlot[data,
           AxesLabel → {"rain\n(mm)", "ΔWL (cm)"},
           PlotStyle → PointSize[0.015],
           PlotRange → {{0, 40}, {0, 6}}]
```

From In[25]:=

ΔWL (cm)

rain (mm)

Out[25]= -Graphics-

Our objective is to determine the straight line $y = c_1 + c_2x$ that best fits the data. It is obvious that there will be no single straight line that passes through all of the data, so we will have to develop a criterion to define what we mean by the best fit. In this example, we will assume that rainfall is the independent variable (x) and water level change is the dependent variable (y) measured with error. This is an important distinction in regression analysis, because standard methods assume that the independent variable is known without error and the dependent variable can be measured only with some experimental error. Special techniques, which will be described further on in this chapter, must be used if both variables contain errors.

The most common approach to linear regression is based on the minimization of the squares of errors between the regression line and the dependent variable, hence the name **least squares**. In order to illustrate the general approach, it will be helpful to use two symbolic arrays representing the rainfall (**x[i]**) and water level (**y[i]**) values. Our criterion will be that the best-fitting line minimizes the sum of squared errors between the line and the data, $\epsilon^2 = \sum_{i=1}^{n}(y_i - \hat{y}_i)^2$ where y_i are the observed data and $\hat{y}_i = c_1 + c_2x_i$ is the equation of the line that we are fitting. We define the sum of the squared errors as

In[26]:= $\epsilon 2 = \sum_{i=1}^{n} (y[i] - c1 - c2\, x[i])^2$

Out[26]= $\sum_{i=1}^{n} (-c1 - c2\, x[i] + y[i])^2$

The sums of squares of the errors can be minimized by taking the derivative of ϵ^2 with respect to c_1 and c_2 and setting the result equal to zero. In this case the variables of interest are not x and y, which are both known at each point, but instead the unknown coefficients c_1 and c_2. First, find the derivatives of ϵ^2

```
In[27]:= Simplify[∂c1 ϵ2 ]
```

$$Out[27]= \sum_{i=1}^{n} -2\ (-c1 - c2\ x[i] + y[i])$$

Next, collect terms to put the result into a more easily understandable form

```
In[28]:= Collect[%, {c1, c2}]
```

$$Out[28]= \sum_{i=1}^{n} -2\ (-c1 - c2\ x[i] + y[i])$$

Inspection shows that the result above is of the form $2\ n\ c_1 + 2\ \sum_{i=1}^{len} x_i\ -\ 2\ \sum_{i=1}^{len} y_i$. Unfortunately, *Mathematica* will not solve equations involving sums with symbolic limits such as n, and the result would be exceptionally messy if we let $n = 14$ to correspond to the number of data (although that method does work). As shorthand, we will use **Sx** and **Sy** to represent the two sums, and **n** to represent the total number of data without specifically using the value of 13. Thus, the results we obtain will be applicable to data sets of any length. Setting the result above equal to zero and dividing through by 2, we get:

```
In[29]:= eq1 = c1 n + c2 Sx - Sy == 0;
```

> **Computer Note:** Use *Mathematica* replacement rules to replace the sums with shorthand variables such as **Sx** and **Sy** instead of manually entering the new equations.

The same procedure can be repeated for the second constant.

```
In[30]:= Simplify[∂c2 ϵ2]
```

$$Out[30]= \sum_{i=1}^{n} -2\ x[i]\ (-c1 - c2\ x[i] + y[i])$$

```
In[31]:= Collect[%, {c1, c2}]
```

$$Out[31]= \sum_{i=1}^{n} -2\ x[i]\ (-c1 - c2\ x[i] + y[i])$$

A quick inspection shows that this result contains three sums and is of the form $2\ c_1\ \sum_{i=1}^{len} x_i + 2\ c_2 \sum_{i=1}^{len} x_i^{\ 2} - 2\ \sum_{i=1}^{len} x_i y_i$. As above, we can rewrite it using shorthand terms for the sums and set the result equal to zero.

```
In[32]:= eq2 = c1 Sx + c2 Sx2 - Sxy == 0;
```

Now that the equations have been assembled, they can be solved to determine the two constants.

```
In[33]:= Simplify[Solve[{eq1, eq2}, {c1, c2}]]
```

$$Out[33]=\ \left\{\left\{c1 \to \frac{Sx\,Sxy - Sx2\,Sy}{Sx^2 - n\,Sx2},\ c2 \to \frac{-n\,Sxy + Sx\,Sy}{Sx^2 - n\,Sx2}\right\}\right\}$$

As usual, the constants can be extracted from the list of replacement rules and assigned to a variable name for future use.

```
In[34]:= constants = %[[1]]
```

$$Out[34]=\ \left\{c1 \to \frac{Sx\,Sxy - Sx2\,Sy}{Sx^2 - n\,Sx2},\ c2 \to \frac{-n\,Sxy + Sx\,Sy}{Sx^2 - n\,Sx2}\right\}$$

The equation for the regression line is in general

```
In[35]:= c1 + c2 x /. constants
```

$$Out[35]=\ \frac{Sx\,Sxy - Sx2\,Sy}{Sx^2 - n\,Sx2} + \frac{(-n\,Sxy + Sx\,Sy)\,x}{Sx^2 - n\,Sx2}$$

and is in this particular case

```
In[36]:= regressionline =
```

$$\%\ /.\ \left\{n \to 14,\ Sx\text{-}> \sum_{i=1}^{len} data[[i,1]],\ Sx2\text{-}> \sum_{i=1}^{len} data[[i,1]]^2,\ Sxy\text{-}> \sum_{i=1}^{len} data[[i,1]]\,data[[i,2]],\ Sy\text{-}> \sum_{i=1}^{len} data[[i,2]]\right\}$$

```
Out[36]= 1.26602 + 0.13763 x
```

The goodness-of-fit can be visually evaluated by superimposing the regression line on the data plot.

```
In[37]:= plot1 = Plot[%, {x, 0, 40},
           AxesLabel → {"rain\n(mm)", "ΔWL (cm)"},
           DisplayFunction → Identity];
         Show[%, dataplot, DisplayFunction → $DisplayFunction]
```

From In[37]:=

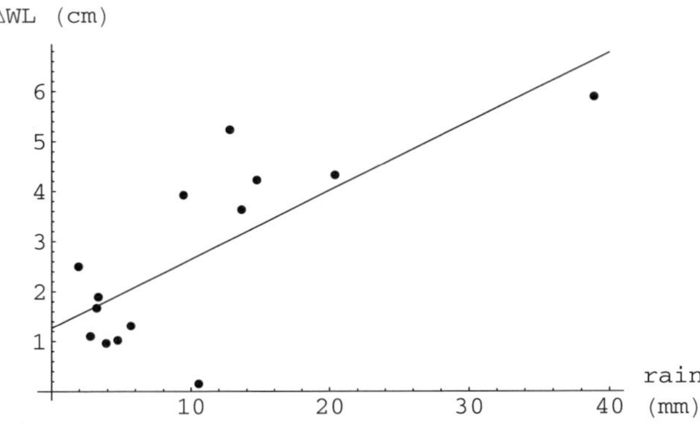


```
Out[37]= -Graphics-
```

There appears to be a reasonably good fit between the data and the regression line. One aspect that needs to be addressed is the physical significance of the regression line, which has a non-zero y intercept. According to the regression line, water level will incease by about 12 cm even if no rain falls. We will accept this dilemma for now and raise the issue again when we discuss the fitting of curves other than straight lines. The next three sub-sections will also explore the issue of goodness-of-fit in more detail.

6.4.2 Residuals

Residuals are the differences between the data and the regression line, and can be easily calculated in *Mathematica* if we first construct tables of the predicted and observed values.

```
In[38]:= regressionline /. x → data[[3, 1]]

Out[38]= 1.70919

In[39]:= predicted =
           Table[regressionline /. x → data[[i, 1]], {i, len}]

Out[39]= {1.53302, 1.72433, 1.70919, 2.04638, 1.91563, 1.8014,
           1.64863, 2.71939, 2.56525, 3.02493, 3.29193, 3.13916,
           4.0723, 6.61845}
```

The preceeding statement simply evaluates the expression **regressionline** at each of the x values. The next statement creates a table of observed values. While not strictly necessary (we could have used the original **data** table), it puts the predicted and observed values into lists of the same format that will be useful as our regression analysis progresses.

```
In[40]:= observed = data[[All, 2]]

Out[40]= {2.5, 1.89, 1.67, 1.31, 1.02, 0.96,
           1.1, 0.15, 3.92, 5.23, 4.22, 3.63, 4.32, 5.89}
```

The statement **data[[All, 2]]** is a quick way to extract a single column, in this case the second column, from a table. The same result could have been obtained by looping through each row of data.

```
In[41]:= Table[data[[i, 2]], {i, len}]

Out[41]= {2.5, 1.89, 1.67, 1.31, 1.02, 0.96,
           1.1, 0.15, 3.92, 5.23, 4.22, 3.63, 4.32, 5.89}
```

Because of the way that *Mathematica* handles lists, there is no need to subtract terms one-by-one. Instead, one list is subracted from the other.

```
In[42]:= residuals = predicted - observed

Out[42]= {-0.966978, -0.165673, 0.0391882, 0.736382, 0.895633,
          0.8414, 0.548631, 2.56939, -1.35475, -2.20507,
          -0.928066, -0.490836, -0.247704, 0.728452}
```

If the residuals are due solely to random experimental error, we would expect them to be normally distributed around a mean value of zero. In fact, the existence of normally distributed errors is one of the underlying assumptions of the linear regression method. The mean in this example is indeed very close to zero.

```
In[43]:= Mean[residuals]
```

Out[43]= 6.34413×10^{-17}

The null hypothesis that there is no significant difference between the residual distribution and a normal distribution can be evaluated using a Kolmogorov-Smirnov test as discussed in Chapter 4. Below is a cumulative plot of the residuals (solid line) and a normal distribution having the sample mean and variance of the residuals (dashed line).

```
In[44]:= KSOneListPlot[residuals, -4, 4, AxesOrigin → {-4, 0},
          AxesLabel → {"residual\n(cm)", "cum. prob."}]

From In[44]:=
```



```
Out[44]= -Graphics-
```

The K-S statistic is

```
In[45]:= KSOneList[residuals]

Out[45]= 0.153801
```

and the K-S probability is

```
In[46]:= KSProb[%, len]

Out[46]= 0.866329
```

Therefore, there is an 87% chance of committing a Type I error if we reject the null hypothesis that the residuals are normally distributed. A stem plot showing the residuals as a function of rainfall provides more insight into their distribution.

```
In[47]:= ListStemPlot[Table[{data[[i, 1]], residuals[[i]]},
           {i, len}], 0.02, AxesLabel → {"rain\n(mm)", "residual"},
           PlotRange → All, AxesOrigin → {1, 0}]
```

From In[47]:=

residual

2

1

5 10 15 20 25 30 35

rain
(mm)

-1

-2

Out[47]= -Graphics-

It appears that, although they are normally distributed around a mean of zero, the residuals may be clustered to some degree. This may indicate that a more complicated polynomial equation would provide a better fit, but for now we will accept the results and proceed.

6.4.3 Goodness-of-Fit and the Correlation Coefficient

The goodness-of-fit of a regression line can be quantitatively evaluated by calculating a correlation coefficient. To do so, first calculate the mean value of the groundwater level measurements.

```
In[48]:= meanval = Mean[data[[All, 2]]]
```

Out[48]= 2.70071

> **Computer Note:** **Mean** is a built-in function in *Mathematica* 5.0, but an add-on function in earlier versions. If you are using an earlier version, before using **Mean** you will have to load the standard package Statistics 'DescriptiveStatistics' using either **Needs** or **<<**. Refer to the *Mathematica* documentation for more information about loading packages.

The goodness-of-fit is defined as the ratio of the sum of squared deviations from the mean (often referred to simply as the sum of squares) of the observed values (total sum of squares, **SST**) and the predicted values (sum of squares due to regression, **SSR**). Both can be calculated using the **predicted** and **observed** lists created above.

```
In[49]:= SSR = Sum[(predicted[[i]] - meanval)^2, {i, 1, len}]
```

$$\texttt{SSR} = \sum_{i=1}^{len} (\texttt{predicted[[i]]} - \texttt{meanval})^2$$

```
Out[49]= 24.1556
```

```
In[50]:= SST = Sum[(data[[i, 2]] - meanval)^2, {i, 1, len}]
```

$$\texttt{SST} = \sum_{i=1}^{len} (\texttt{data[[i, 2]]} - \texttt{meanval})^2$$

```
Out[50]= 42.4667
```

The goodness-of-fit is thus

```
In[51]:= SSR/SST

Out[51]= 0.568813
```

Values of the goodness-of-fit can range from 0 to 1. Comparing the two sum of squares equations, it can be seen that the two will be equal if and only if the predicted values are exactly the same as the observed values, which will yield a value of 1. The goodness-of-fit is conventionally written as r^2or R^2, and its square root is the correlation coefficient r.

```
In[52]:= √SSR/SST

Out[52]= 0.754197
```

6.4.4 Significance of Regression Results: ANOVA

Goodness-of-fit and correlation coefficient values provide simple quantitative estimates of the adequacy of a regression line. It is obvious that $r = 0.99$ is a much better fit than, say, $r = 0.50$. We can also take another step and ask whether the regression line is statistically significant at a specified confidence level. To do this, we will employ a method known as analysis of variance (ANOVA). We will first postulate that the total sum of squares is the sum of two components: the sum of squares due to regression and the sum of squares due to errors between the observed and predicted values. Furthermore, we will proceed on the assumption that the statistical significance of the regression equation is somehow related to the two component sums of squares and the number of data used to obtain the regression curve.

Two of the three sums have already been calculated as **SST** and **SSR**, so the third can be found by subtraction.

```
In[53]:= SSE = SST - SSR

Out[53]= 18.3111
```

This is the same result that is obtained by summing the squares of the differences between predicted and observed values.

In[54]:= $\sum_{i=1}^{len}$ (predicted[[i]] - observed[[i]])^2

Out[54]= 18.3111

One seemingly obvious way to incorporate the number of data into the process is to calculate mean values for the sums of squares. In this case, however, calculation of mean sums of squares is not quite as simple as dividing the sums by the number of data. This is because mean values were used to calculate both **SSR** and **SST**, and those mean values were in turn calculated from a sample of an underlying data set rather than from the complete (and infinite) population of possible rainfall and groundwater level values. To account for this fact, the degrees of freedom associated with **SST** are reduced from **len** to **len - 1**. This is the same logic that is used when distinguishing between sample and population standard deviations (see Chapter 4). Upon first consideration, it might seem that there would be 2 degrees of freedom associated with **SSR** because the regression line has two variables: its slope and *y* intercept. It turns out, however, that a line fitted to data using least squares methods will always pass through the mean of the dependent variable and, as above, the use of a sample mean requires that the degrees of freedom associated with **SSR** be reduced from 2 to 1. The number of degrees of freedom associated with **SSE** is the difference between those associated with **SST** and **SSR**, or **len - 2**. All that having been written, we can now calculate the means of the sums of squares due to regression and error.

In[55]:= $\overline{\textbf{SSR}}$ = **SSR/1.**

Out[55]= 24.1556

In[56]:= $\overline{\textbf{SSE}}$ = **SSE/(len - 2)**

Out[56]= 1.52592

The ratio of mean sums of squares can now be tested against the null hypothesis that the ratio is 1 (*i.e.*, there is no difference between the values) at a specified significance level using an *F* ratio test. *F* ratio tests are used to compare calculated variances from two data sets of populations, and variances are mean sums of squared deviations (see the definition in Chapter 4 or refer to a statistics text). If the null hypothesis is true, then the regression **is not** significant because its mean sum of squares is no different from that of the errors. If the null hypothesis is rejected, however, then the regression **is** significant at the specified level. We will use the common significance level of 0.05.

In[57]:= **FRatioPValue[$\overline{\textbf{SSR}}$/$\overline{\textbf{SSE}}$, 1, len - 2,
SignificanceLevel → 0.05]**

Out[57]= {OneSidedPValue → 0.00183053,
Reject null hypothesis at significance level → 0.05}

Would you have predicted that the regression line is statistically significant if you had plotted the results, and perhaps even calculated a correlation coefficient, but not performed the ANOVA? The statistical significance of a regression line, especially for a small number of data, is not always apparent unless an ANOVA is performed.

6.4.5 Using Fit and Regress

We have done quite a bit of work to estimate a straight line. The same results can be obtained in one step using the **Fit** function.

```
In[58]:= line1 = Fit[data, {1, x}, x]

Out[58]= 1.26602 + 0.13763 x
```

The list **{1, x}** describes the form of the polynomial to be fitted, and **x** identifies the independent variable. An analogous procedure can be followed for higher order polynomials or multi-dimensional surfaces. The correlation coefficient can be quickly calculated using the **Correlation** function in the standard package Statistics`MultiDescriptiveStatistics`

```
In[59]:= Correlation[observed, predicted]

Out[59]= 0.754197
```

The standard package Statistics`LinearRegression` contains the function **Regress**, which performs linear regression and returns the results, including diagnostic statistics such as r^2, estimated errors for each of the regression coefficients, and ANOVA results as a list of replacement rules. It is used the same way as **Fit**, for example **Regress[data, {1, x}, x]**. One drawback to **Regress**, however, is that the table it produces is too wide to be shown on a printed page if the type is large enough to be legible. Therefore, it is not shown here.

Computer Note: Use **Regress** to calculate a regression line for the rainfall and water level data set, then compare the results to those obtained in the step-by-step demonstration above.

Computer Note: *Mathematica* 5.0 includes the new function **FindFit**, which performs both linear and nonlinear regression. The syntax of its use is slightly different than that of **Fit** and **Regress**, and its main advantage is that it will automatically decided whether linear or nonlinear regression is appropriate.

Fitting Polynomials of Arbitrary Form

The **Fit** and **Regress** functions can be used to fit any linear polynomial to data. In the discussion above, it was noted that the pattern of residuals suggests that a different curve might yield a better fit to the rainfall and groundwater level data. We will try a curve of the form $y = \sqrt{x}$ for two reasons. First, the data seem to follow a concave-downward arc that might be better approximated by a $\sqrt{x}$ curve than a straight line. One physical justification for a concave-downward curve is that large storms might cause the infiltration capacity of the soil to be exceeded and generate runoff, so only a portion of the rain contributes to the water level change during heavy storms. A contrasting statistical explanation might be that the lone 40 mm rainfall value is an anomaly that exerts a disproportionately large influence on the shape of the trend. Remove that outlier, and the data follow a much stronger straight line trend. There is no way of knowing how the trend would appear if there were more data from storms in the 20 to 40 mm range. Therefore, the apparent concavity may or may not represent the actual hydrologic behavior of the hillside. Field observations are the only way to fill the gap. Second, a $\sqrt{x}$ curve passes through the origin, thereby eliminating the problem of a non-zero y intercept and producing a more physically plausible relationship than the straight line with a non-zero y intercept obtained above.

```
In[60]:= line2 = Fit[data, {√x}, x]

Out[60]= 0.93216 √x

In[61]:= plot2 = Plot[line2, {x, 0, 40},
           AxesLabel → {"rain\n(mm)", "ΔWL (cm)"},
           DisplayFunction → Identity];
         Show[%, dataplot, DisplayFunction → $DisplayFunction]

From In[61]:=
```

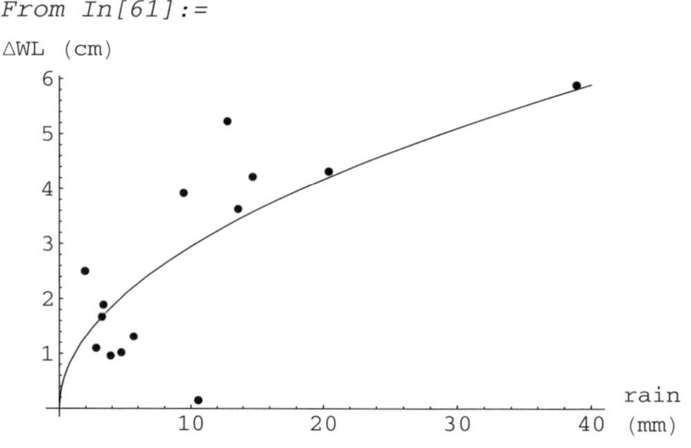


```
Out[61]= -Graphics-
```

To calculate a correlation coefficient, we need to generate a new list of predicted values for the $\sqrt{x}$ curve.

```
In[62]:= predicted = Table[line2 /. x → data[[i, 1]], {i, len}]

Out[62]= {1.29835, 1.70103, 1.6727, 2.21964, 2.02517, 1.83851,
           1.55422, 3.02916, 2.86402, 3.33239, 3.57639, 3.4389,
           4.2092, 5.81312}
```

It turns out that the correlation coefficient is just about the same as that calculated for the straight line (0.75 *vs*. 0.77). Therefore, the $\sqrt{x}$curve does not do an appreciably better job of explaining the data than the straight line even though we can make a physical argument for the $\sqrt{x}$ curve.

```
In[63]:= Correlation[observed, predicted]

Out[63]= 0.766548
```

> **Computer Note:** Perform an ANOVA for the $\sqrt{x}$ regression line to determine if it is more or less significant than the straight line. With regard to the problem of a non-zero y intercept, does the $\sqrt{x}$ provide a better or worse fit than a line of the form $y = c_1 x$?

6.4.6 Can I Solve for the Independent Variable?

The short answer is that yes, you can, but in general no, you should not! At least, you should not do it unless you have thought about your intended actions and are confident that you know what you are doing. It can be tempting to rearrange a regression equation to express the independent variable in terms of the dependent variable. A common reason is that someone has published a regression equation giving y in terms of x, but you want to the opposite: calculate x in terms of y. Except for special cases in which $r^2 = 1$, however, the results obtained by rearranging a regression equation to solve for x in terms of y will be different than those obtained by performing a new regression with x as the dependent variable.

To show what happens when the change in water level is used as the independent variable, switch the values in each element of the rainfall and water level data set.

```
In[64]:= switcheddata = Table[{data[[i, 2]], data[[i, 1]]},
           {i, Length[data]}];
```

Then, use **Fit** to find the regression line for rainfall in terms of water level change.

```
In[65]:= Fit[switcheddata, {1, y}, y]

Out[65]= -0.737532 + 4.13291 y
```

In order to plot this result on the same set of axes as the first regression line, we need to solve for y

```
In[66]:= Simplify[Solve[% == x, y]]

Out[66]= {{y → 0.178453 + 0.24196 x}}

In[67]:= line3 = y /. %[[1]]

Out[67]= 0.178453 + 0.24196 x
```

Now, plot but do not display the second regression line.

```
In[68]:= plot3 = Plot[line3, {x, 0, 40},
           PlotStyle → Dashing[{0.01}],
           DisplayFunction → Identity]

Out[68]= -Graphics-
```

and combine it with the data plot and previous straight-line regression plot to show how both lines are related to the data. The dashed line represents the regression line calculated using water level change as the independent variable, and the solid line represents the line calculated using rainfall as the independent variable.

```
In[69]:= Show[dataplot, plot1, plot3,
           DisplayFunction → $DisplayFunction]

From In[69]:=
```

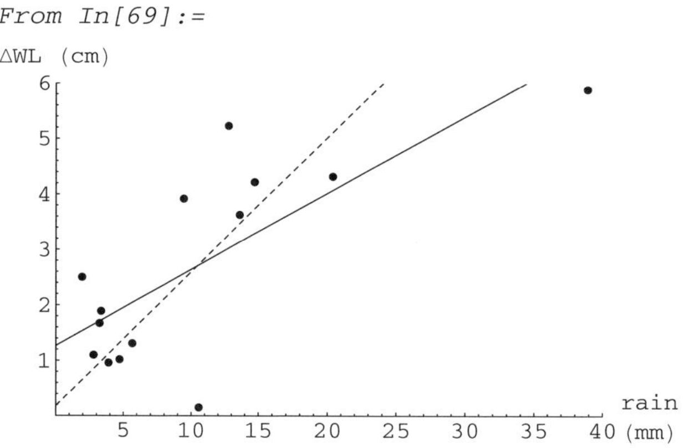


```
Out[69]= -Graphics-
```

The point at which the two lines intersect is given by the means of the rainfall and water level data. This can be demonstrated by calculating the mean values of each. The specification **data[[All, *i*]]** returns all of the values in the *i*th column of **data.**

```
In[70]:= rain = Mean[data[[All, 1]]]

Out[70]= 10.4243

In[71]:= WL = Mean[data[[All, 2]]]

Out[71]= 2.70071
```

Although they are clearly different, both regression lines appear to fit the data equally well. In fact, they both have the same *P* value and are therefore equally significant from a purely statistical perspective. Which one is correct? The answer depends on whether rainfall or change in water level is the dependent variable for which the sum of errors is minimized. In this case, it seems clear that rainfall is the independent variable, because it is unlikely that rainfall is dependent on the change in water level.

If your objective is to estimate a value for the independent variable in a situation where you have only a value for the dependent variable, then it might be permissible to rearrange the original regression equation. For example, the rain gauge was broken for one storm but you know the water level change and want to estimate the magnitude of the storm. If, however, you are asserting that what was the dependent variable should become the independent variable, then the correct approach is to calculate a new regression line. The point to remember is this: solving a regression equation for the independent variable is **not** the same as switching the variables and performing a new regression unless $r^2 = 1$.

Most discussions of linear regression assume that the dependent variable is the one associated with error, but this is something of an artificial distinction. It is also possible to imagine situations in which measurements of the independent variable have larger errors than measurements of the dependent variable. For example, there might be less error in measurements of water level change than in rainfall, even though rainfall is clearly the independent variable in terms of cause and effect. In such a situation it might be appropriate to perform linear regression by treating rainfall as if it were the dependent variable. A more common geoscientific situation is to have errors associated with both variables or two dependent variables, in which case one appropriate technique is **reduced major axis regression**.

Fitting Fourier Series to Data

Another application of **`Fit`** and **`Regress`** is the least-squares fitting of Fourier series to geoscientific data sets such as elevation profiles. This can be useful when there is a need to represent a complicated profile with a continuous function of distance (or time) rather than a set of discrete points. Haneberg (1999), for example, used Fourier series to simulate the effects of real topography on the subsurface state of stress in an aquifer system beneath the Rio Grande valley of central New Mexico. Hamming (1973) discusses the mathematical details of this process, and we will illustrate its application. Least-squares fitting of Fourier series can produce results similar to those obtained using numerical Fourier transforms, which are introduced in Chapter 8. One important difference is that least-squares methods can use profiles in which the data are irregularly spaced.

The underlying principle of Fourier series is that complicated curves can be approximated by adding together a series of sine or cosine curves. For example, if the curves $2 \sin(2\pi x)$, $5 \sin(3\pi x)$, and $3 \sin(4\pi x)$ are added together, the result is:

```
In[72]:= Plot[2 Sin[2 π x] + 5 Sin[3 π x] + 3 Sin[4 π x], {x, 0, 1}]
```

From In[72]:=

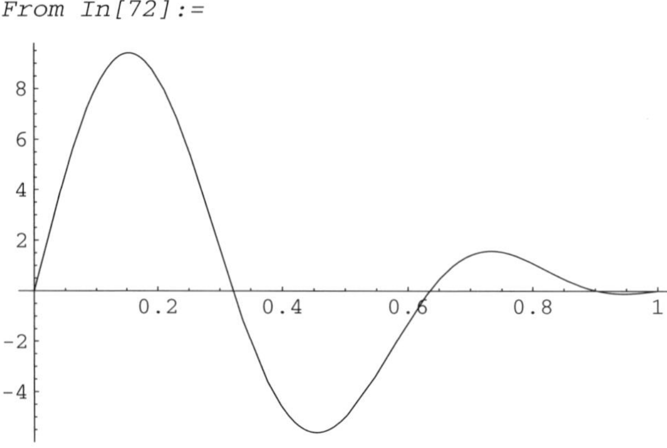

Out[72]= -Graphics-

Expanding upon this simple illustration, you might imagine how a large number of curves might be added together to approximate a complicated curve like a topographic profile. Our task will be to determine the amplitudes of the sine curves that are necessary to replicate a particular profile. This would be virtually impossible to accomplish by trial-and-error estimation, but it is very easy to do using least squares methods. To begin, read in a set of elevation values.

```
In[73]:= ReadList[
           "/Users/bill/Mathematica_Book/elevations.dat",
           Number];
```

> **Computer Note:** You will have to change the file path above to reflect the directory in which you have placed the file **elevation.dat**.

Because this data set consists of only one column, whereas **Import** assumes the general case of a multi-column table, it is just as easy to import the data using **ReadList** as it is to use **Import** and remove the extra set of brackets. The data are from a digital elevation model with 10 m spacing, however, so it is easy to create a new table that assigns horizontal coordinates starting with the first elevation value.

```
In[74]:= elev = Table[{(i - 1) * 10., %[[i]]}, {i, Length[%]}];
```

Here is a plot of **elev**:

```
In[75]:= topodataplot = ListPlot[elev,
           AxesLabel → {"dist", "elev"},
           PlotStyle → {GrayLevel[0.7], PointSize[0.01]}]
```

From In[75]:=

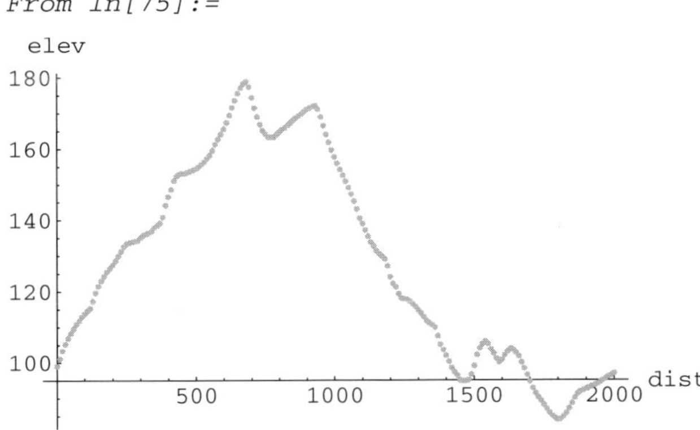

Out[75]= -Graphics-

We will begin by fitting a simple sine curve to the elevatation data

```
In[76]:= Fit[elev, {1, Sin[2 π x/2000]}, x]
```

Out[76]= $128.157 + 30.2979\ \mathrm{Sin}\left[\frac{\pi x}{1000}\right]$

and plotting the regression curve along with the data.

```
In[77]:= Plot[%, {x, 0, 2000}, AxesLabel → {"dist", "elev"},
           DisplayFunction → Identity];
         Show[topodataplot, %,
           DisplayFunction → $DisplayFunction]
```

From In[77]:=

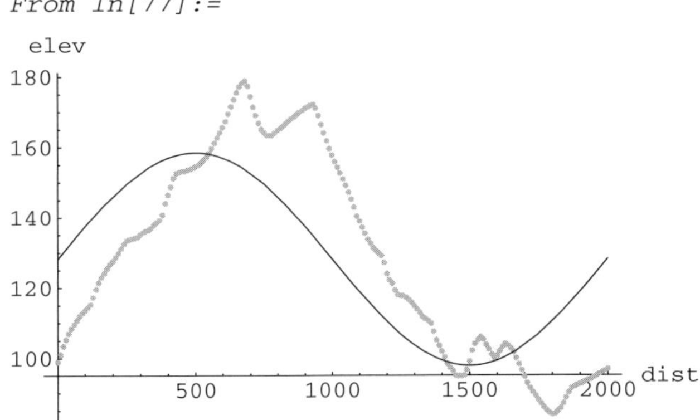

Out[77]= -Graphics-

Although the correlation is crude, it is easy to see how sine or cosine curves might be useful tools. To fit a Fourier series to data, first generate a table of the sine or cosine curves that will be used.

```
In[78]:= terms = Table[Sin[n π x/2000] , {n, 1, 10}]
```

Out[78]= $\left\{\text{Sin}\left[\frac{\pi x}{2000}\right], \text{Sin}\left[\frac{\pi x}{1000}\right], \text{Sin}\left[\frac{3\pi x}{2000}\right], \text{Sin}\left[\frac{\pi x}{500}\right], \text{Sin}\left[\frac{\pi x}{400}\right], \text{Sin}\left[\frac{3\pi x}{1000}\right], \text{Sin}\left[\frac{7\pi x}{2000}\right], \text{Sin}\left[\frac{\pi x}{250}\right], \text{Sin}\left[\frac{9\pi x}{2000}\right], \text{Sin}\left[\frac{\pi x}{200}\right]\right\}$

and then add the constant term using **Prepend**. Its compliment, **Append**, could have been just as easily used to add a term to the end of the list without affecting the regression results.

```
In[79]:= terms = Prepend[terms, 1]
```

Out[79]= $\left\{1, \text{Sin}\left[\frac{\pi x}{2000}\right], \text{Sin}\left[\frac{\pi x}{1000}\right], \text{Sin}\left[\frac{3\pi x}{2000}\right], \text{Sin}\left[\frac{\pi x}{500}\right], \text{Sin}\left[\frac{\pi x}{400}\right], \text{Sin}\left[\frac{3\pi x}{1000}\right], \text{Sin}\left[\frac{7\pi x}{2000}\right], \text{Sin}\left[\frac{\pi x}{250}\right], \text{Sin}\left[\frac{9\pi x}{2000}\right], \text{Sin}\left[\frac{\pi x}{200}\right]\right\}$

Now, use **Fit** in the usual way. Because **sinterms** is already a list with its own brackets, it does not have to be enclosed in another set. Doing so will return an incorrect result.

```
In[80]:= Fit[elev, terms, x]
```

Out[80]= $99.5485 + 47.7102\ \text{Sin}\left[\frac{\pi x}{2000}\right] + 30.2979\ \text{Sin}\left[\frac{\pi x}{1000}\right] - 7.16466\ \text{Sin}\left[\frac{3\pi x}{2000}\right] - 3.39745\ \text{Sin}\left[\frac{\pi x}{500}\right] + 0.782245\ \text{Sin}\left[\frac{\pi x}{400}\right] + 3.34785\ \text{Sin}\left[\frac{3\pi x}{1000}\right] - 2.26662\ \text{Sin}\left[\frac{7\pi x}{2000}\right] + 2.57889\ \text{Sin}\left[\frac{\pi x}{250}\right] + 0.0724944\ \text{Sin}\left[\frac{9\pi x}{2000}\right] + 1.22961\ \text{Sin}\left[\frac{\pi x}{200}\right]$

As usual, we can plot the regression curve along with the data.

```
In[81]:= Plot[%, {x, 0, 2000}, AxesLabel → {"dist", "elev"},
           DisplayFunction → Identity];
         Show[topodataplot, %,
           DisplayFunction → $DisplayFunction]
```

From In[81]:=

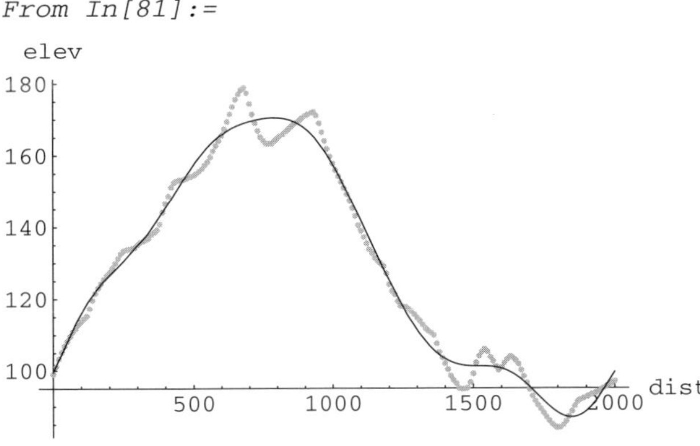

Out[81]= -Graphics-

The Fourier series represents many aspects of the topography well, but has trouble with the details. This problem can be addressed by adding some higher frequency (shorter wavelength) terms to deal with small-scale elevation changes. Because the expressions are long, we will use semicolons to suppress the output and display only the final plot of the data and Fourier series. Also notice that **dataplot** is listed before **%** in the **Show** statement in order to place the gray points in the background. Reversing the order would obscure much of the regression line with the points.

```
In[82]:= terms = Table[Sin[n π x/2000], {n, 1, 50}];

In[83]:= terms = Prepend[terms, 1];

In[84]:= line = Fit[elev, terms, x];

In[85]:= Plot[%, {x, 0, 2000}, AxesLabel → {"dist", "elev"},
           DisplayFunction → Identity];
         Show[topodataplot, %,
           DisplayFunction → $DisplayFunction]
```

From In[85]:=

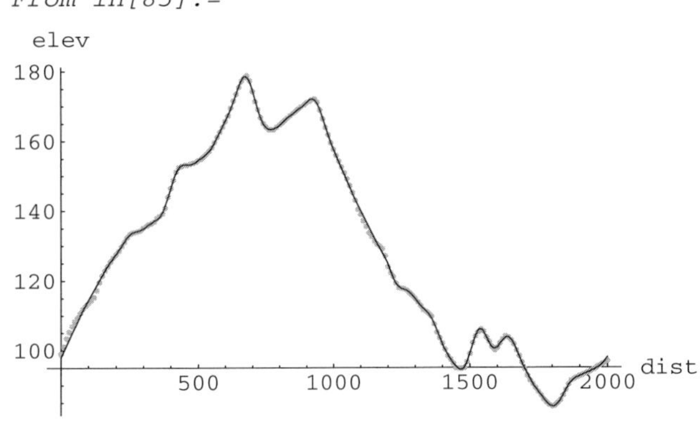

Out[85]= -Graphics-

A 51 term Fourier series reproduces the topography almost exactly. In most geoscientific applications, a few tens of terms provide an adequate representation of data. The relative contribution of each waveform can be illustrated by plotting the absolute value of its amplitude. This is conventionally done in terms of frequency, *f*, where $f = n/2$ is the number of cycles per 2000 m. The sum of terms in **line** can be treated as a list, so **line[[4]]** returns the fourth term of the summation.

```
In[86]:= line[[4]]

Out[86]= -6.51467 Sin[3 π x / 2000]
```

The coefficient can be isolated by using a replacement rule that sets the sin term to unity

```
In[87]:= line[[4]] /. Sin[__] → 1

Out[87]= -6.51467
```

This logic can be extended to create a table of frequencies and coefficients known as an amplitude spectrum. Notice that $i - 1$ is used to skip the first term of **line**, which is the constant.

```
In[88]:= Table[{(i - 1)/2, Abs[Part[line, i]]/. Sin[__] → 1},
           {i, 2, Length[line]}];
         ListStemPlot[%, 0.02, PlotRange → All,
           AxesOrigin → {0, 0}, AxesLabel → {"f", |Ai|}]

From In[88]:=
```

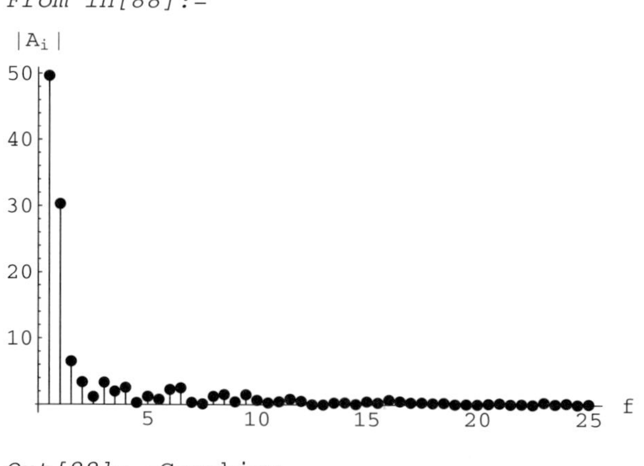


```
Out[88]= -Graphics-
```

6.4.7 Reduced Major Axis Regression: Two Variables with Error

Standard least-squares methods are not designed for situations in which there are errors associated with both variables, because they determine best-fit lines by minizing the error between the regression line and one of the variables (usually understood to be the dependent variable). A more complicated technique known as **reduced major axis regression** fits lines by minimizing the error along lines that are perpendicular to the regression line and pass through the data. That is to say, reduced major axis regression minimizes the product of the x and y errors, or$\sum_{i=1}^{n}(x_i - X_i)(y_i - Y_i)$ instead of just the sum of the squares of y errors, or $\sum_{i=1}^{n}(y_i - \hat{y})^2$. Carr (2002) describes in some detail a related but more complicated method known as **major axis regression**. Despite the fact that reduced major axis regression would seem to be the method of choice in many scientific problems where both variables have appreciable amounts of error, it is generally not discussed in statistics textbooks or included in many statstistics software packages.

Solution of the reduced major axis problem is more difficult than the ordinary least squares problem, and the derivation below is based on notes written by Wessel (2000). It uses **Lagrange multipliers** (*e.g.*, Arfken, 1985) to create a new function f by adding the function to be minimized to the constraints imposed when the function is minimized. In this case, the constraint is that the result must be a line of the form $c_1 + c_2 x - y = 0$. The new function is thus

$$In[89]:=\ \mathbf{f} = \sum_{i=1}^{n}((\mathbf{x[i]} - \mathbf{X[i]})(\mathbf{y[i]} - \mathbf{Y[i]})) + \sum_{i=1}^{n}(\lambda\mathbf{[i]}(\mathbf{c1} + \mathbf{c2\,x[i]} - \mathbf{y[i]}));$$

The upper-case **X** and **Y** variables represent the known data, whereas the lower-case **x**and **y** represent the corresponding point on the regression line yet to be determined. We differentiate f and set the results equal to zero in order to minimize the function. This time, however, there are four variables: x, y, c_1, and c_2.

$$In[90]:=\ \partial_{\mathbf{x[i]}}\,\mathbf{f}$$

$$Out[90]=\ \sum_{i=1}^{n}(y[i] - Y[i]) + \sum_{i=1}^{n} c2\,\lambda[i]$$

$$In[91]:=\ \partial_{\mathbf{y[i]}}\,\mathbf{f}$$

$$Out[91]=\ \sum_{i=1}^{n}(x[i] - X[i]) + \sum_{i=1}^{n} -\lambda[i]$$

$$In[92]:=\ \partial_{\mathbf{c1}}\,\mathbf{f}$$

$$Out[92]=\ \sum_{i=1}^{n}\lambda[i]$$

$$In[93]:=\ \partial_{\mathbf{c2}}\,\mathbf{f}$$

$$Out[93]=\ \sum_{i=1}^{n}x[i]\,\lambda[i]$$

As in the least squares example, we will use shorthand notation for the summations. Solving the rewritten forms of the first two derivatives above, we get

```
In[94]:= Solve[(Sy - SY) + c2 Sλ == 0, Sy]
Out[94]= {{Sy → SY - c2 Sλ}}
In[95]:= Solve[(Sx - SX) - Sλ == 0, Sx]
Out[95]= {{Sx → SX + Sλ}}
```

Next, substitute the two preceeding solutions into the straight-line equation

```
In[96]:= Sy == n c1 + c2 Sx /. %[[1]] /. %%[[1]]
Out[96]= SY - c2 Sλ == c1 n + c2 (SX + Sλ)
In[97]:= Solve[%, Sλ]
```

$$\text{Out[97]= } \left\{\left\{S\lambda \to \frac{-c1\,n - c2\,SX + SY}{2\,c2}\right\}\right\}$$

The third derivative above shows that $\sum_{i=1}^{n} \lambda[i] = 0$, so

```
In[98]:= Sλ == 0 /. %[[1]]
```

$$\text{Out[98]= } \frac{-c1\,n - c2\,SX + SY}{2\,c2} == 0$$

and, solving this result for c_1, we find that

```
In[99]:= Solve[%, c1]
```

$$\text{Out[99]= } \left\{\left\{c1 \to \frac{-c2\,SX + SY}{n}\right\}\right\}$$

The sums of X and Y divided by n are mean values, so in more traditional format $c_1 = \overline{Y} - c_2\overline{X}$, where the overbars denote mean values. The next task is to determine c_2. From the results obtained above we know that $\lambda = \frac{Y - c_2 X - c_1}{2\,c_2}$ and $x = X + \lambda$, and can therefore rewrite the fourth of the derivatives as

```
In[100]:= xλ == 0/.x → X + λ/.λ → (Y - c2X - c1)/(2c2)/.c1 → Ȳ - c2X̄
```

$$\text{Out[100]= } \frac{\left(X + \frac{-c2\,X + Y + c2\,\overline{X} - \overline{Y}}{2\,c2}\right)\left(-c2\,X + Y + c2\,\overline{X} - \overline{Y}\right)}{2\,c2} == 0$$

which is one equation with one unknown. In this case the order of the replacement rules is critical, and switching them will produce an incorrect result. Solving for c_2, we find

```
In[101]:= Solve[%, c2]
```

$$\text{Out[101]= } \left\{\left\{c2 \to \frac{Y - \overline{Y}}{X - \overline{X}}\right\}, \left\{c2 \to \frac{-Y + \overline{Y}}{X + \overline{X}}\right\}\right\}$$

Now it is time for one last application of creative algebraic visualization to discern the simple patterns in this result. If we square the numerators and denominators, apply summation operators to the results, divide through by n/n, and then take the square root, we get

$$c_2 = \sqrt{\frac{\frac{1}{n}\sum_{i=1}^{n}(Y - \overline{Y})^2}{\frac{1}{n}\sum_{i=1}^{n}(X - \overline{X})^2}} = \frac{s_y}{s_x}$$

It may seem tricky, but the result above is algebraically equivalent to the expression with which we started: $(Y - \overline{Y})/(X - \overline{X})$. It is just in a more convenient form, which is the ratio of the standard deviation of y to the standard deviation of x.

We still have to compare our newly found reduced major axis regression line to the water level data set and the two least squares lines. To perform the necessary calculations, first break **data** into two separate lists.

```
In[102]:= rain = data[[All, 1]]

Out[102]= {1.94, 3.33, 3.22, 5.67, 4.72, 3.89, 2.78,
           10.56, 9.44, 12.78, 14.72, 13.61, 20.39, 38.89}

In[103]:= WL = data[[All, 2]]

Out[103]= {2.5, 1.89, 1.67, 1.31, 1.02, 0.96,
           1.1, 0.15, 3.92, 5.23, 4.22, 3.63, 4.32, 5.89}
```

The slope of the reduced major axis line is the ratio of the standard deviations

```
                 StandardDeviation[WL]
In[104]:= c2 = ─────────────────────────
                StandardDeviation[rain]

Out[104]= 0.182486
```

and, now that the slope has been calculated, the y intercept is found by combining it with the two mean values.

```
In[105]:= c1 = Mean[WL] - c2 Mean[rain]

Out[105]= 0.798433
```

As it has been many times before, the next step is combine a plot of the reduced major axis line with plots of the data set and the two least squares lines. The reduced major axis line is the thick line bisecting the two least squares lines, illustrating that it does indeed do a good job of selecting a line that simultaneously minimizes errors in both the x and y directions.

```
In[106]:= Plot[c1 + c2 x, {x, 0, 40}, PlotStyle → Thickness[0.007],
            DisplayFunction → Identity];
          Show[dataplot, plot1, plot3, %,
            DisplayFunction → $DisplayFunction]
```

From In[106]:=

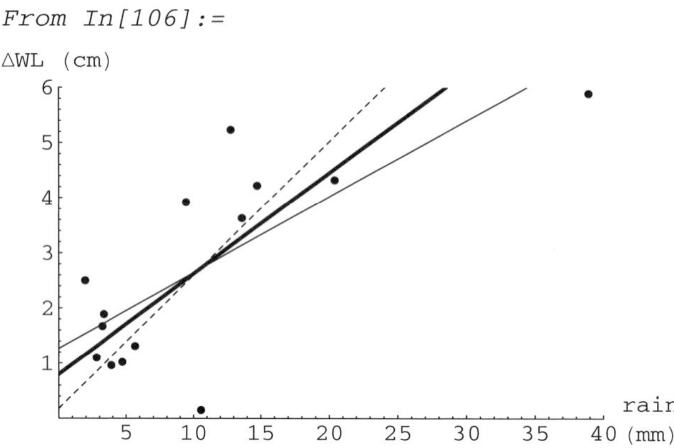

`Out[106]= -Graphics-`

6.5 Nonlinear Regression

Because there is no unique way to minimize the errors between data and a nonlinear function of the dependent variable(s), nonlinear regression involves iterative numerical methods that are significantly more complicated than linear regression. *Mathematica* contains a standard package, Statistics`NonlinearFit`, that contains the functions **NonlinearFit** and **NonlinearRegress**. As with **Fit** and **Regress**, **NonlinearFit** returns only an expression for the best-fit curve whereas **NonlinearRegress** returns a list of replacement rules containing the regression results and diagnostics. Both use iterative least squares algorithms that require initial estimates of the coefficients, which must be chosen carefully. In essence, iterative least squares methods work by continually adjusting values of the coefficients until it appears that a minimum value has been reached. Initial coefficient estimates that are too far from the true values sought may result in erroneous results because the iterative error minimization algorithm may become stuck on a local rather than global minimum as it works its way through the range of possible coefficient values. The default method of minimization is the Levenberg-Marquardt algorithm, although other options are available (see the *Mathematica* documentation in the online Help Browser for more details). Several other options allow users to specify the maximum number of iterations, the required precision of the iterations, and other variables. As illustrated in the Logistic Regression sub-section below, weighted nonlinear regression is also allowed. Because of the possibility that the results of **NonlinearFit** and **NonlinearRegress** may represent a local rather than global error minimum, it is importat to treat any results with caution and plot them to ensure that they provide a good representation of the data to which the curve is being fit.

6.5.1 Nonlinear Least Squares

Sediment compaction curves, in which porosity is plotted as a function of depth, provide an excellent example of a geoscientific problem that can be solved by the use of nonlinear least squares methods. Compaction curves can be used to help understand the geologic evolution of sedimentary basins and aid in petroleum exploration (*e.g.*, Athy, 1930; Dickinson, 1953; Hart *et al.*, 1995) and predict the land subsidence potential of aquifer systems with heavy groundwater pumping (Haneberg, 1995; Helm, 1984). Athy (1930), based on his study of Paleozoic shales buried to different depths, proposed that porosity decreases with depth according to the function $n = n_0 \exp(-z/z_0)$, where n is porosity, z is depth, n_0 is the porosity at $z = 0$, and z_0 is a reference depth at which the porosity is n_0/e (or approximately 0.37 n_0). His original estimates were $n_0 = 0.48$ (which is the porosity of a cubic packed array of uniformly sized spheres; see Bear, 1972) and $z_0 = 714$ m.

To illustrate the fitting of nonlinear sediment compaction curves, first import some depth and density porosity geophysical log data from a deep exploratory water well in the Albuquerque basin, New Mexico (Haneberg, 1995).

```
In[107]:= data = Import[
            "/Users/bill/Mathematica_Book/porosity.dat"];
```

> **Computer Note:** You will have to change the file path above to reflect the directory into which you have placed the file **porosity.dat**.

Its length is

```
In[108]:= len = Length[data]
Out[108]= 6312
```

As always, plot the data. This is especially important in nonlinear regression because it can help to provide initial estimates for the coefficients.

```
In[109]:= dataplot = ListPlot[data, PlotStyle → GrayLevel[0.7],
            PlotRange → All, AxesLabel → {"depth", "porosity"},
            AxesOrigin → {0, 0.2}]
From In[109]:=
```

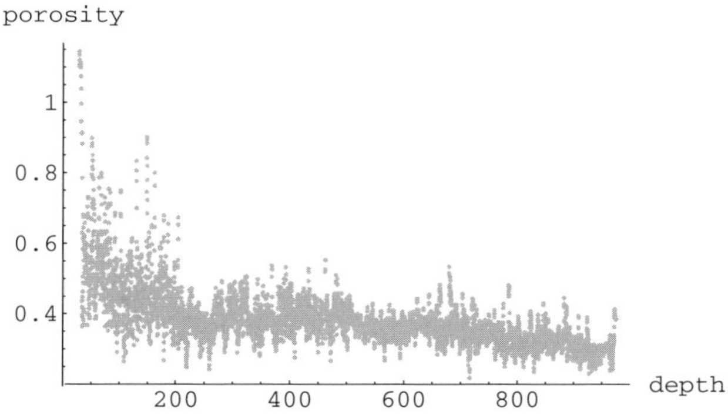

porosity

1
0.8
0.6
0.4

200 400 600 800 depth

```
Out[109]= -Graphics-
```

There are a few erroneous porosity values approaching or exceeding 1 at very shallow depths where the well is cased and above the static water level. In general, though, the high porosity measurements near the surface that range between 0.5 and 1.0 are not unreasonable because sediments in shallow subsurface include pumice gravels deposited by the ancestral Rio Grande. We will ignore them in this example because, with more than 6300 data, their influence will be minor.

Next, perform the regression. If no starting values are specified for the coefficients, *Mathematica* assigns default values of 1. In this case, the result is

```
In[110]:= compactioncurve = NonlinearFit[data, n0 Exp[-z/z0],
            {z}, {n0, z0}]

Out[110]= 0.502093 e^(-0.000552994 z)
```

> **Computer Note: `FindFit`**, which is new in *Mathematica* 5.0, can perform both linear and nonlinear regression. The syntax of its use is slightly different than that of **`Fit`** and **`Regress`**, and its main advantage is that it will automatically decided whether linear or nonlinear regression is appropriate.

This is an unacceptable result that shows what can happen when the iterative algorithm converges on a local rather than global minimum error. The calculated values for the coefficients are n_0= 21 (not 0.21) and z_0= 1/0.58 = 1.7 m. The physically possible range for the porosity term is only $0 \le n_0 \le 1$, and the physically realistic range is more likely to be something like $0 \le n_0 \le 0.5$. The $z_0$term is also physically unrealistic because it suggests that the observed porosity should decrease to 37% of its original value at a depth of 1.7 m (in this case 0.37 × 21 = 7.8). The results can be improved by using physically plausible initial estimates for the coefficients. Looking at **`dataplot`**, an initial estimate of n_0= 0.5 seems realistic. The observed porosity averages about 0.30 at a depth approaching 1000 m, so a good first estimate for $z_0$might found using the ratio of Athy equations using values inferred from the data plot.

```
In[111]:= Solve[ (0.5 Exp[z0]) / (0.5 Exp[-1000/z0]) == 0.5/0.3, z0]

Out[111]= {{z0 → 0.255413 - 31.6217 i}, {z0 → 0.255413 + 31.6217 i}}
```

Using these more enlightened initial estimates, **`NonlinearFit`** returns strikingly different results.

```
In[112]:= compactioncurve = NonlinearFit[data, n0 Exp[-z/z0],
              {z}, {{n0, 0.5}, {z0, 2000.}}]

Out[112]= 0.502093 e^(-0.000552994 z)
```

The porosity value, although high, is not unrealistic for poorly compacted Cenozoic sediments near the surface. The reference depth is 1/0.00055, or about 1800 m. It, too, is consistent with the measured porosity data. Finally, a plot of the porosity data and the Athy compaction curve show that the agreement is reasonably good.

Using **NonlinearRegress** for the same problem will return an extensive table of diagnostics.

```
In[113]:= Plot[compactioncurve, {z, 0, 1000},
            DisplayFunction → Identity];
          Show[dataplot, %, DisplayFunction → $DisplayFunction]
```

From In[113]:=

porosity

1
0.8
0.6
0.4
200 400 600 800 1000 depth

```
Out[113]= -Graphics-
```

Our results show that, while n_0 is not appreciably different between the Albuquerque basin sediments and the shales studied by Athy, the Albuquerque basin sediments are much less compressible than the muds that lithified to form Athy's shales. This is to be expected, because our data set included porosity measurements from a combination of sandy, silty, and clayey strata. Even so, the Albuquerque basin sediments are compressible enough that heavy groundwater pumping could cause compaction of up to 10 cm for every meter the water level decreases once a threshold is exceeded (Haneberg, 1995).

> **Computer Note:** Dickinson (1953) and Helm (1984) proposed a compaction curve with three coefficients: $n = n_0 + c \log (z/z_0)$. The curve that represents Dickinson's data from the Gulf of Mexico has the coefficients $n_0 = 0.05$, $c = -0.103$, and $z_0 = 10{,}000$ m. The significance of the Dickinson-Helm coefficients is much different than that of the Athy coefficients. In the Dickinson-Helm model, n_0 is the limiting value of porosity at great depth as $z/z_0 \to 1$ and $n \to n_0$. Use **NonlinearFit** or **NonlinearRegress** to fit a Dickinson-Helm compaction curve to the Albuquerque basin data set, taking care to estimate reasonable initial coefficient values from the data plot. Which curve provides a better fit?

6.5.2 Logistic Regression

Logistic regression is appropriate in cases where the dependent variable y is binary. For example, $y = 1$ if a landslide exists (or at least is mapped) at a point and $y = 0$ if one does not (*e.g.*, Bernknopf *et al.*, 1988; Jäger and Wieczorek, 1994; Ohlmacher and Davis, 2003). Or, $y = 1$ if a contaminant is present above some threshold amount and $y = 0$ if it is not (*e.g.*, Focazio *et al.*, 2002; Tesoriero and Voss, 1997). In such a situation, the dependent variable is actually an estimate of the probability that an event has occurred or will occur. This section includes a example developed around a simple univariate relationship between landslide occurrence and slope, but more complicated problems can be addressed using multivariate logistic regression. The extension from one variable or dimension to several should be obvious.

At this point is seems logical to ask why standard least squares or reduced major axis regression is not appropriate in the situations where logistic regression is used. The primary reason is that they can yield predicted values less than 0 or greater than 1, which make no sense if the dependent variable is a probability. Other problems include the fact that if the dependent variable can take on values of only 0 and 1, the error between the observations and the regression line will not be normally distributed and will depend on the indpendent variable (recall the discussion of normally distributed residuals earlier in this chapter). These shortcomings can be addressed by fitting a line that resembles the logistic population curve introduced in Chapter 3, hence the name logistic regression.

We will use a data set of landslide occurrence and slope angles in a 2×3 km area near Wheeling, West Virginia to illustrate a geoscientific application of logistic regression. Most landslides in the area develop in colluvium that covers hillsides and is underlain by a variety of nearly flat lying clastic and carbonate sedimentary rocks. Water levels in the colluvium can approach or reach the ground surface during the late winter and early spring of unusually wet years, so landslides are common. Landslide occurrence was taken from a slope stability map prepared by Davies *et al.* (1978) and recoded so that active landslides, dormant landslides, and areas deemed susceptible to landsliding were assigned values of 1. All other areas were assigned a value of 0. The map was digitized using 30×30 m cells corresponding to the resolution of the U.S. Geological Survey digital elevation model of the area, which was used to calculate a maximum slope angle for each cell using a method that will be described in Chapter 7. The result is contained in the file **logisticdata.dat**, which is imported by the next statement.

```
In[114]:= data = Import[
            "/Users/bill/Mathematica_Book/logisticdata.dat"];
```

> **Computer Note:** You will have to change the file path above to reflect the directory into which you have placed the file **porosity.dat**.

The data file is large, containing more than 6800 values.

```
In[115]:= len = Length[data]

Out[115]= 6868
```

Here is a plot of **logisticdata**.

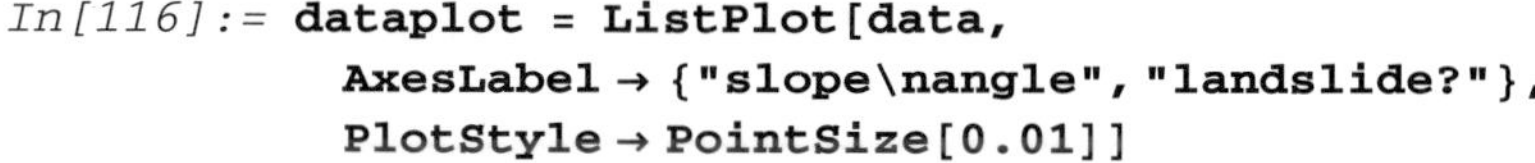

```
In[116]:= dataplot = ListPlot[data,
            AxesLabel → {"slope\nangle", "landslide?"},
            PlotStyle → PointSize[0.01]]
```

```
From In[116]:=
```

landslide?

slope angle

```
Out[116]= -Graphics-
```

Although there is a great deal of overlap, notice that the slope angles at which landslides were mapped or inferred ($y = 1$) extend to higher values than slope angles for which landslides were note mapped or inferred ($y = 0$). Also, notice that there seem to be more no-landslide slope angles than landslide slope angles at the low end of the range. Because there are so many data, however, it is difficult to discern details about the distributions of slope angles for which landslides were or were not mapped.

One way to make the data more comprehensible is to calculate the percentage of cells in which landslides were mapped for different slope angle increments, which is also the probability of landslide occurrence as a function of slope angle. First create a new table to accumulate to results.

```
In[117]:= slideangles = Table[Null, {0}];
```

Now loop through **data**. If a landslide was mapped (*i.e.*, **data[[i, 2]] == 1** is true), then add that slope angle to **slideangles**.

```
In[118]:= Do[If[data[[i, 2]] == 1,
           AppendTo[slideangles, data[[i, 1]]]], {i, len}]
```

Count the number of cells in which landslides were mapped for each 1° slope angle increment from 0° to 30°. *Mathematica* increments the **BinCounts** indices such that a number at the lower edge of a bin is included in that bin, meaning that a number at the upper edge of the bin is not (it is put into the next larger bin). To include the 0° slope angles, therefore, a starting value of –0.5 is used in the following counts.

```
In[119]:= BinCounts[slideangles, {-0.5, 30.5, 1}]

Out[119]= {4, 43, 53, 113, 98, 143, 223, 213, 236, 270, 263, 350, 184,
    235, 170, 206, 191, 186, 177, 118, 129, 102, 67, 70, 73,
    27, 19, 13, 11, 11, 1}
```

and similarly count the total number of cells in each 1° slope angle increment.

```
In[120]:= BinCounts[data[[All, 1]], {-0.5, 30.5, 1}]

Out[120]= {111, 361, 296, 344, 241, 279, 359, 333, 380, 398, 384,
    493, 269, 338, 242, 297, 253, 253, 269, 187, 183, 152,
    118, 131, 94, 36, 25, 18, 12, 11, 1}
```

and take their ratio to find the probability

```
In[121]:= %%/% //N

Out[121]= {0.036036, 0.119114, 0.179054, 0.328488, 0.406639,
    0.512545, 0.62117, 0.63964, 0.621053, 0.678392,
    0.684896, 0.709939, 0.684015, 0.695266, 0.702479,
    0.693603, 0.754941, 0.735178, 0.657993, 0.631016,
    0.704918, 0.671053, 0.567797, 0.534351, 0.776596,
    0.75, 0.76, 0.722222, 0.916667, 1., 1.}
```

The probability values will be more useful if we associate them with their slope angles, which are this case the midpoints of each slope angle increment.

```
In[122]:= probability = Table[{1. (i-1), %[[i]]}, {i, Length[%]}]

Out[122]= {{0, 0.036036}, {1., 0.119114}, {2., 0.179054},
    {3., 0.328488}, {4., 0.406639}, {5., 0.512545},
    {6., 0.62117}, {7., 0.63964}, {8., 0.621053},
    {9., 0.678392}, {10., 0.684896}, {11., 0.709939},
    {12., 0.684015}, {13., 0.695266}, {14., 0.702479},
    {15., 0.693603}, {16., 0.754941}, {17., 0.735178},
    {18., 0.657993}, {19., 0.631016}, {20., 0.704918},
    {21., 0.671053}, {22., 0.567797}, {23., 0.534351},
    {24., 0.776596}, {25., 0.75}, {26., 0.76},
    {27., 0.722222}, {28., 0.916667}, {29., 1.},
    {30., 1.}}
```

Finally, plot the probability of finding a landslide in a cell as a function of slope angle. Although the relationship isn't ideal, it does illustrate a general correspondence between landslide occurrence and slope angle in the study area.

```
In[123]:= probplot = ListPlot[probability,
    AxesLabel → {"slope\nangle", "probability"},
    PlotStyle → PointSize[0.018]]
```

From In[123]:=

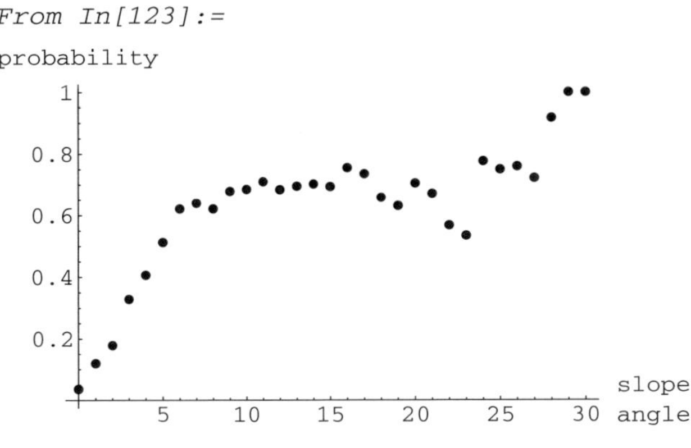

Out[123]= -Graphics-

Now that we have a feel for the data, we can move on to the regression. Instead of fitting a line of the form $y = c_1 + c_2x$, in logistic regression one of the form log $(p/(1-p)) = c_1 + c_2x$ or $p/(1-p) = e^{c_1} e^{c_2x}$, where p is the probability of occurrence is used. Areas in which landslides were mapped have $p = 1$ because a landslide has occurred, and areas in which landslides were not mapped have $p = 0$. The utility of a logistic regression equation is that it will allow the estimation of the probability that a landslide will be found or occur in an area that has not yet been mapped (but which has similar geologic and topographic characteristics). The term log $(p/(1-p))$ is known as the **log odds ratio** or **logit** and $p/(1-p)$ is, logically enough, known as simply the **odds ratio**. Examination of the odds ratio shows that it is simply the ratio of the probability that a landslide occurs (p) to the probability that it does not occur $(1-p)$ at a specific location.

To fit a logistic curve, we will first define the logit function and solve it for p. We've already assigned numerical values to **c1** and **c2**, so the first step is to clear the existing values.

```
In[124]:= Clear[c1, c2]

In[125]:= Simplify[Solve[Log[p/(1 - p)] == c1 + c2 x, p]]
```

Out[125]= $\left\{\left\{p \to \frac{e^{c1+c2\,x}}{1 + e^{c1+c2\,x}}\right\}\right\}$

```
In[126]:= psltn = p /. %[[1]]
```

Out[126]= $\frac{e^{c1+c2\,x}}{1 + e^{c1+c2\,x}}$

We will use the result in **NonlinearFit** and assign the result to **logcurve**. Many descriptions of logistic regression are intimately tied to a method of solution known as maximum likelihood estimation (MLE), and it is easy to falsely conclude that MLE is the only way to solve logistic regression problems. But, it is not. Nonlinear least squares will work, and the problem can even be transformed and solved by linear least squares. Lowry (2003) explains the linear least squares approach, and suggests that insistence on MLE is an example of Alfred North Whitehead's fallacy of misplaced concreteness. He argues that MLE may not be worth the extra effort in most practical cases, and this is probably true.

Because each of the probability values is based on a different number of observations, we can assign weights to each value. In this case the simplest way to weight the values is to construct a table of the number of values in each slope angle increment (we did this before, but did not give the result a name, so we will repeat the operation).

```
In[127]:= wts = BinCounts[data[[All, 1]], {-0.5, 30.5, 1}] //N

Out[127]= {111., 361., 296., 344., 241., 279., 359., 333., 380.,
            398., 384., 493., 269., 338., 242., 297., 253., 253.,
            269., 187., 183., 152., 118., 131., 94., 36., 25., 18.,
            12., 11., 1.}
```

The weights are incorporated using the **Weights** option. The default value is **Weights → Automatic**, which gives each point equal weight.

```
In[128]:= results = NonlinearFit[probability, psltn , {x},
            {c1, c2}, Weights → wts]
```

Out[128]= $\dfrac{e^{-0.860731+0.135306x}}{1+e^{-0.860731+0.135306x}}$

```
In[129]:= logcurve = %
```

Out[129]= $\dfrac{e^{-0.860731+0.135306x}}{1+e^{-0.860731+0.135306x}}$

To see the results, plot **logcurve** over the observed range of slope angles and superimpose the result with the probability values that were used in **NonlinearFit**.

```
In[130]:= logplot = Plot[logcurve, {x, 0, 30},
            DisplayFunction → Identity];
          Show[logplot, probplot,
            AxesLabel → {"slope\nangle", "probability"},
            DisplayFunction → $DisplayFunction]
From In[130]:=
```



```
Out[130]= -Graphics-
```

The agreement is not perfect, but the plot does show a definite correspondence between the regression line and the probability data.

You may also be thinking that the regression line does not look much like the logistic population curves obtained in Chapter 3, but that is because of the data set and range of slope angles we have chosen. This can be demonstrated by plotting the curve over a much wider (but physically unrealistic) range of angles. The 0° to 30° range of slope angles used in this example is denoted by the gray rectangle.

```
In[131]:= Plot[logcurve, {x, -90, 90},
            DisplayFunction → Identity];
          Show[Graphics[{GrayLevel[0.8],
            Rectangle[{2, 0.01}, {39.9, 1.}]}],
            DisplayFunction → Identity, Axes → True
          ];
          Show[%, %%, probplot,
            DisplayFunction → $DisplayFunction]
```

From In[131]:=

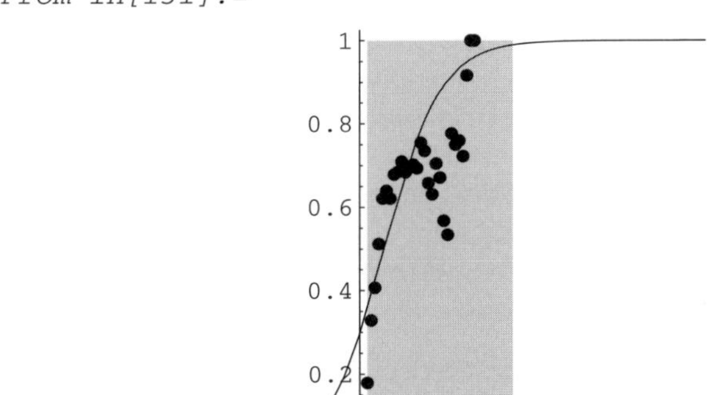

Out[131]= -Graphics-

Ohlmacher and Davis (2003) also obtained wide logistic regression curves in their study of landslide hazards in Kansas, some of which predicted $p < 1$ for slopes as high as 90°. Although the results in this example are much messier than the simple examples typically used in textbook logistic regression tutorials, they are representative of the results obtained from observations of complicated natural systems.

Computer Note: The non-randomness of the residuals suggests that a better curve can be found. Modify the logit function to include a higher order polynomial, for example:

$$\log\left(\frac{p}{1 - p}\right) = c_1 + c_2 x + c_3 x^2 + c_4 x^3$$

(Continued.) The visible agreement the regression curve to the probability data will be much better, but will now include a small interval in which the probability of observing a landslide decreases as the slope angle increases. The geoscientific question is, why? Should the probability of landsliding always increase as slope angle increases? A histogram of slope angles in the study area and an understanding of the infinite slope stability equation used in Chapter 5 may help to explain the relationship. You may assume that the angle of internal friction for hillside soils and colluvium in the area is about 25°, cohesive strength is negligible, and the area receives enough rainfall to saturate the slopes nearly to the ground surface from time to time. What kinds of field evidence would you look for to test any hypotheses that you develop?

In logistic regression, there are no statistics analogous to the r^2 goodness-of-fit used for linear regression models. This is because the concept of an r^2 value is postulated on the existence of normally distributed errors, which does not occur with binary data. Most of the goodness-of-fit statistics that do exist for logistic regression results, moreover, are difficult to apply and some are closely tied to the method of solution (such as maximum likelihood estimation). One measure of goodness-of-fit that is easy to calculate is a simple percentage of correct predictions. To calculate the percentage of correct predictions, assume that a value of 0.5 is the cutoff between the occurrence or non-occurrence of a landslide (or whatever other dependent variable is being studied). This is, after all, what we really want to know: will a landslide occur at a given place or not? Then, use the cutoff to transform **probability** into a list of binary "yes-no" variables. This is most easily accomplished using **Round**.

```
In[132]:= list1 = Round[probability[[All, 2]]]

Out[132]= {0, 0, 0, 0, 0, 1, 1, 1, 1, 1, 1, 1, 1,
           1, 1, 1, 1, 1, 1, 1, 1, 1, 1, 1, 1, 1, 1, 1, 1, 1, 1}
```

Create a corresponding table of predicted values by evaluating **logcurve** for each of the slope angles, then apply **Round** to recast the results as binary variables.

```
In[133]:= list2 = Round[Table[logcurve, {x, 0, 30}]]

Out[133]= {0, 0, 0, 0, 0, 0, 0, 1, 1, 1, 1, 1, 1,
           1, 1, 1, 1, 1, 1, 1, 1, 1, 1, 1, 1, 1, 1, 1, 1, 1, 1}
```

The next step is to determine what percentage of the elements in **list1** agrees with the corresponding elements in **list2**. There are different ways to accomplish this, one of which is to add the two lists together.

```
In[134]:= list3 = list1 + list2

Out[134]= {0, 0, 0, 0, 0, 1, 1, 2, 2, 2, 2, 2, 2,
           2, 2, 2, 2, 2, 2, 2, 2, 2, 2, 2, 2, 2, 2, 2, 2, 2, 2}
```

Values of 0 indicate slope angle categories for which both **list1** and **list2** agree that no landslide exists or should not exist. Values of 2 indicate categories for which both lists agree that a landslide does or should exist. The remaining values of 1 indicate categories where the observations and predictions differ. **Select** extracts the elements of **list3** that are either 0 or 2 and returns the result as a list, the length of which is determined by **Length**. The two lengths are then added together and divided by the total number of slope angle increments to calculate the percentage of agreement. The **//N** operator is necessary to obtain a numerical result from calculations involving only integers.

```
In[135]:= (Length[Select[list3, # == 0&]]+
            Length[Select[list3, # == 2&]])/Length[list3] //N

Out[135]= 0.935484
```

In this very simple measure of its ability to predict whether the observed probability of landslide occurrence is less or greater than 50% for a given slope angle increment, therefore, the logistic regression model appears to work well. Consult a logistic regression reference such as Kleinbaum and Klein (2002) or Menard (2001) for information about more sophisticated assessments and the maximum likelihood estimation method of solution.

6.6 References and Recommended Reading

Arfken, G., 1985, *Mathematical Methods for Physicists (3rd ed.)*: Academic Press.

Athy, L.F., 1930, Density, porosity, and the compaction of sediments: *American Association of Petroleum Geologists Bulletin*, v. 14 p. 1–24.

Bear, J., 1972, *Dynamics of Fluids in Porous Media*: American Elsevier (reissued by Dover Publications).

Bernknopf, R.L., Campbell, R.H., Brookshire, D.S., and Shapiro, C.D., 1988, A probabilistic approach to landslide hazard mapping in Cincinnati, Ohio, with applications for economic evaluation: *Bulletin of the Association of Engineering Geologists*, v. 25, p. 39–56.

Carr, J.R., 2002, *Data Visualization in the Geosciences*: Prentice Hall.

Davis, J.C., 2002, *Statistics and Data Analysis in Geology (3d ed.)*: John Wiley & Sons.

Davies, W.E., Pomeroy, J.S., and Ohlmacher, G.C. , 1978, *Landslides and Related Features, Ohio, West Virginia, and Pennsylvania Canton 1 × 2 Degree Sheet*: U.S. Geological Survery Open-File Map 78-1057.

Dickinson, G., 1953, Geological aspects of abnormal reservoir pressure in Gulf Coast Louisiana:*American Association of Petroleum Geologists Bulletin*, v. 37, p. 410–432.

Focazio, M.J., Reilly, T.E., Rupert, M.G., and Helsel, D.R., 2002, *Assessing Ground-Water Vulnerability to Contamination: Providing Scientifically Defensible Information for Decision Makers*: U.S. Geological Survey Circular 1224.

Hamming, R.W., 1973, *Numerical Methods for Scientists and Engineers (2d ed.)*: McGraw Hill (reissued by Dover Publications).

Haneberg, W.C., 1995, Depth-porosity relationships and virgin specific storage estimates for the upper Santa Fe Group aquifer system, central Albuquerque Basin, New Mexico: *New Mexico Geology*, v. 17, p. 62–71.

Haneberg, W.C., 1999, Effects of valley incision on the subsurface state of stress – theory and application to the Rio Grande valley near Albuquerque, New Mexico: *Environmental & Engineering Geoscience*, v. 5, p. 117–131.

Haneberg, W.C. and Gökce, A.Ö., 1994, *Rapid Water-Level Fluctuations in a Thin Colluvium Landslide West of Cinncinati, Ohio*: U.S. Geological Survey Bulletin 2059-C.

Hart, B.E., Flemings, P.B., and Deshpande, A., 1995, Porosity and pressure: role of compaction disequilibrium in the development of geopressures in a Gulf Coast Pleistocene basin: *Geology*, v. 23, p. 45–48.

Helm, D. C., 1984, Field-based computational techniques for predicting subsidence due to fluid withdrawal, *in* T.L. Holzer, editor, *Man-Induced Land Subsidence*: Geological Society of America Reviews in Engineering Geology, v. 6, p. 1–22.

Jäger, S. and Wieczorek, G.F., 1994, *Landslide Susceptibililty in the Tully Valley Area, Finger Lakes Region, New York*: U.S. Geological Survey Open-File Report 94-615 (online version at http://pubs.usgs.gov/of/1994/ofr-94-0615/tvstudy.htm).

Kleinbaum, D.G. and Klein, M., 2002, *Logistic Regression (2d ed.)*: Springer Verlag.

Lowry, R., 2003, *Simple Logistic Regression*: web page located at http://faculty.vassar.edu/lowry/logreg1.html.

Menard, S., 2001, *Applied Logistic Regression Analysis*: Sage Publications.

Middleton, G.V., 2000, *Data Analysis in the Earth Sciences Using Matlab*: Prentice Hall.

Ohlmacher, G.C. and Davis, J.C., 2003, Using multiple logistic regression and GIS technology to predict landslide hazard in northeast Kansas, USA: *Engineering Geology*, v. 69, p. 331–343.

Swan, A.R.H. and Sandilands, 1995, *Introduction to Geological Data Analysis*: Blackwell Science.

Tesoriero, A.J., and Voss, F.D., 1997, Predicting the probability of elevated nitrate concentrations in the Puget Sound Basin-Implications for aquifer susceptbility and vulnerability: *Ground Water*, v.35, p.1029-1039.

Wessel, P., 2000, *Geologic Data Analysis*: http://www.higp.hawaii.edu/~cecily/courses/gg313/DA_book/index.html.

7 Visualizing and Analyzing Surfaces

7.1 *Mathematica* Packages You Will Need

Be sure to execute the following statement to ensure that you will have available all of the add-on and book-specific *Mathematica* functions used in this chapter.

```
In[1]:= Needs["Graphics`Graphics`"]
        Needs["Graphics`Graphics3D`"]
        Needs["Graphics`Legend`"]
        Needs["Statistics`DescriptiveStatistics`"]
        Needs["Statistics`MultiDescriptiveStatistics`"]
        Needs["Statistics`DataManipulation`"]
        Needs["NumericalMath`ListIntegrate`"]
        Needs["CompGeosci`"]
```

Computer Note: The CompGeosci package will load correctly only if it is located in one of the directories in *Mathematica*'s standard file path. Execute the statement **`$Path`** to see a list of the default paths on your computer and place the file CompGeosci.m in one of those directories. The specific file paths may differ from one operating system to another. See Chapter 1 for more information about installing the CompGeosci package.

7.2 Gridded Data

Data representing surfaces of interest to geologists comes in two basic forms: gridded and ungridded. Gridded data are arranged along regularly spaced rows and columns, and are therefore relatively easy to analyze. One of the most ubiquitous forms of gridded data used by geologists is the digital elevation model, or DEM, which consists of rows and columns of elevation data. The results of finite difference groundwater flow models are also in gridded form, although the grid spacing may be variable, and digital image files can also be thought of as gridded data. Ungridded data are randomly or irregularly located, for example at locations where a drilling rig could be set up or a sample was collected from an outcrop. Before irregularly spaced data can be contoured, analyzed, and in many cases even plotted, they must generally be gridded using two dimensional equivalents the interpolation methods discussed in Chapter 6.

7.2.1 Digital Elevation Models

Digital elevations models, often referred to as DEMs, are available in a variety of sizes and resolutions. They are freely available through the U.S. Geological Survey, state GIS clearinghouses, and sources such as www.gisdatadepot.com. The example used in this section is a 201 x 201 grid of elevation values, listed to the nearest decimeter, from the USGS digital elevation model of the Bremerton West, Washington, 7.5' quadrangle. This is a 10 m digital elevation model, meaning that the rows and columns of elevation data are separated by 10 m. *Mathematica* 4.2 and higher can import Spatial Data Transfer Standard (SDTS) digital elevation models, which are now being used for U.S. Geological Survey 7.5' DEMs, and there are some public domain *Mathematica* functions that will read earlier ASCII (also known as DEM format) models. Geographic information system (GIS) or specialized DEM reading software can also be used to read the DEM files, isolate portions of interest, and save them as a tab delimited ASCII or text file containing only the elevation values. The east-west and north-south coordinates, given either as Universal Transverse Mercator (UTM) grid coordinates or latitude and longitude, can also be exported but will result in a much larger file.

The following statement reads in the elevation data:

```
In[2]:= temp = Import[
          "/Users/bill/Mathematica_Book/bremerton.dat"];
```

> **Computer Note:** You will have to change the file pathway to locate the bremerton.txt file on your computer. The easiest way to do this is to use the Get File Path... item on the Input menu.

The next step is to reorganize the elevation data, because *Mathematica* plots grids of data with row 1 at the bottom of the plot whereas digital elevation models typically have row 1 at the top (northern edge). Skipping this step would produce maps with North at the bottom of the map. *Mathematica* stores matrices as lists of lists, so it is easy to write a one line routine to take lists from **temp** starting at the end and put them into **elev** starting at the beginning. The iteration statement simply places values from **temp** into **elev** starting with the last row of **temp** and working towards the first row.

```
In[3]:= elev = Table[temp[[i]], {i, Length[temp], 1, -1}];
```

Now we can calculate some basic values that will be useful in subsequent calculations. Notice that there are six different variables defined in one input, and that the results of the calculations are returned as six separate output lines.

```
In[4]:= nrows = Length[elev]
        ncols = Length[elev[[1]]]
        aratio = nrows/ncols
        minval = Min[elev]
        maxval = Max[elev]
        relief = maxval - minval

Out[4]= 201

Out[4]= 201

Out[4]= 1

Out[4]= 53.4

Out[4]= 223.

Out[4]= 169.6
```

7.2.2 Importing SDTS DEM Files

SDTS DEMs consist of directories or folders containing several files, and are usually obtained as compressed tape archives (as indicated by .gz and .tar suffixes). The software necessary for uncompressing and unarchiving the files, which must be done before they can be imported into *Mathematica*, will vary from computer to computer. Most of the files in the DEM directory contain metadata such as information about the location of the DEM, its accuracy, and the data format. Only one of the files, which has a name of the form *nnnn*CELO.DDF, contains the elevation data. The following statement imports the elevation data file of an SDTS DEM for the Wheeling, West Virginia, area from a directory cryptically named 1632825.DEM.SDTS.TAR.GZ.Folder:

```
In[5]:= Import[
          "/Users/bill/Mathematica_Book/
          1632825.DEM.SDTS.TAR.GZ.Folder/8671CEL0.DDF"];
```

Computer Note: Readers following these examples on their own computers will have to change the file path shown above to correspond to the location of the DEM file on their computers. The full uncompressed and unarchived SDTS directory is provided on the CD accompanying this book.

The **Import** function accepts two conversion options for SDTS files. The first, "**ElementSkip**" → *i*, specifies that every *i*th row and column of data should be imported. The default, "**ElementSkip**" → 1, reads all of the elevation values. The second option, "**FillValue**" → *i*, specifies an integer value that should be assigned to points with no numerical values. Its default value is –10000. Points with no numerical values occur along the edges of some DEMs because the alignment of the DEM grids for U.S. Geological Survey quadrangle maps is not exactly parallel to the quadrangle boundaries, as illustrated by the difference between True North and Grid North shown on paper copies of the maps. Once an SDTS DEM has been imported, it can be manipulated and plotted in the same was as the Bremerton example data set used in this chapter.

Computer Note: The **Import** function was modified between versions 4.2 and 5.0 of *Mathematica,* requiring conversion options to be specified differently in each version. In version 5.0, adding a conversion option to the **Import** statement above would change it to:

```
Import["/Users/bill/Mathematica_Book/
1632825.DEM.SDTS.TAR.GZ.Folder/8671CEL0.DDF",
ConversionOptions → {"ElementSkip" → 2}];
```

In version 4.2, however, the analogous **Import** statement would be:

```
Import["/Users/bill/Mathematica_Book/
1632825.DEM.SDTS.TAR.GZ.Folder/8671CEL0.DDF",
ElementSkip → 2];
```

7.2.3 Contour Plots

The first choice of geoscientists working with surfaces is often a contour plot, in this case made using the **ListContourPlot** function. The option **AspectRatio** controls the proportions of the plot, in this case the ratio of rows to columns. The default value of 1 produces a square plot that distorts any data set that does not contain equal numbers of row and columns. You will need to change the aspect ratio if you are working with a grid that does not contain the same number of rows and columns. **Frame→False** produces a plot without any borders; the default of **Frame→True** would have produced a plot with a rectangular frame and ticks showing the column and row numbers. A final detail is the specification of contours. The default is 10 contour levels calculated automatically by *Mathematica*. The number of equally spaced contours can be specified using **Contours→{*n*}** or a list of specific contours with any spacing can be given as an option. For example, **Contours→{50,200,300}** would produce a plot with contours at the three values in the list. The option used below creates a table filled with the variable *c*, which is allowed to range from 50 to 400 in increments of 10.001. The 0.001 is added because *Mathematica* sometimes produces strange contours when an integer value is used.

```
In[6]:= shadedtopomap = ListContourPlot[elev,
          AspectRatio → aratio, Frame → False,
          ColorFunction → Function[z, GrayLevel[(0.3 + 0.7z)]],
          Contours → Table[c, {c, 50., 220., 10.001}]]
```

From In[6]:=

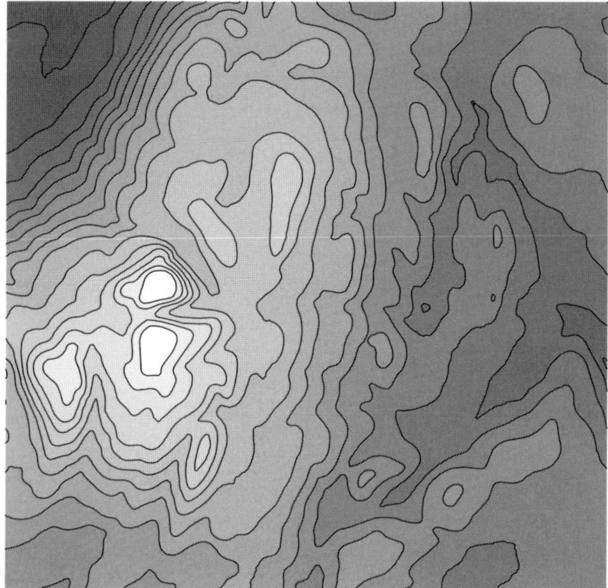

Out[6]= -ContourGraphics-

Computer Note: Experiment with different color functions to see what effects they produce. For example

```
ColorFunction →
  Function[z, RGBColor[0.2 + 0.8 z^2, 0.6 - 1.6 z + 2 z^2,
  0.2 z + 0.8 z^2]]
```

produces an atlas-like range of colors ranging from bright green at the lowest elevations through brown and on to white at the highest elevations. Each of the three polynomials given as arguments in `RGBColor` describes how the red, green, or blue value changes as values of the plotted function range from their minimum to their maximum. See Appendix B for more detailed instructions.

One drawback to contour plots made with *Mathematica* is that the contour lines are not labeled. An option that works for shaded contour and density plots is to use the **`ShowLegend`** function contained in Graphics`Legend`. In the case of the shaded contour plot, **`ShowLegend`** is invoked as follows:

```
In[7]:= ShowLegend[shadedtopomap,
          {Function[z, GrayLevel[0.3 + 0.7 z]], 16, "53", "223 m",
            LegendShadow → None, LegendPosition → {1.1, -0.9},
            LegendSize → {0.5, 1.8}}]
```

From In[7]:=

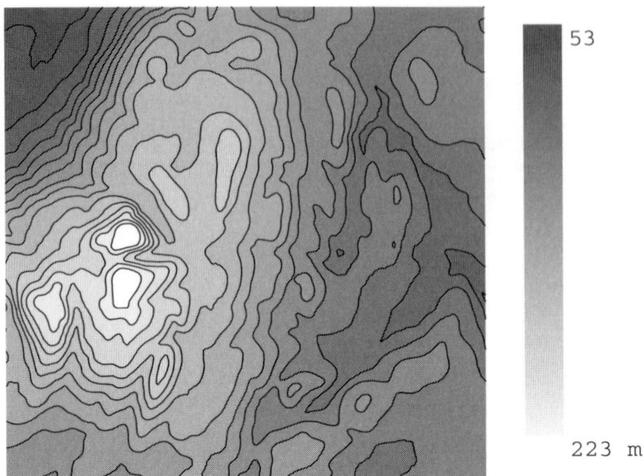

Out[7]= -Graphics-

The list of variables required by **ShowLegend** includes a color function (or, in this case, a gray level function) that should be identical to that used in the original plot, the number of divisions within the legend, and the labels for the low and high ends of the elevation range. The topographic map contains 17 contour lines, meaning that there will be 16 filled intervals between the contours. Optional variables include **LegendShadow**, **LegendPosition**, and **LegendSize**. The default for **LegendShadow** is true, which puts a black drop-shadow behind the legend. The dimensions for **LegendLocation** and **LegendSize** are given relative to the size of the plot, which is centered at 0,0 and ranges from –1 to 1 in each direction, and **LegendLocation** refers to the *x*, *y* coordinates of the lower left-hand corner of the legend. The documentation for Graphics`Legend` describes other options that can be used to control the exact appearance of the legend.

> **Computer Note:** Modify the options in **ShowLegend** so that the lightest color and highest elevation label are at the top, rather than the bottom, of the legend.

The default for **ContourPlot** and **ListContourPlot** is, as illustrated above, to shade each contour interval with a different shade of gray. Using the option **ColorFunction→Hue**will produce a rainbow of colors, although both the highest and lowest values will be colored red. This is not always helpful so, as described in Appendix B, the argument given to **ColorFunction** can be scaled to avoid the problem. **ColorFunction** can also be customized, for example to shade topography ranging from green at low elevations to white at high elevations. Below is the same data set plotted with **ContourShading→False**. The option **ContourShading→False** is used to produce black contour lines set against a white background.

```
In[8]:= topomap = ListContourPlot[elev, ContourShading → False,
           AspectRatio → aratio, Frame → False,
           Contours → Table[c, {c, 50., 220., 10.001}]]
```

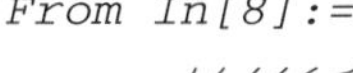

```
From In[8]:=
```

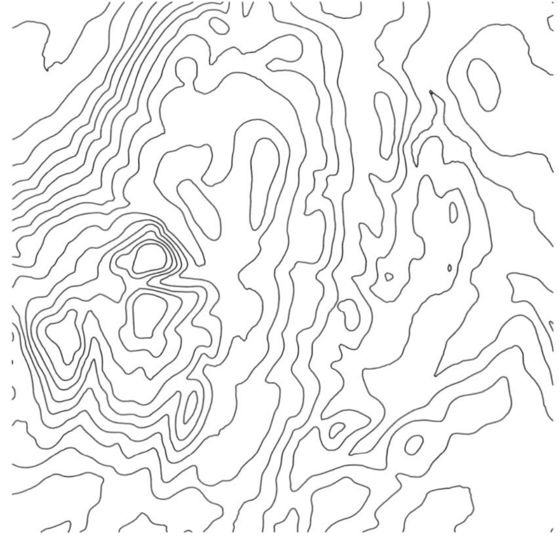

```
Out[8]= -ContourGraphics-
```

> **Computer Note:** Download and install the contour labeling package described in Appendix B, then use it to label contour plots of the Bremerton data set.

> **Computer Note:** Contour maps are often shown with every fifth contour as a thicker or darker line. Make a contour plot using a 25 m contour interval, use the **`ContourStyle`** option to increase the thickness of the contours, and then use **`Show`** to superimpose the result on a contour map with a 5 m contour interval.

7.2.4 Density Plots

Surfaces can also be visualized using density plots in which low values are assigned dark shades and high values are assigned light shades. The default is a series of gray scale values, but this can be overridden by specifying **`ColorFunction→Hue`** or any other user-defined color function. By default, density plots have a mesh showing the rows and columns. For a data set of this size, however, the mesh lines would be so closely spaced that they would obscure the shading. So, the option **`Mesh→False`** was used to suppress the mesh.

```
In[9]:= ListDensityPlot[elev, AspectRatio → aratio,
          Frame → False, Mesh → False,
          ColorFunction- > Function[z, GrayLevel[0.3 + 0.7 z]]]
```

From In[9]:=

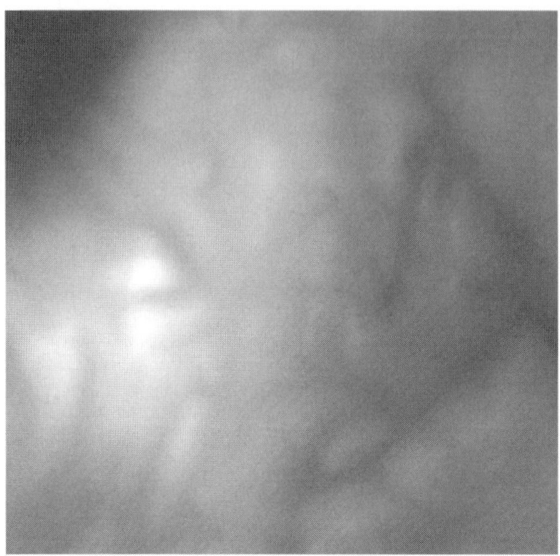

Out[9]= -DensityGraphics-

The contour and density plots can be superimposed using the **Show** function.

```
In[10]:= Show[%, topomap]
```

From In[10]:=

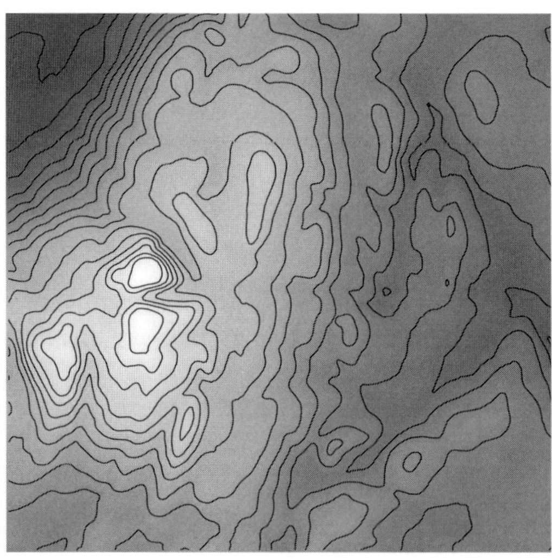

Out[10]= -Graphics-

7.2.5 Three Dimensional Surface Plots

Three dimensional surface or wireframe plots are also simple to construct although, as in the density plot example, the number of mesh lines in a wireframe plot would produce a solid black surface. To alleviate this problem, **elev** can be resampled to include every fifth value in both directions. This is done by creating a new table and filling it with values obtained by iterating through the original data set in steps of 5.

```
In[11]:= sampledelev = Table[elev[[r, c]], {r, 1, nrows, 5},
            {c, 1, ncols, 5}];
```

As illustrated below, the resampled data makes a reasonably representative wireframe plot. Because three dimensional plots are used for qualitative visualization rather than as the basis for calculations, resampling does not create any problems. In this case the **Boxed** option was not overridden, although it could have been by using **Boxed→False**.

```
In[12]:= ListPlot3D[sampledelev, Mesh → True, Shading → False,
            BoxRatios → {50 * ncols, 50 * nrows, 10 * relief}]
```

From In[12]:=

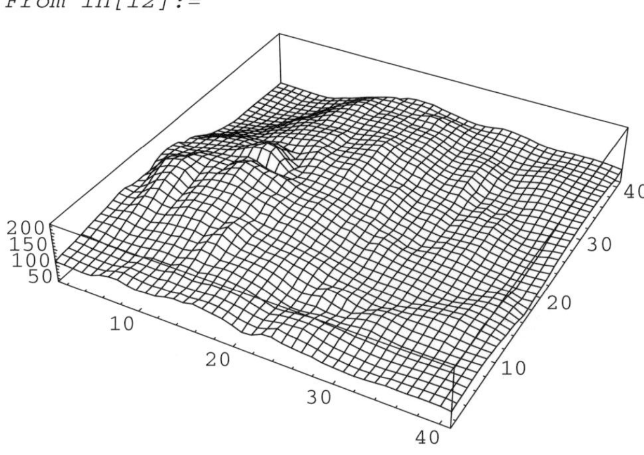

Out[12]= -SurfaceGraphics-

The **BoxRatios** option in the example above functions similarly to the **AspectRatio** option for two dimensional plots. In the case of **ListPlot3D**, the horizontal coordinates are the row and column numbers; therefore, multiplying them by the 50 m spacing of the resampled elevation data converts the two horizontal box dimensions to meters. The third value uses the actual elevation values multiplied by a vertical exaggeration factor of 10. Here is the same plot with the bounding box and axes removed:

```
In[13]:= ListPlot3D[sampledelev, Mesh → True, Shading → False,
            BoxRatios → {50 * ncols, 50 * nrows, 10 * relief},
            Boxed → False, Axes → None]
```

From In[13]:=

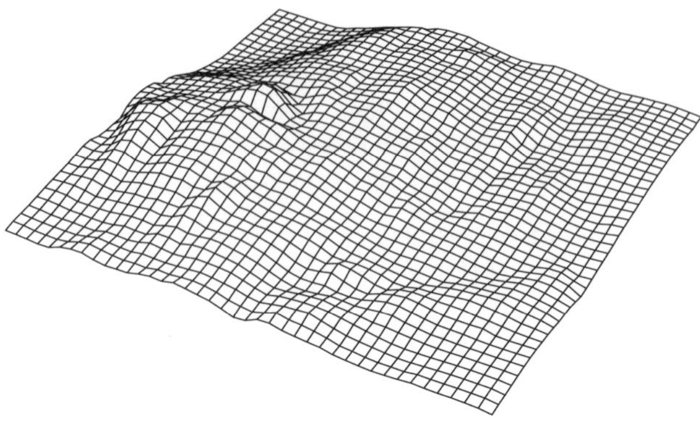

Out[13]= -SurfaceGraphics-

with the surface meshed and shaded:

```
In[14]:= ListPlot3D[sampledelev,
           ColorFunction → Function[z, GrayLevel[0.3 + 0.7 z]],
           BoxRatios → {50 * ncols, 50 * nrows, 10 * relief},
           Boxed → False, Axes → None]
```

From In[14]:=

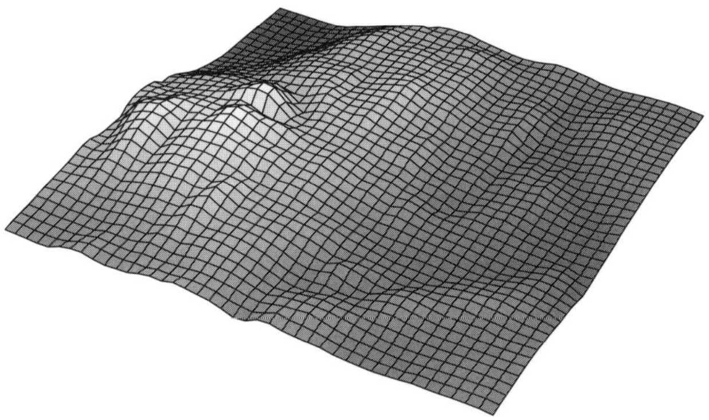

Out[14]= -SurfaceGraphics-

and with the surface shaded but not meshed:

```
In[15]:= ListPlot3D[sampledelev,
           ColorFunction → Function[z, GrayLevel[0.3 + 0.7 z]],
           Mesh → False,
           BoxRatios → {50 * ncols, 50 * nrows, 10 * relief},
           Boxed → False, Axes → None]
```

From In[15]:=

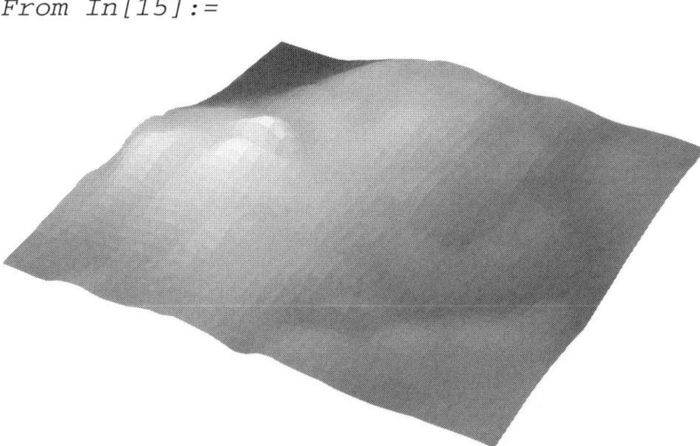

Out[15]= -SurfaceGraphics-

Mathematica allows complete control over surface shading and coloration, but these features are cannot be adequately illustrated in a black and white textbook. Some examples of surface coloration are included in Appendix B on the companion CD.

The viewpoint for three dimensional graphics can be changed using the **ViewPoint** option, which is not interactive in the standard version of *Mathematica*. The easiest way to change the viewpoint is to select the 3DViewPointSelector item from the Input menu, which brings up an interactive dialog box in which a cube can be rotated and zoomed. Clicking on the Paste button will paste the new viewpoint coordinates at the location of the cursor in the notebook. The plot below was created by typing **Show[%,** and leaving the cursor at the end of the line. Then, the viewpoint selector dialog box was used to rotate the graphics cube and paste a viewpoint value at the cursor location. Finally, the input line was completed by typing the final **]** and pressing the Enter key.

In[16]:= **Show[%%, ViewPoint- > {-6.463, 3.17, 2.979}]**

From In[16]:=

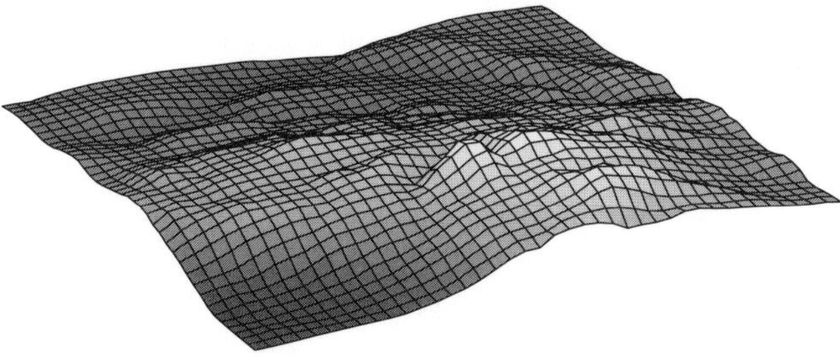

Out[16]= -SurfaceGraphics-

7.2.6 Quantitative Terrain Analysis

Hypsometric Integral

The grid of elevation values that constitutes a digital elevation model is ripe for quantitative analysis. The branch of science concerned with the study of absolute or relative elevations is known as hypsometry, and the simplest form of hypsometric analysis is an elevation histogram, which is easily constructed using the **Histogram** function from the standard Graphics`Graphics` package. To do this, however, the matrix of elevation values must first be flattened into a single list using **Flatten**. Here is the result for the resampled Bremerton data set:

```
In[17]:= Histogram[Flatten[elev],HistogramCategories → 20,
           BarStyle → GrayLevel[0.6],
           AxesLabel → {"elev.", "number"}]
```

From In[17]:=

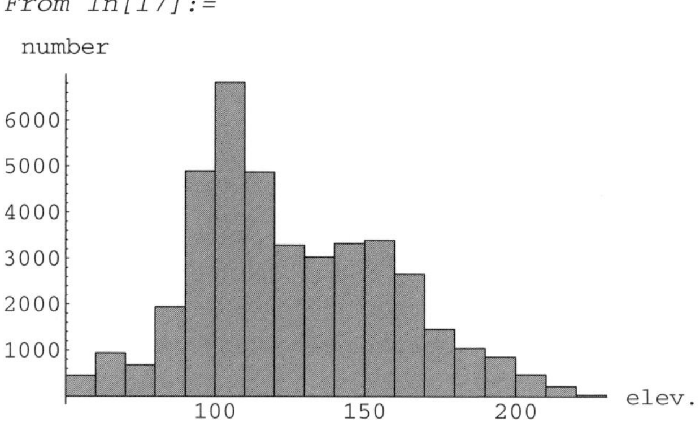


```
Out[17]= -Graphics-
```

> **Computer Note:** Because the Bremerton elevation data set **elev** consists of 40,401 values, readers following the examples in this section and using computers that are slow or have limited memory may wish to substitute the resampled data set **sampledelev**. The results will be similar regardless of which data set is used.

Elevation data can also be summarized using the **CumFreqs** or **CumFreqPlot** functions in the *Mathematica* package included with this book. **CumFreqPlot** takes three arguments: a list of data (which must be flattened if it is a matrix of values), the minimum plot value, and the maximum plot value. **CumFreqs** returns a list of cumulative frequencies without a plot. The cumulative distribution of relative elevations in the resampled Bremerton data set is shown in the graph below, with the relative elevation on the horizontal axis and the cumulative proportion of elevations less than each relative elevation shown on the vertical axis. It is the emprical equivalent of the cumulative distribution function

(CDF) plots introduced in Chapter 4. The elevations are normalized relative to the relief so that they range from 0 to 1, which allows curves from different areas to be easily compared even if their absolute elevations are different.

```
In[18]:= CumFreqPlot[(Flatten[elev] - minval)/relief, 0., 1.,
           AxesLabel → {"Relat.\nElev.", "Cum.\nFreq."}]
```

From In[18]:=

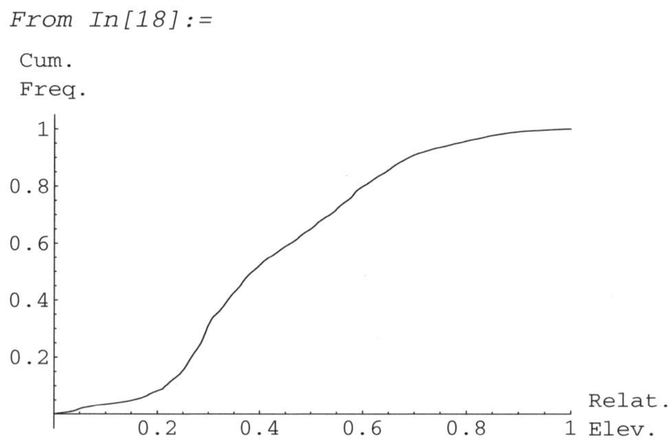

Out[18]= -Graphics-

This cumulative plot of relative elevations is very similar to the **hypsometric curve** used in many geomorphological studies, except that the traditional hypsometric curve shows cumulative proportion on the horizontal axis and elevation on the vertical axis. The traditional hypsometric curve can be plotted by obtaining the cumulative frequencies of the elevation data

```
In[19]:= elevfreqs = CumFreqs[(Flatten[elev] - minval)/relief];
```

putting them into a new data table in which the two columns are interchanged.

```
In[20]:= len = Length[elevfreqs]
```

Out[20]= 40401

```
In[21]:= Table[{elevfreqs[[i, 2]], elevfreqs[[i, 1]]}, {i, len}];
```

and then plotting the new data table

```
In[22]:= ListPlot[%,
           AxesLabel → {"Cum.\nProb.", "Relat.\nElev."}]
```

From In[22]:=

Relat.
Elev.

1
0.8
0.6
0.4
0.2

0.2 0.4 0.6 0.8 1 Cum. Prob.

Out[22]= -Graphics-

The area beneath the hypsometric curve is the **hypsometric integral**, which can be used as a scalar reflection of the degree of incision. Values of the hypsometric integral can range from 0 to 1. Large values indicate high plateaus that are incised by a few narrow valleys, whereas small values indicate flat plains interrupted by only a few hills or hummocks. The hypsometric integral can be calculated from the swapped cumulative frequency list using the **`ListIntegrate`** function contained in the standard NumericalMath`ListIntegrate` package.

```
In[23]:= ListIntegrate[%%]

Out[23]= 0.425987
```

A variable known as the **elevation-relief ratio** was introduced shortly after the concept of the hypsometric integral was developed, and was later shown to produce a value virtually indistinguishable from the hypsometric integral (Scheidegger, 1991). The elevation-relief ratio for the resampled Bremerton data set is

In[24]:= $\frac{\text{Mean[Flatten[sampledelev]] - minval}}{\text{maxval - minval}}$ **//N**

Out[24]= 0.420897

which is, indeed, virtually indistinguishable.

Slope Angle or First Derivative Maps

Another useful geomorphic attribute is slope angle, which can be a factor in slope stability, erosion and sedimentation, and land use restrictions. Maps showing the slope of non-topographic surfaces, for example gravity or magnetic data, are usually referred to as first derivative maps because the slope of a line or plane is its first derivative. One way to calculate slope angles from digital elevation data is to fit a plane or higher order surface to each elevation point and its immediate neighbors using least-squares methods and then take the derivative of the fitted surface to find the slope. An alternative is to use finite difference approximations, in which derivatives are approximated as elevation changes over finite distances, for example the elevation difference between two adjacent values divided by the horizontal distance between them. In practice, finite difference methods are implemented using either the four or eight neighbors of each elevation point. A finite difference approximation can be illustrated using a set of nine elevation values taken from the Bremerton elevation data set. First, fill a table with a subset of nine values from the full data set. The choice of rows and columns is arbitrary.

```
In[25]:= data = Table[elev[[r, c]], {r, 100, 102}, {c, 50, 52}]

Out[25]= {{215.1, 213.8, 212.2}, {219.2, 218., 217.1},
          {221.9, 221.1, 220.6}}
```

The **TableForm** option can be used to display the elevations in rows and columns, recalling that the northernmost row is at the bottom of the table because we reversed the row order at the beginning of this section.

```
In[26]:= data //TableForm

         215.1 213.8 212.2
Out[26]= 219.2 218.  217.1
         221.9 221.1 220.6
```

The north-south and east-west components of slope are calculated separately. Notice that the elevation differences are divided by twice the elevation grid spacing because the center point itself is not used in this calculation.

In[27]:= $NS = \frac{data[[3, 2]] - data[[1, 2]]}{20.}$

Out[27]= 0.365

In[28]:= $EW = \frac{data[[2, 3]] - data[[2, 1]]}{20.}$

Out[28]= -0.105

Each of the two slope components is a vector quantity, so the resultant maximum downward slope at row 101 and column 51 of the Bremerton elevation data set is calculated as the square root of the sum of their squares, or

In[29]:= $\sqrt{NS^2 + EW^2}$

Out[29]= 0.379803

The value calculated above is the**slope gradient**, which is the tangent of the slope angle and therefore dimensionless. It can be converted into a **slope angle** using the **ArcTan** function as shown below.

In[30]:= $\frac{1}{\circ} ArcTan\left[\sqrt{NS^2 + EW^2}\right]$

Out[30]= 20.7969

Mathematica, like other computer programs, calculates angles in radians rather than degrees and **Degree** is a built-in conversion factor. Values given in radians are divided by **Degree** to obtain degrees, and those given in degrees are multiplied by **Degree** to obtain radians. If the elevation data set represented a structural geologic surface, for example the top of a petroleum reservoir or aquifer, the slope angle would be the dip of the surface.

Slope angles for an entire table of values can be calculated by combining the previous four steps into a single equation and then using that equation to produce a new table filled with the slope angles at each data point. Because the slope angle calculation method that we are using is based on values of neighboring data points, however, it cannot calculate slopes for points around the edges of the data set. Therefore, the resulting tables will have two fewer rows and two fewer columns than **elev**.

```
In[31]:= Δ = 10.;
         slopes =
          1/°
           ArcTan[
            Table[Sqrt[((elev[[r + 1, c]] - elev[[r - 1, c]])/(2 Δ))^2 +
               ((elev[[r, c + 1]] - elev[[r, c - 1]])/(2 Δ))^2],
             {r, 2, nrows - 1}, {c, 2, ncols - 1}
            ]
           ];
```

> **Computer Note:** Write a *Mathematica* function that will take an entire table of gridded elevation values and their grid spacing as input and produce a table of slope angles as output. The usage might be something like **SlopeAngle[elev, Δ]**.

> **Computer Note:** Develop a method for calculating slope angles that will allow values to be calculated along the edges of the data set. It may help to read about the treatment of boundary conditions in finite difference simulations of groundwater or heat flow.

Contour plots of slope angles can be difficult to interpret, and the best visualization choice is often a density plot that shows a continuous range of tones or colors. A **MeshRange** specification is included in the density plot below so that it can be centered beneath a topographic map of **elev**, which has two more rows and columns of data.

```
In[32]:= ListDensityPlot[slopes, AspectRatio → aratio,
           Frame → False,
           ColorFunction → Function[z, GrayLevel[0.3 + 0.7 z]],
           Mesh→False, MeshRange→ {{2, nrows - 1}, {2, ncols - 1}}]
```

From In[32]:=

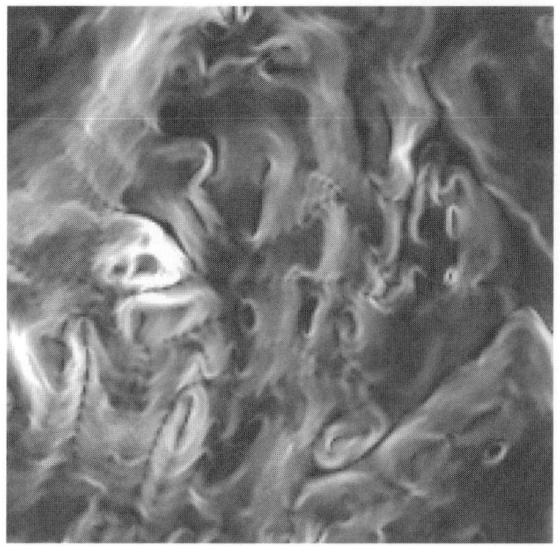

Out[32]= -DensityGraphics-

The white and black banding on the slope map is an artifact of the digital elevation model, which rounds the elevation data to the nearest decimeter. The information in the slope map can be tied to the landscape by overlaying it with a topographic map. Because most of the slope plot is gray to black, the usual black contour lines would not show up well. Therefore, the first step will be to make another contour map with white contours and a 20 m contour interval so that the slope information is not obscurred by the contours. **DisplayFunction→Identity** is used to suppress output of the contour map, which would be invisible against a white background.

```
In[33]:= whitetopomap = ListContourPlot[elev,
           ContourShading → False,
           AspectRatio → aratio, Frame → False,
           Contours → Table[c, {c, 50, 400, 20.0001}],
           ContourStyle → {Thickness[0.005], GrayLevel[1]},
           DisplayFunction → Identity]
```

Out[33]= -ContourGraphics-

Next, Show is used to place the contour map over the slope angle map. **DisplayFunction → $DisplayFunction** is used to make both maps visible.

```
In[34]:= Show[%%, %, DisplayFunction → $DisplayFunction]
```

From In[34]:=

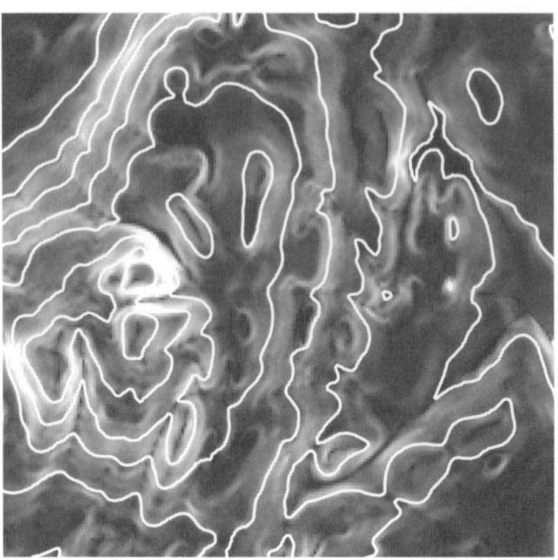

```
Out[34]= -Graphics-
```

> **Computer Note:** Experiment with different color functions to help visualize the slope angle distribution. Using **ColorFunction- > Function [z, RGBColor[z, 1 - z, 0]]** in **DensityPlot** or **ContourPlot** produces plots that range from bright green for low values to bright red for high values.

> **Computer Note:** Overlay a gray scale density plot with a colored contour plot. What combinations of colors and map styles best convey the information about slope angles and topography?

> **Computer Note:** Make a density plot that shows slope angles above a certain threshold, say 20°, in red and all other values in green.

> **Computer Note:** Create a new density plot without using the **MeshRange** specification, then overlay it with a contour plot containing black or colored lines. This will illustrate the mismatch that occurs if the two missing rows and columns of slope angles are not taken into account.

Another option for visualizing slope angles is the use of vector plots such as those used to illustrate groundwater flow directions in Chapter 2. In the case of a 201 × 201 grid of elevation data, however, the vectors would be too crowded to read.

Slope Aspect Maps

Slope aspect is the azimuth of the maximum slope angle, and is equivalent to the the azimuth of the dipline of a structural geologic surface such as a tilted bed. It can range in value from 0 to 360°. As you might imagine, slope angle and aspect are therefore closely related. The slope aspect azimuth for **data** is calculated from the two previously calculated components of slope gradient, **EW** and **NS**, and is.

```
In[35]:= 1/° ArcTan[-NS, -EW]
Out[35]= 163.951
```

The elevation points in **data** thus define a slope that is facing about 16° east of south (164°) and dipping about 21° in that direction. Notice that the arc tangent function used above is different than the one used to calculate the slope angle. Because the possible range of slope angles occupies only one quadrant (0 to 90°), the arc tangent could be calculated from the simple ratio of the slope components. Aspect, however, can range through all four quadrants (0 to 360°) and the sign of each component must therefore be considered. The four-quadrant arc tangent of *y/x* is calculated using **ArcTan[x, y]**.

The relationship between the east-west slope gradient, the north-south slope gradient, the slope angle, and the slope aspect can be illustrated by drawing a simple vector diagram. The slope gradient will be the resultant of the two orthogonal slope gradient components, and the aspect is the supplement of the angle measured clockwise-positive from North to the resultant. The angle itself gives the maximum upslope gradient, hence 180° must be added or subtracted in order to obtain the direction of the maximum downslope gradient. Multiplying the two gradients by –1 in the expression above has the same effect as adding or subtracting 180°.

The orientation of the surface defined by **data** can be visualized with a surface plot of its nine elevations, as shown below.

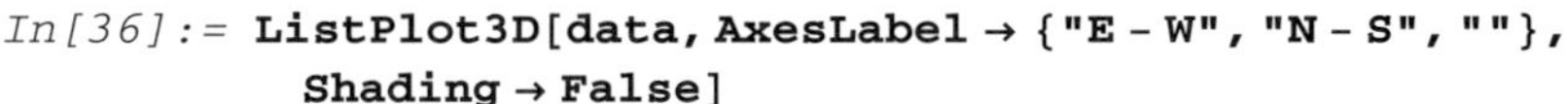

```
In[36]:= ListPlot3D[data, AxesLabel → {"E - W", "N - S", ""},
           Shading → False]
```

From In[36]:=

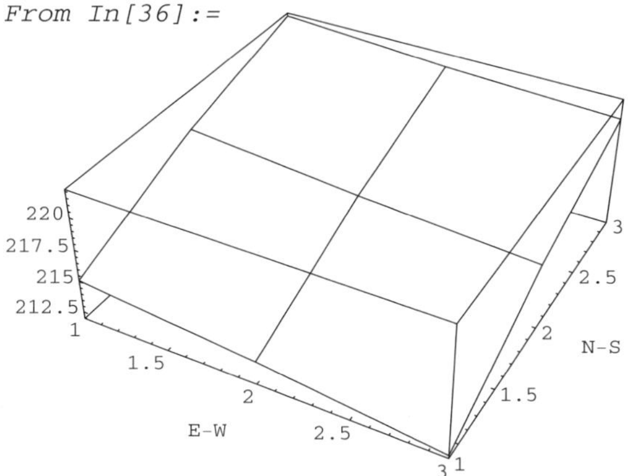


```
Out[36]= -SurfaceGraphics-
```

Recall from structural geology that the strike of a surface is the compass direction of its intersection with an imaginary horizontal plane. Picking an elevation within the range of **data**, plotting it but supressing the output, and then combining it with the dipping surface plot produces the figure below.

```
In[37]:= Plot3D[218, {x, 1, 3}, {y, 1, 3}, PlotPoints → 25,
           DisplayFunction → Identity];
         Show[%%, %, DisplayFunction → $DisplayFunction]
```

From In[37]:=

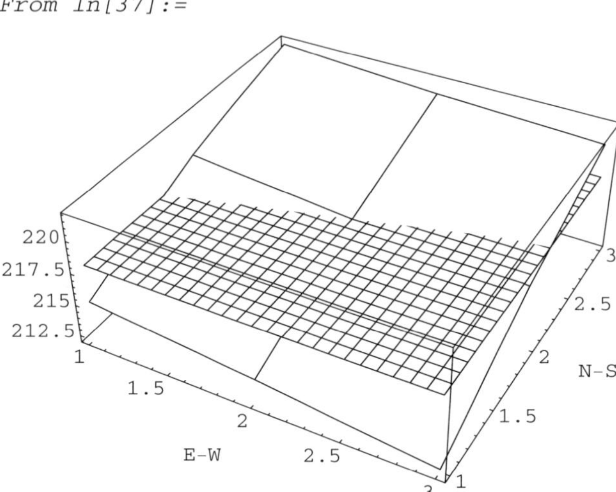

Out[37]= -Graphics3D-

The direction of the strike line is the aspect ±90°, or 164° - 90° = 74°. You can verify this by picking any three points that do not lie along the same line from **data** and using the three-point interpolation method developed in Chapter 2.

Slope aspect azimuths for the entire **elev** data set can be calculated by making just a few changes to the slope angle calculation. Remember to correctly order the two gradients in the arc tangent function and multiply them by –1. An error will occur if any of the east-west gradients is zero and *Mathematica* will not calculate a value for that point. This potential problem can be alleviated by adding a very small quantity (say, 0.0000001) to the east-west gradient to ensure that there will be no divide by zero errors

```
In[38]:= Δ = 10.;
         aspect =
          1/° Table[ArcTan[-((elev[[r + 1, c]] - elev[[r - 1, c]])/(2 Δ)),
              -((elev[[r, c + 1]] - elev[[r, c - 1]])/(2 Δ)) + 1. 10^-7],
            {r, 2, nrows - 1}, {c, 2, ncols - 1}
           ];
```

A density plot of the slope aspect angle, shown below, looks something like a shaded relief map, but not quite. The problem is the existence of unnatural looking black and white patches throughout the plot. These are produced because *Mathematica* scales the gray levels linearly between the lowest and highest azimuth values but, in reality, azimuths are continuously distributed. The result is that an azimuth of 001° would be plotted as black whereas an azimuth of 359° would be plotted as white. The second fact that contributes to the unusual appearance of the map below is that *Mathematica* returns arc tangent values in a range between –180° and +180°, another mathematical convention. When working with maps, however, it is much more convenient to have azimuths between 0° and 360°. Both of these problems can be easily fixed.

```
In[39]:= ListDensityPlot[aspect, Mesh → False, Frame → False,
           MeshRange → {{2, nrows - 1}, {2, ncols - 1}}]
```

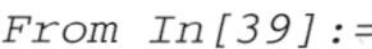

```
From In[39]:=
```

```
Out[39]= -DensityGraphics-
```

To make a more realistic looking shaded relief map, we will need to come up with way to avoid the discontiuity that occurs where the high and low ends of the scale meet. Because the aspect azimuth data vary continuously over a range of 0° to 360°, the logical choice is a trigonometric function, such as a sine or cosine curve, that likewise varies continuously over the same range. One such solution is illustrated below. A cosine curve will have its largest values, and therefore lightest shades on the density map, for aspect azimuths near 0° and its smallest values, and therefore darkest shades, for azimuths near 180°.

```
In[40]:= ListDensityPlot[Cos[(aspect)°], Mesh → False,
           MeshRange → {{2, nrows - 1}, {2, ncols - 1}},
           Frame → False]
```

From In[40]:=

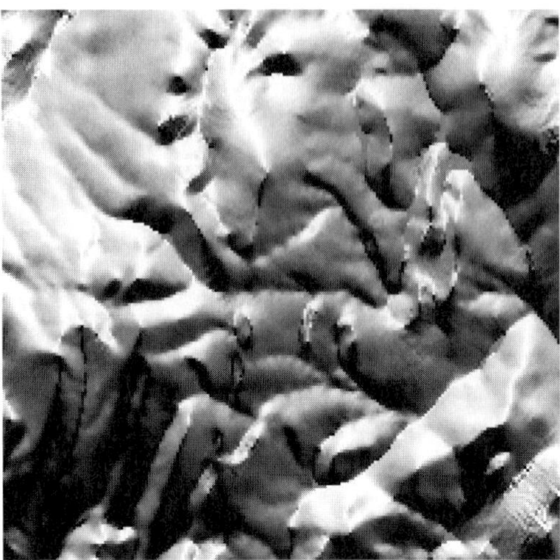

```
Out[40]= -DensityGraphics-
```

The simulated lighting can be adjusted by shifting the cosine curve. For example, the plot below has lighting from a direction of 045°. It also scales the **GrayLevel** option so as to remove the darkest values from the image.

```
In[41]:= ListDensityPlot[Cos[(aspect - 45.)°], Mesh → False,
           ColorFunction → Function[z, GrayLevel[0.2 + 0.8 z]],
           MeshRange → {{2, nrows - 1}, {2, ncols - 1}},
           Frame → False]
```

From In[41]:=

```
Out[41]= -DensityGraphics-
```

As before, adding topographic contours helps to visualize the topography.

```
In[42]:= Show[%, topomap]
```

```
From In[42]:=
```

```
Out[42]= -Graphics-
```

Much more sophisticated shaded relief maps can be constructed by specifying the degree of reflectance as a function of the angle between the topography and the light source.

> **Computer Note:** Use **ListPlot3D** to plot the surface and explore the effects of changing lighting on three dimensional shaded relief plots by varying the **LightSources**, **AmbientLight**, and **Lighting** options. Consult the *Mathematica* documentation for more information about these options.

> **Computer Note:** Generate a series of aspect plots with different lighting angles and then animate them. The *Mathematica* documentation contains information about animating a series of plots.

Curvature or Second Derivative Maps

Curvature maps are used to delineate areas of concave, planar, and convex topography. When constructed from geophysical or structural data, curvature maps are often referred to as second derivative maps. The **SlopeCurvature** function included in the *Mathematica* package accompanying this book uses a second order accurate finite difference expression to calculate total curvature using the elevation at a point and its four nearest neighbors. Again using the nine values in data, the curvature at row 101 and column 51 is calculated as

```
In[43]:= 1/10^2 (data[[1, 2]] + data[[3, 2]] + data[[2, 1]] +
             data[[2, 3]] - 4. data[[2, 2]])

Out[43]= -0.00799988
```

Curvature, as can be shown be examining the units in the expression above, has units of reciprocal length (meters in this case). Positive values of curvature indicate concave-upwards topography (for example, valleys), whereas negative values indicate convex-upwards topography (for example, ridges). Some geomorphologists further distinguish between plan curvature (the curvature of contour lines shown in map view) and profile curvature (the curvature measured down a slope such as the axis of a valley), and equations to calculate those two variations can be found in geomorphology or GIS books such as Burrough and McDonnell (1998).

```
In[44]:= curvature = SlopeCurvature[elev, 10.];

In[45]:= ListDensityPlot[curvature, AspectRatio → aratio,
           Frame → False, Mesh → False,
           MeshRange → {{2, nrows - 1}, {2, ncols - 1}}]

From In[45]:=
```

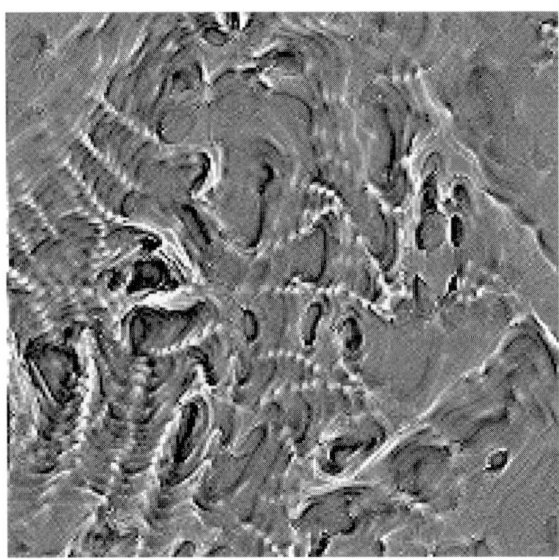

```
Out[45]= -DensityGraphics-
```

The lightest colors in the map above are areas of concave slopes such as unchanneled hollows or stream valleys. The darkest colors are convex areas such as hilltops and ridges. Curvature maps have been combined with slope maps to identify debris flow source areas and runout paths, and may be useful for identifying topographically subtle features such as the scarps and toes of dormant landslides. As before, superimposing a topographic map can help to show the significance of slope curvature. Because most of the values are light gray to white, a black line (or colored) contour map will be more useful than the white map used in *Out[34]*.

```
In[46]:= Show[%, topomap]
```

From In[46]:=

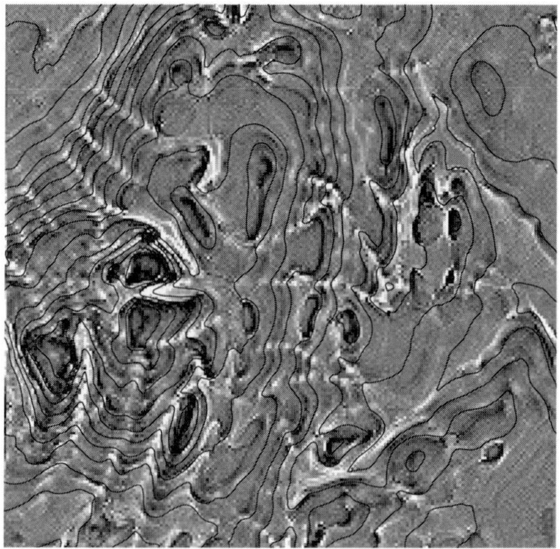

```
Out[46]= -Graphics-
```

7.2.7 Composite Geomorphic Maps

The quantities calculated so far in this chapter have been purely descriptive geometric attributes such as slope angle, aspect, and curvature. They can be used as the basis for process-based models of phenomena such as soil erosion, landslide potential, and flooding or linked into composite models that allow several empirical data layers to be examined in concert. The potential for shallow translational landsliding, for example, can be expressed in its simplest form using the **infinite slope** model, which is the ratio of driving forces to resisting forces acting in a planar slope of infinite extent. No real slope exactly satisfies these conditions, of course, but the infinite slope model is a reasonable first-order approximation that can be used to quickly screen large areas and identify potentially unstable zones that merit more detailed field investigations. The simplest form of infinite slope model, which applies to unsaturated granular soils, is the ratio of the tangents of the angle of internal friction (which reflects the shear strength of the soil) and the slope angle. The result is a factor of safety against sliding, for which values greater than 1 indicate stability. The factor of safety for a saturated slope is about one-half that of an unsaturated slope, so slopes with dry factors of safety between 1 and 2 are conditionally stable depending on their degree of saturation. Slopes with factors of safety greater than 2 should be unconditionally stable.

Assume that the angle of internal friction is constant over the entire map area and is 30°, or

```
In[47]:= tanφ = Tan[25. °]

Out[47]= 0.466308
```

If more data were available, for example a map showing different formations with different angles of internal friction, they could have been incorporated. For now, though, we will assume that the value is a constant. Factor of safety values can now be calculated and stored in a new table named **FS**

```
In[48]:= FS = Table[tanφ / Tan[(slopes[[r, c]] + 0.00001) °],
            {r, 1, nrows - 2}, {c, 1, ncols - 2}];
```

and then plotted and overlain with a contour map. The plot is of the quantity $1 - FS$, so light values indicate low factors of safety and are more susceptible to landsliding.

```
In[49]:= ListDensityPlot[1 - FS, Mesh- > False,
           MeshRange → {{2, nrows - 1}, {2, ncols - 1}},
           Frame → False,
           ColorFunction → Function[z, GrayLevel[0.3 + 0.7 z]],
           DisplayFunction → Identity];
         Show[%, topomap, DisplayFunction → $DisplayFunction]

From In[49]:=
```

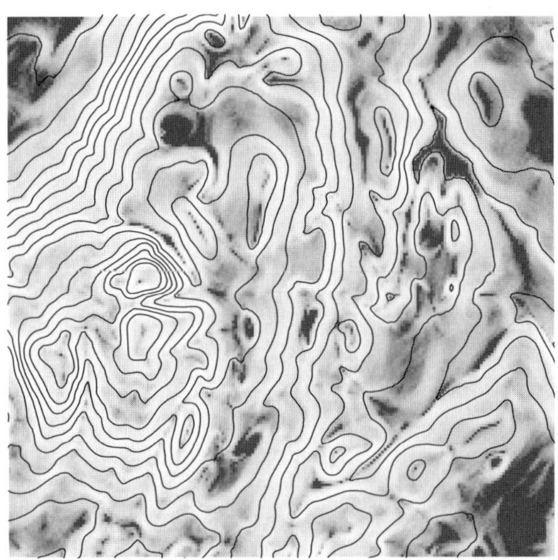

```
Out[49]= -Graphics-
```

Some geomorphologists argue, however, that other factors contribute to landslide potential. Concave portions of slopes, for example, can contain thicker and wetter soils than planar or convex slopes. How can we combine this qualitative information with the quantitative factor of safety? One way is to produce a composite map that contains three different categories; concave slopes with factors of safety less than 2 (the most hazardous), planar or convex slopes with factors of safety less than 2 (moderately hazardous), and slopes with factors of safety greater than 2 (the least hazardous).

The following set of *Mathematica* statements creates a table named **landslide** and sets all of its values to 2. Then, the lines within the **Do** statement check the **FS** and **curvature** values for each data point and assign a value of 0 or 1.5 depending on the result. These values were chosen so that areas with the lowest hazard will appear as white and those with moderate hazard will appear as gray in a grayscale density map.

```
In[50]:= landslide = Table[2., {nrows - 2}, {ncols - 2}];
         Do[
           Block[{},
             If[curvature[[r, c]] ≤ 0. && FS[[r, c]] < 2.,
               landslide[[r, c]] = 1.5];
             If[curvature[[r, c]] > 0. && FS[[r, c]] < 2.,
               landslide[[r, c]] = 0.]
           ], {r, 1, nrows - 2}, {c, ncols - 2}
         ]
```

Here is a plot of the **landslide** table overlain with a topographic map.

```
In[51]:= ListDensityPlot[landslide, Mesh- > False,
           MeshRange → {{2, nrows - 1}, {2, ncols - 1}},
           Frame → False, DisplayFunction → Identity];
         Show[%, topomap, DisplayFunction → $DisplayFunction]
```

```
From In[51]:=
```

```
Out[51]= -Graphics-
```

It cannot be overemphasized that this is a very simple approach to a very complicated problem, and that there are other factors that control landslide potential. They include other components of shear strength, the hydrologic and mechanical effects of vegetation, seismic effects, the previous occurrence of landslides, and the magnitude and frequency of rainstorms that may trigger landslides. Not all of them are well understood or easily modeled, and calculations should never be used as a replacement for field observations. Nonetheless, the simple model developed above illustrates how easily data sets and their derivative products can be combined with geologic inference to produce reconnaissance level screening tools that can be used in conjunction with field and laboratory investigations.

7.3 Irregularly Spaced Data

More often than not, the spatial data of interest to geologists are not collected on a regular grid. Geophysical surveys, for example, may be conducted along roads or flight lines that do not form a rectangular grid. Borehole information such as depth to water, formation thickness, the elevation of formation tops, or measurements obtained from geophysical logs may be clustered or spaced at irregular intervals. This poses a problem because many contouring and plotting alorithms require regularly gridded data. The same is true of the slope angle, curvature, and aspect routines developed in the previous section. One exception is methods based on networks of triangles connecting irregularly spaced data points (triangulated irregular networks, or TINs). *Mathematica* includes add-on functions for calculating TINs and plotting three dimensional surfaces defined by TINs, but it is not an easy matter to perform mathematical operations such as slope angle or curvature calculations on TINs. This section describes three different methods of gridding or interpolating irregularly spaced data so that it can be displayed or analyzed using the methods developed above for gridded data.

Our approach will be to examine the results of different gridding methods by comparing them to a surface for which values are known everywhere. Therefore, we will start by generating a surface in with the dependent variable is given by the function

In[52]:= $\mathbf{f = 0.02\,x + 0.02\,y + 1.\,10^{-6}\,x\,y + 100\,Sin\left[2\,\pi\frac{x}{8000}\right]\,Sin\left[2\,\pi\frac{y + 1000}{25000}\right]}$

Out[52]= $0.02x+0.02y+1.\times10^{-6}xy+100\,\mathrm{Sin}\left[\frac{\pi x}{4000}\right]\,\mathrm{Sin}\left[\frac{\pi\,(1000+y)}{12500}\right]$

The surface-generating function was obtained by trial and error, adjusting values to produce a surface that might reasonably represent a series of antiforms and synforms superimposed on a regionally dipping surface. We'll consider a map area that ranges over $0 \leq x \leq 10{,}000$ and $0 \leq y \leq 10{,}000$ units. In map view, the surface looks like this:

```
In[53]:= truesurfacemap=ContourPlot[f,{x,0,10000},{y,0,10000},
          AspectRatio → 1, PlotPoints → 40, Frame → False,
          ColorFunction → Function[z, GrayLevel[(0.3 + z)/1.3]],
          Contours → Table[c, {c, 50., 550., 50.0001}]]
```

From In[53]:=

```
Out[53]= -ContourGraphics-
```

The **GrayLevel** specification was used to scale the values so that the darkest shade is dark gray rather than black, which would have obscurred any data points later plotted on the map.

Now, generate a series of randomly located points at which the surface is to be sampled. The statement **Random[Real, {0, 10000}]** selects a real number between 0 and 10,000 at random, so the table below consists of 25 pairs of random x and y coordinates. Random number generation is discussed in much more detail in Chapter 4.

```
In[54]:= SeedRandom[6];

In[55]:= locs = Table[{Random[Real, {0, 10000}],
           Random[Real, {0, 10000}]}, {25}]

Out[55]=
    {{3605.14, 9447.96}, {5545.13, 6210.91},
      {5589.31, 9487.48}, {3358.64, 9733.6}, {9580.38, 8719.48},
      {9357.68, 1640.3}, {9201.46, 2841.85}, {7602.61, 5907.51},
      {277.95, 4364.93}, {1620.37, 5073.31}, {1425.84, 4317.25},
      {8547.71, 5196.8}, {7820.71, 4869.3}, {3002.59, 8985.89},
      {2231.39, 5381.82}, {9643.95, 9252.29},
      {2651.02, 6662.33}, {286.274, 7611.99}, {3449.56, 3820.48},
      {2683.66, 1704.48}, {3171.61, 9455.56}, {1063.29, 6631.16},
      {1745.77, 5138.3}, {2515.58, 1434.37}, {3925.06, 269.01}}
```

Once the random coordinates have been generated, fill a table with values of **f** calculated only at the **locs** points. Each triplet in the table below contains an east-west coordinate, a north-south coordinate, and a z value for those coordinates.

```
In[56]:= surfdata = Table[{locs[[i, 1]], locs[[i, 2]],
           f /.{x- > locs[[i, 1]], y- > locs[[i, 2]]}},
           {i, Length[locs]}]

Out[56]=
  {{3605.14, 9447.96, 310.174},
    {5545.13, 6210.91, 178.594}, {5589.31, 9487.48, 308.612},
    {3358.64, 9733.6, 315.268}, {9580.38, 8719.48, 510.402},
    {9357.68, 1640.3, 289.23}, {9201.46, 2841.85, 333.602},
    {7602.61, 5907.51, 284.827}, {277.95, 4364.93, 115.194},
    {1620.37, 5073.31, 237.588}, {1425.84, 4317.25, 208.56},
    {8547.71, 5196.8, 361.01}, {7820.71, 4869.3, 277.91},
    {3002.59, 8985.89, 308.431}, {2231.39, 5381.82, 262.572},
    {9643.95, 9252.29, 518.608}, {2651.02, 6662.33, 285.703},
    {286.274, 7611.99, 178.625}, {3449.56, 3820.48, 197.802},
    {2683.66, 1704.48, 146.348}, {3171.61, 9455.56, 312.302},
    {1063.29, 6631.16, 230.655}, {1745.77, 5138.3, 244.626},
    {2515.58, 1434.37, 135.398}, {3925.06, 269.01, 86.7817}}
```

The locations of the data points can be illusrated using **ListPlot**

```
In[57]:= wellmap = ListPlot[locs, PlotStyle → PointSize[0.02],
           AspectRatio → 1, AxesLabel → {"E - W", "N - S"},
           PlotRange → {{0, 10000}, {0, 10000}}]

From In[57]:=
```

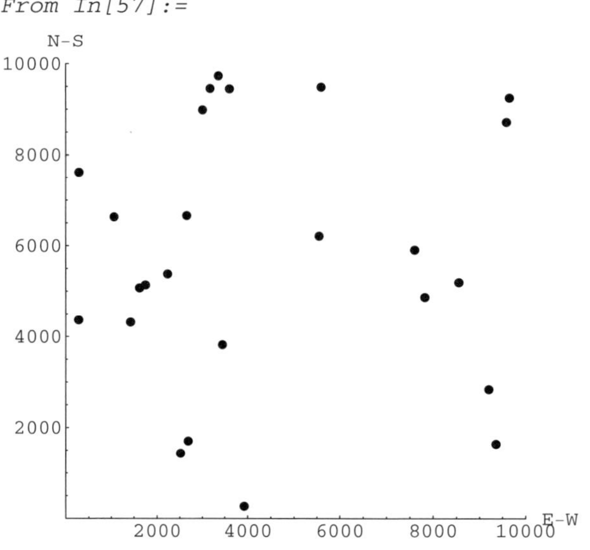


```
Out[57]= -Graphics-
```

and superimposed on the true surface map

```
In[58]:= Show[truesurfacemap, wellmap]
```

From In[58]:=

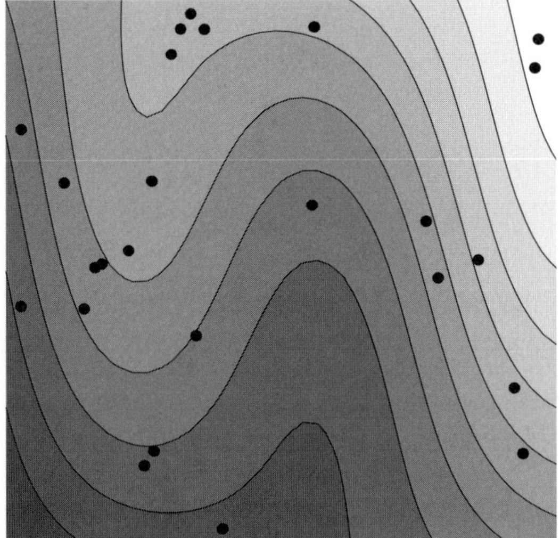

```
Out[58]= -Graphics-
```

Nearest Neighbor Gridding

The simplest approach to gridding is to formulate a grid and then assign each grid point the value of the nearest data point. Below is a *Mathematica* function that takes as input a table of x, y, z values and two lists containing information about the size and resolution of the grid to be filled. In these examples, the x axis is assumed to represent the east-west direction and the y axis is assumed to represent the north-south direction. The **xvals** list includes the minimum x value to be gridded, the maximum x value to be gridded, and the number of grid lines in the x direction. The **yvals** list contains the analogous data for the y direction.

```
In[59]:= NearestNeighborGrid[indata_, xvals_, yvals_] :=
           Block[{i, j, k, d, len, Δx, Δy, zvals, nearest},
             xmin = xvals[[1]];
             xmax = xvals[[2]];
             nx = xvals[[3]];
             ymin = yvals[[1]];
             ymax = yvals[[2]];
             ny = yvals[[3]];
             len = Length[indata];

             d = Table[{0., k}, {k, len}];
             zvals = Table[0., {nx}, {ny}];
```

```
Δx = (xmax - xmin)/(nx - 1);
Δy = (ymax - ymin)/(ny - 1);

Do[
  Block[{x, y, nearest},
    x = xmin + (i - 1) Δx;
    y = ymin + (j - 1) Δy;

    Do[
      d[[k, 1]] =
        N[Sqrt[(indata[[k, 1]] - x)^2 + (indata[[k, 2]] - y)^2]],
      {k, len}
    ];

    nearest = Sort[d];

    zvals[[j, i]] = indata[[nearest[[1, 2]], 3]];

  ], {i, 1, nx}, {j, 1, ny}
];
Return[zvals]
]
```

NearestNeighborGrid is executed using the following syntax, in this case to produce a 21 by 21 grid of data ranging from 0 to 10,000 units in each of the coordinate directions.

```
In[60]:= neighborresults = NearestNeighborGrid[surfdata,
           {0, 10000, 21}, {0, 10000, 21}];
```

The results of **NearestNeighbor**, along with the data point locations, are shown in the contour map below.

```
In[61]:= Show[
           ListContourPlot[neighborresults,
             ColorFunction → Function[z, GrayLevel[0.3 + 0.7 z]],
             Contours → Table[c, {c, 50, 500, 50.0001}],
             MeshRange → {{0, 10000}, {0, 10000}}, Frame → False,
             DisplayFunction → Identity],
           wellmap, DisplayFunction → $DisplayFunction
         ]
```

From In[61]:=

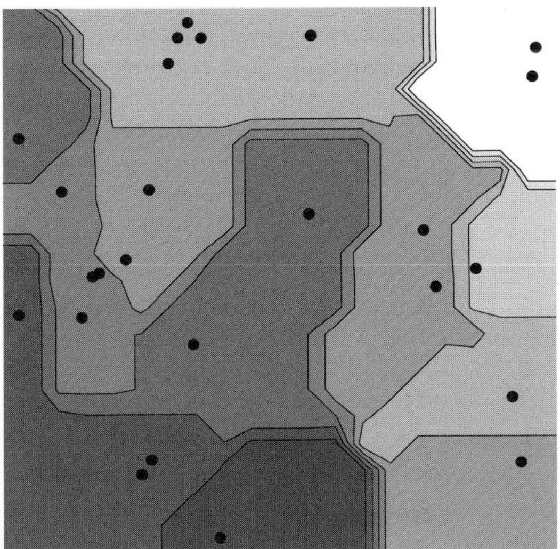

Out[61]= -Graphics-

Although the general pattern of two antiforms and a synform can be discerned with some imagination, a knowledge of the underlying surface, and an appreciation of cubism, the surface is not very realistic. Its stairstep nature can be emphasized by making a three dimensional surface plot.

```
In[62]:= ListPlot3D[neighborresults, MeshRange → {{0, 10000},
          {0, 10000}},
          ColorFunction → Function[z, GrayLevel[0.3 + 0.7 z]],
          AxesLabel → {"E - W", "N - S", " "}]
```

From In[62]:=

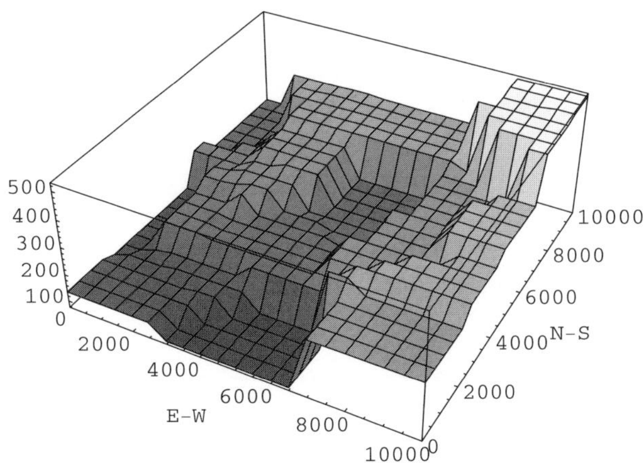

Out[62]= -SurfaceGraphics-

7.3.1 Reciprocal Distance Gridding

Another commonly used method is interpolation based on a weighted average of the N nearest neighbors, with the weights being the reciprocal of the distance between the known and unknown points raised to some power. The interpolated values of z are given by

$$\hat{z}_{x,y} = \frac{\sum_{i=1}^{N} \frac{z_i}{d_i^n}}{\sum_{i=1}^{N} \frac{1}{d_i^n}}$$

where $\hat{z}_{x,y}$ is the interpolated or gridded value, z_i are the nearest N data points, and $d_i$are the distances between the interpolated value and the nearest N data points. The function **InverseDistanceGrid** takes as input a list of x, y, and z values such as **sur fdata**; lists consisting the minimum value, maximum value, and number of grid points in the x and y dimensions; the power to which the distance is raised; and the number of neighbors to be included in each interpolation. Its output is a table containing **ny** rows and **nx** columns of interpolated values.

```
In[63]:=
ReciprocalDistanceGrid[
      indata_, xvals_, yvals_, power_, neighbors_] :=
  Block[{len, Δx, Δy, zvals},
    xmin = xvals[[1]];
    xmax = xvals[[2]];
    nx = xvals[[3]];
    ymin = yvals[[1]];
    ymax = yvals[[2]];
    ny = yvals[[3]];
    len = Length[indata];
    Δx = (xmax - xmin)/(nx - 1);
    Δy = (ymax - ymin)/(ny - 1);
    zvals = Table[0., {nx}, {ny}];
    d = Table[{0., k}, {k, len}];
    Do[
      Block[{x, y, k, m, mind, nearest},
        x = xmin + (i - 1) Δx;
        y = ymin + (j - 1) Δy;
        Do[d[[k, 1]] = Sqrt[(indata[[k, 1]] - x)^2 + (indata[[k, 2]] - y)^2],
          {k, len}];
        nearest = Take[Sort[d], neighbors];
        zvals[[j, i]] = Sum[indata[[nearest[[k,2]],3]]/nearest[[k,1]]^2, {k, 1, neighbors}] /
                        Sum[1./nearest[[k,1]]^2, {k, 1, neighbors}];
      ], {i, 1, nx}, {j, 1, ny}
    ];
    Return[zvals]
  ]
```

The following statement creates the table **reciprocalresults** and fills it with a 21×21 grid of interpolated values using an exponent of 2 and the 15 nearest neighbors to each point.

```
In[64]:= reciprocalresults =
           ReciprocalDistanceGrid[surfdata, {0, 10000, 21},
             {0, 10000, 21}, 2, 15];
```

Here is a contour plot of the results:

```
In[65]:= Show[
           ListContourPlot[reciprocalresults,
             ColorFunction→Function[z,GrayLevel[(z+0.3)/1.3]],
             MeshRange → {{0, 10000}, {0, 10000}},
             Contours → Table[c, {c, 50, 500, 50.0001}],
             Frame → False, DisplayFunction → Identity],
           wellmap, DisplayFunction → $DisplayFunction
         ]
```

From In[65]:=

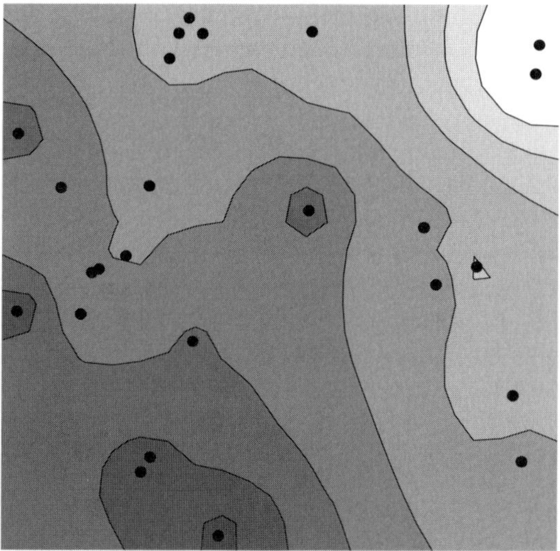

Out[65]= -Graphics-

The result appears more realistic than the stair-step surface generated by the nearest neighbor approach, but the underlying antiforms and synforms are still very difficult to discern and the contours seem unrealistically jagged. The surfaces generated by the reciprocal distance method will, of course, also be influenced by the location of the data points and you will see different results if you select a different set of randomly located data points.

Computer Note: As written, the user-defined function **`ReciprocalDistanceGrid`** cannot account for data points that are located exactly on a grid point. The distance to the nearest neighbor would be zero in such a case, which would produce an infinite value for the weight of that neighbor. Modify **`ReciprocalDistanceGrid`** to check the distances to the *N* nearest neighbors and, if one of them is zero, assign the value of the data point to the interpolated grid point.

Computer Note: Use **`GraphicsArray`** to create a 2 x 2 array of graphs showing contour plots produced for values of **`power`** = 1, 2, 3, and 4.

7.3.2 Thin Plate Spline Gridding

A third commonly used method is thin plate spline interpolation or, as it is commonly known among geophysicists, minimum curvature gridding. Splines are flexible strips of wood or plastic once commonly used for drawing smooth curved lines through a series of points. A drafter would use weights to hold the spline in place and bend it to pass through all of the points, adding more weights as necessary, and then trace the smooth curve onto the paper with his or her pencil. The same task can be accomplished digitally using the equations that described the flexure of a thin elastic beam in two dimensions or a thin elastic plate in three dimensions. Hence the name thin plate spline interpolation. This method is sometimes referred to as the minimum curvature method because an elastic beam or plate passing through a series of points will follow the curve or surface of minimum curvature.

Thin plate spline gridding involves the iterative solution of the **biharmonic equation**, a differential equation that describes two dimensional elastic deformation. Think of it as a two dimensional extension of the elastic beam equation used in Chapter 3 to simulate deformation above laccoliths. The form of the biharmonic equation is

$$\frac{\partial^4 z}{\partial x^4} + 2\frac{\partial^4 z}{\partial x^2 \partial x^2} + \frac{\partial^4 z}{\partial y^4} = 0$$

The biharmonic equation has some known analytical solutions, but none of them apply to interpolation problems in which the value of the dependent variable is specified at various locations throughout the problem domain. Therefore, numerical methods such as finite difference approximations must be used to iteratively solve the equation (*e.g.*, Timoshenko and Goodier, 1970).

The function **`ThinPlateGrid`** in the *Mathematica* package accompanying this book takes as input an array of x, y, and z values such as **`surfdata`**, lists of minimum and maximum x and y values, a grid spacing value that, for simplicity, must be the same in both dimensions, and a numerical tolerance value. The tolerance value is the threshold to which the solution is iterated, and is specified as some fraction of the range of z values in the input. For example, an input data set with values ranging over $0 \leq z \leq 500$ and a tolerance of 0.001 would cause the solution to be iterated until the maximum difference between iterations is 0.5 units of measurement. The input data points will not necessarily fall on any of the grid points where interpolations will be performed. Indeed, it would be surprising if any of them fell exactly on a grid point. To account for this discrepancy, **ThinPlateGrid** assigns each known data point values to the nearest grid point and holds the value constant throughout the iterations. The actual function is not shown below because of its length, but it can be examined by opening the accompanying *Mathematica* package as a notebook or with a text editor. In general, the spacing of the interpolated grid should be smaller than the spacing of the sampled grid. Otherwise, the function may produce undesired results because it will try to assign more than one known value to some interpolation grid points.

Here is an example of **ThinPlateGrid** using the same data and ranges as the previous examples, and with a numerical tolerance of 0.001.

```
In[66]:= thinplateresults = ThinPlateGrid[surfdata,{0.,10000.},
           {0., 10000.}, 500, 0.001];
```

Be prepared to wait a couple of minutes if you execute the previous statement on your computer, because several hundred iterations can be required to obtain a solution when the grid spacing and tolerance are small. If you are routinely gridding very large data sets or producing finely meshed grids, it may be worthwhile to used specialized software that is compiled and optimized for speed. The resulting surface, along with the data point locations, is:

```
In[67]:= ListContourPlot[thinplateresults,
           MeshRange → {{0, 10000}, {0, 10000}},
           Contours→Table[c,{c,50.,550.,50.0001}],Frame→False,
           ColorFunction → Function[z, GrayLevel[(z + 0.3)/1.3]],
           DisplayFunction → Identity];
         Show[%, wellmap, DisplayFunction → $DisplayFunction]
```

```
From In[67]:=
```

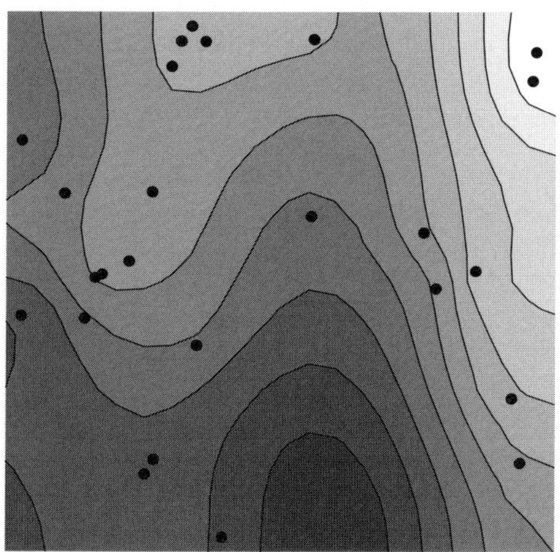

```
Out[67]= -Graphics-
```

This is probably the most natural looking of the three surfaces, and most geologists would probably not hesitate to consider it a successfully interpreted data set. It is generally, although not exactly, similar to the map produced using the reciprocal distance method. As stated above, the nature of the surface obtained by any gridding method will be strongly dependent upon the distribution of data points. More sophisticated variations of the thin plate spline method also include a tension component that lets the user tighten or loosen the imaginary elastic plate being used for interpolation.

7.3.3 A Note About Kriging

Kriging is a sophisticated interpolation technique that incorporates information about the spatial correlation structure of the surface, and could be the subject of an entire course or book. It has many proponents. Kriging can work well and be worth the effort when the number of data points is large and the data satisfy certain conditions. In other cases the surfaces generated by kriging are no better, and can be appreciably worse, than those produced by the methods we have examined. In situations where data too sparse to yield reliable information about their spatial correlation structure, assumptions about their spatial relationships must be made and kriging loses much of its attractiveness. The books by Isaaks and Srivastava (1989), Burrough and McDonnell (1998), Middleton (2000), Carr (2002), and Davis (2002) listed in the Recommended Reading section of this chapter describe the theory and application of kriging methods in various degrees of detail.

7.3.4 Adding Well Locations to Surface Plots

It is relatively straightforward to consruct a three dimensional version of **wellmap** in order to visualize the relationship between a surface and boreholes from which the data were obtained. We know from previous plots that the z values in **surfdata** range from 0 to about 500, so this would be a good vertical range for lines representing boreholes. The statement below constructs a table filled with 25 vertical lines, each representing a well from which an elevation datum was obtained.

```
In[68]:= lines = Table[
           Line[{{locs[[i, 1]], locs[[i, 2]], 0},
             {locs[[i, 1]], locs[[i, 2]], 500}}],
             {i, Length[locs]}
         ];
```

Line[{{x_1, y_1, z_1},{x_2, y_2, z_2}}] creates, but does not display, a line from {x_1, y_1, z_1} to {x_2, y_2, z_2}. The table of lines can be plotted by identifying it as a **Graphics3D** object and then using **Show**. The **Thickness** function controls the thickness of the lines relative to the entire width of the plot and the **GrayLevel** function controls the darkness of the lines. The latter could have been replaced by an **RGBColor** function.

```
In[69]:= wellplot3d = Show[Graphics3D[{Thickness[0.008],
          GrayLevel[0.5], lines}]]
```

From In[69]:=

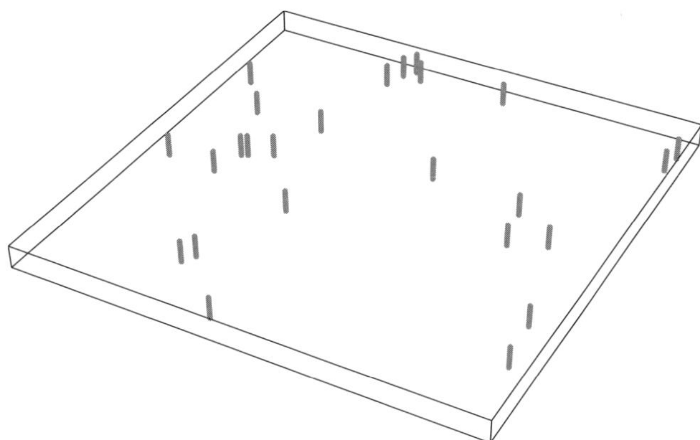

Out[69]= -Graphics3D-

Now that the well locations are plotted, make a three dimensional surface plot of **thinplateresults**

```
In[70]:= thinplateplot3d = ListPlot3D[thinplateresults,
          MeshRange → {{0, 10000}, {0, 10000}}, Shading → False,
          DisplayFunction → $DisplayFunction]
```

From In[70]:=

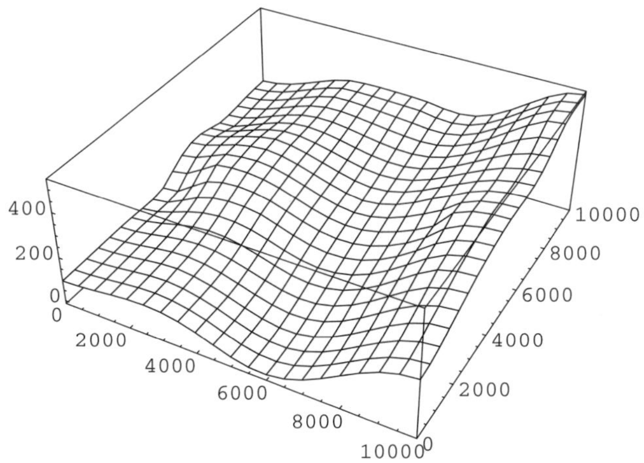

Out[70]= -SurfaceGraphics-

and superimpose the two.

```
In[71]:= Show[thinplateplot3d, wellplot3d, Axes → None]
```

From In[71]:=

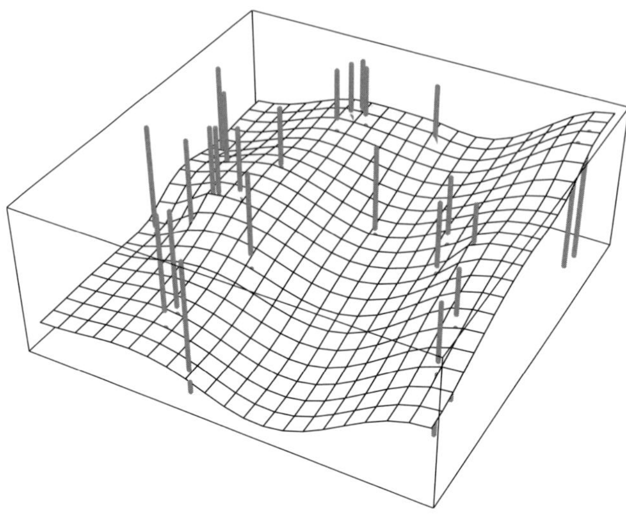

```
Out[71]= -Graphics3D-
```

7.3.5 Comparing Results

In most geologic applications it will be impossible to know how well an interpolated or gridded surface represents the true surface. In this case, however, we know the true surface because it was generated from a mathematical function and then sampled to form our data set. We can therefore compare the true and interpolated surfaces to better understand how the two differ. Although any of the interpolated surfaces from the previous sections can be used, we will take for an example the thin plate spline surface because to the trained eye it appears to be the most geologically realistic of the three.

One way to compare the true and interpolated surfaces is to generate a grid of true surface values that corresponds the the interpolated surface grid and then take their difference. To do so, first fill a table with values of **f** calculated along the same grid as **thinplateresults**.

```
In[72]:= truevals = Table[f, {y, 0, 10000, 500.},
            {x, 0, 10000, 500.}];
```

Then, find the error as the difference between the two tables.

```
In[73]:= errorsurface = thinplateresults - truevals;
```

Mathematica automatically takes into account the fact that **thinplateresults** and **truevals** are both two dimensional tables and not one dimensional vectors or scalars. An error message would have been returned if the the two tables had not contained the same numbers of rows or columns. A three dimensional surface plot is a convenient way to visualize the interpolation errors.

```
In[74]:= ListPlot3D[errorsurface,
           AxesLabel → {"E - W", "N - S", " "},
           ColorFunction→Function[z, GrayLevel[(0.3 + z)/1.3]],
           MeshRange → {{0, 10000}, {0, 10000}}]
```

From In[74]:=

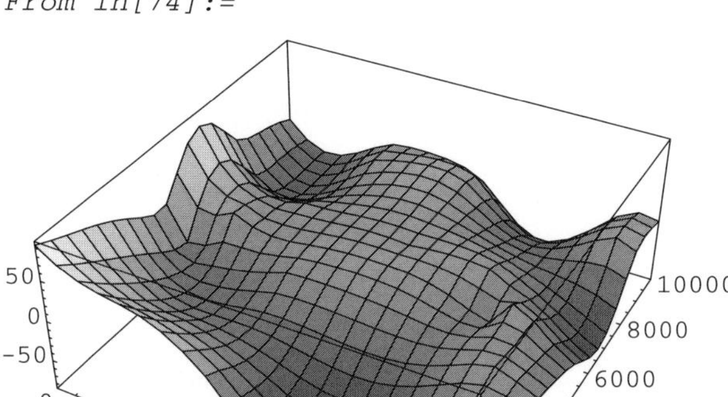


```
Out[74]= -SurfaceGraphics-
```

The surface plot shows that the interpolation errors are quite large along the edges of the grid. Although certainly not desirable, large errors along the edges of the grid are inevitable because they represent an extrapolation beyond the data points rather than an interpolation between data points. Keep this in mind when extrapolating any kind of curve or surface beyond the range of the data!

Another way to represent this tendency for large interpolation errors to occur along the edges of the interpolation grid is to create a table consisting of the distance from the center of the grid and the error at each interpolated point. The table is created by the statement

```
In[75]:= Table[{√((500 (r - 1) - 5000.)^2 + (500 (c - 1) - 5000.)^2),
            Abs[errorsurface[[r, c]]]}, {r, 1, 21}, {c, 1, 21}];
```

and plotted by

```
In[76]:= ListPlot[Flatten[%, 1],
           AxesLabel → {"Distance", "Error"}]
```

From In[76]:=

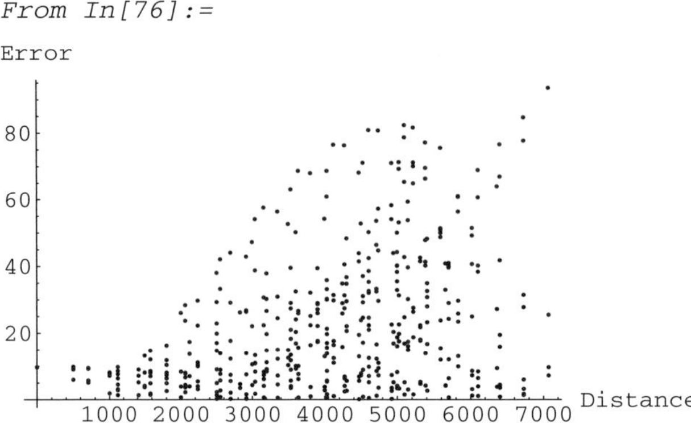

Out[76]= -Graphics-

> **Computer Note:** Generate a new set of "data", this time on a regularly spaced grid that covers the entire area, and compare the results produced by **`ThinPlateGrid`** to the true surface. Is the agreement better or worse?

7.4 Trend Surface Mapping

Another approach to surface analysis is to use least squares methods to fit a surface to the data instead of attempting to interpolate a surface that will pass exactly through each data point. If the data to be analyzed are particularly simple, for example the elevations the top of a homoclinal formation, then a low order polynomial such as $z = a + bx + cy$ may provide a realistic representation of the surface and produce results that are not much different than one might obtain via interpolation. If, however, the surface is more complicated then the low order polynomial will represent a **trend surface**. In geological terms, a trend surface might be the regional dip within a basin or a gradual proximal to distal decrease in average sediment grain size. Differences between the original surface (true or interpolated, although in geological problems it will almost always be the latter) and the trend surface are, just as in other forms of regression, known as **residuals**. A typical application of trend surface analysis might be to subtract the regional dip from structural or geophysical data to highlight smaller scale folds or faults.

Fitting a Trend Surface

Trend surface fitting can be illustrated using the same **`surfdata`** points that were used in the interpolation examples. The equation for the surface itself is found by using *Mathematica*'s **`Fit`** function, although the add-on function **`Regress`** can be used if more detailed statistical output is required. The following statement fits a plane of the form $z = a + bx + cy$ to **`surfdata`**.

```
In[77]:= trendsurface = Fit[surfdata, {1, x, y}, {x, y}]

Out[77]= 42.4842 + 0.022069 x + 0.0218598 y
```

It isn't necessary to restrict trend surfaces to planes. Other low order polynomials can be just as easily used, although there should be some geologic reason for doing so. In general, however, the order of the polynomial is generally much smaller than the number of data points so that the process is one of regression rather than interpolation. The statement below superimposes three dimensional surface plots of **trendsurface** and **thinplateresults** to illustrate the relationship between the two.

```
In[78]:= Plot3D[trendsurface, {x, 0, 10000}, {y, 0, 10000},
           PlotPoints → 50,
           ColorFunction→Function[z, GrayLevel[(0.3 + z)/1.3]],
           DisplayFunction → Identity];

         ListPlot3D[thinplateresults, MeshRange → {{0, 10000},
           {0, 10000}},
           ColorFunction→Function[z, GrayLevel[(0.3 + z)/1.3]],
           DisplayFunction → Identity];

         Show[%, %%, DisplayFunction → $DisplayFunction,
           Axes → None]
```

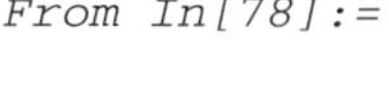

From In[78]:=

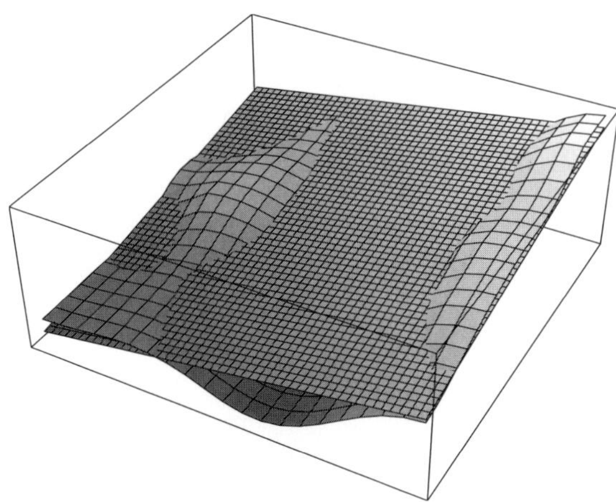

```
Out[78]= -Graphics3D-
```

Calculating Residuals

Residuals have the same definition as in Chapter 6, although in trend surface analysis it is common to concentrate on the residuals and the information they convey rather than trying to minimize them. We'll compare the planar trend surface fitted above to the thin plate spline gridded data, so the next step is to create a table of trend surface values corresponding to the 21 x 21 grid of values in **thinplateresults**.

```
In[79]:= trenddata =
           Round[
             Table[{locs[[i, 1]], locs[[i, 2]],
                 trendsurface /.{x- > locs[[i, 1]],
                 y- > locs[[i, 2]]}},
               {i, Length[locs]}]];
```

The residual is the difference between the third columns of the two tables, or

```
In[80]:= residualdata =
           Round[
             Table[{locs[[i, 1]], locs[[i, 2]],
                 surfdata[[i, 3]] - trenddata[[i, 3]] },
                 {i, Length[locs]}]];
```

Notice that simply executing **surfdata - trenddata** will not provide the answer we want because, in addition to finding the residuals from the third columns, it will subtract the first two columns from each other and set all of the x and y coordinates to zero. Alternatively, we could have simply subtracted the polynomial **trendsurface** from the true surface **f** using the statement

```
In[81]:= f - trendsurface
```

Out[81]= $-42.4842 - 0.00206896\, x - 0.00185985\, y + 1.\times 10^{-6}\, x\, y + 100\, \mathrm{Sin}\left[\frac{\pi x}{4000}\right]\, \mathrm{Sin}\left[\frac{\pi\,(1000 + y)}{12500}\right]$

and evaluated the result using **locs**. In most real world geological problems, though, the true underlying surface **f** is unknown and can only be estimated from a finite number of data points. The minimum and maximum residual values, which will be useful for defining contour intervals, are

```
In[82]:= Min[Column[residualdata, 3]]

Out[82]= -122

In[83]:= Max[Column[residualdata, 3]]

Out[83]= 65
```

As above, we cannot apply **Min** and **Max** to the entire **residualdata** table because it includes x and y coordinates along with the residual values. **Column**, which is an add on function in Statistics`DataManipulation`, solves the problem by isolating the column containing the residuals. With these minimum and maximum values in mind, a contour map of the residuals can be superimposed with a map of the randomly selected data points using the statement

```
In[84]:= residualgrid = ThinPlateGrid[residualdata, {0, 10000},
            {0, 10000}, 500., 0.001];

In[85]:= ListContourPlot[residualgrid,
           MeshRange → {{0, 10000}, {0, 10000}},
           Contours → Table[c, {c, -200, 100, 25.0001}],
           Frame → False,
           ColorFunction → Function[z, GrayLevel[(z + 0.3)/1.3]],
           DisplayFunction → Identity];

         Show[%, wellmap, DisplayFunction → $DisplayFunction]

From In[85]:=
```

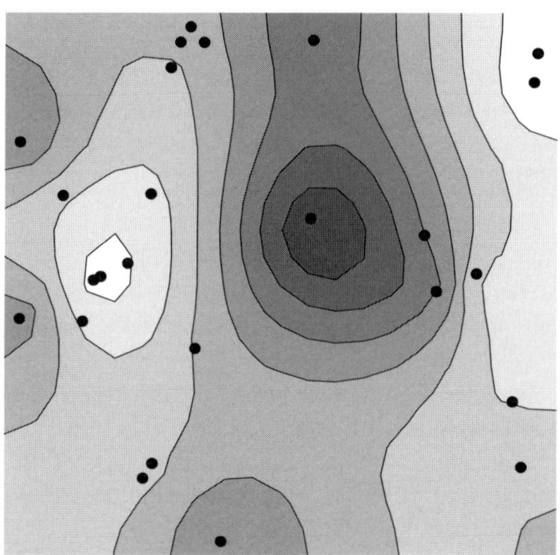

```
Out[85]= -Graphics-
```

How does this compare to the residual map produced from the true surface? We can easily produce one by subtracting the best-fit function **trendsurface** from the true surface **f**, as shown below.

```
In[86]:= ContourPlot[Release[f - trendsurface], {x, 0, 10000},
            {y, 0, 10000},
          Contours → Table[c, {c, -200, 100, 25.0001}],
          Frame → False, PlotPoints → 25,
          ColorFunction → Function[z, GrayLevel[(z + 0.3)/1.3]],
          DisplayFunction → Identity];

        Show[%, wellmap, DisplayFunction → $DisplayFunction]
```

From In[86]:=

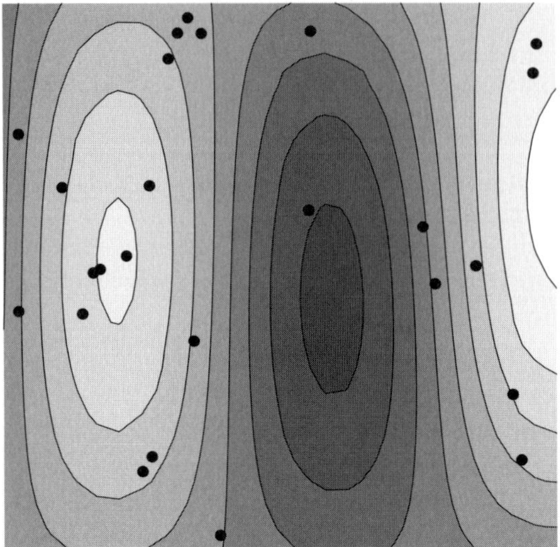

Out[86]= -Graphics-

As with the interpolation examples, there is a general correspondence between the two preceding plots although there are significant differences in the details. Are the differences geologically significant? Would the two residual maps produce different strategies if you were exploring for structural petroleum traps or LNAPL (light non-aqueous phase liquid) accumulations associated with structural highs? What about localized perched water tables that might accumulate in structural lows along a paleotographic surface now covered by surficial deposits? Keep the nature of these differences in mind when interpreting any map based on gridded data.

Goodness-of-Fit

The goodness-of-fit of a trend surface can be evaluated by calculating an r^2 value analogous to that used to characterize a fitted line. In the case of trend surfaces, the residuals commonly represent real differences such as small-scale structures rather than experimental errors, and the objective is not to obtain a particularly good fit between the trend surface and the data. Instead, the objective is to simply remove the low-order trend so as to make the residuals more apparent. In other cases, for example spatially distributed chemical concentrations, the residuals may represent a combination of experimental errors and natural variability. Therefore, it may or may not make much sense to test for the statistical significance of the trend surface. It may still be useful, however, to use a goodness-of-fit value to quantify the relative influence of the regional trend and the local perturbations on the form of the surface. This can be done using the **`Correlation`** function contained in the standard package Statistics`MultiDescriptiveStatistics`.

```
In[87]:= r2 = Correlation[Column[trenddata, 3],
            Column[surfdata, 3]]^2
```

```
Out[87]= 0.810102
```

Therefore, the linear regional trend accounts for about 81% of the variability of the z values contained in **surfdata**.

Derivative Maps

The same slope and curvature mapping tools that we developed for topographic surfaces can be applied to any gridded surface. In this case, we'll assume that **surfdata** represents the top of a petroleum reservoir or aquifer. Therfore, a contour map of the gridded **surfdata** values is a structural contour map. Although structural contour maps can be interpreted as-is, it can sometimes be helpful to produce first derivative (slope) and second derivative (curvature) maps to aid in their interpretation. For example, the elastic beam theory used in Chapter 3 to analyze deformation above laccoliths suggests that faults should occur where the shearing force (which is proportional to the slope of the surface) is greatest and that joints should occur where the fiber stress (which is proportional to the curvature) is greatest. Therefore, slope and curvature maps of a folded surface may help to identify areas that may contain faults that impede fluid flow or fractures that increase porosity and permeability (*e.g.*, Fischer and Wilkerson, 2000; Stewart and Wynn, 2000). The results returned by **SlopeAngle** have units of degrees, but can be converted to dimensionless gradients (vertical/horizontal) by taking their tangents. As long as the results are used for visualization and interpretation rather than calculations, however, the choice is a matter of personal preference. Below is a contour plot showing the first derivative of **thinplateresults** along with the data point locations from which that table was interpolated.

```
In[88]:= ListContourPlot[SlopeAngle[residualgrid, 500.],
           MeshRange → {{0, 10000}, {0, 10000}}, Frame → False,
           ColorFunction → Function[z, GrayLevel[(z + 0.3)/1.3]],
           DisplayFunction → Identity];
         Show[%, wellmap, DisplayFunction → $DisplayFunction]
```

From In[88]:=

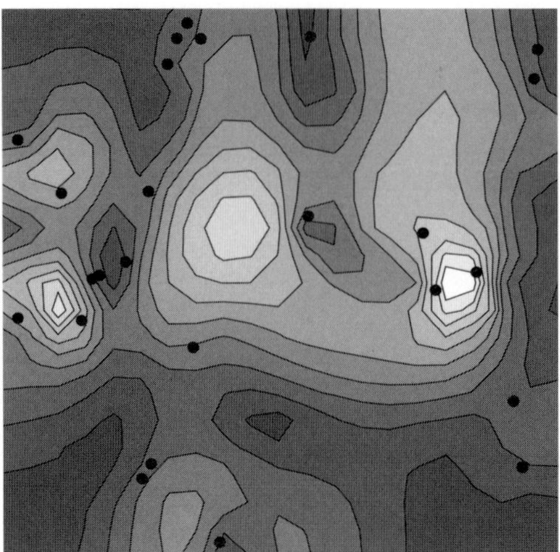

Out[88]= -Graphics-

Second derivative maps of interpolated surfaces can just as easily be produced using the function **SlopeCurvature**. Here is one such map made from **thinplateresults**.

```
In[89]:= ListContourPlot[SlopeCurvature[residualgrid, 500.],
           MeshRange → {{0, 10000}, {0, 10000}}, Frame → False,
           ColorFunction → Function[z, GrayLevel[(z + 0.3)/1.3]],
           DisplayFunction → Identity];
         Show[%, wellmap, DisplayFunction → $DisplayFunction]
```

From In[89]:=

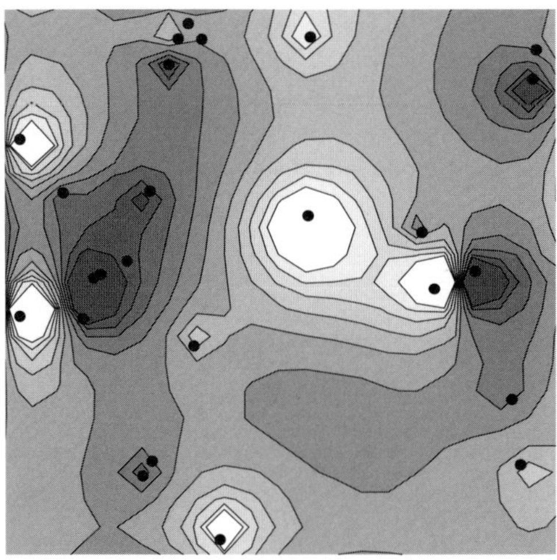

Out[89]= -Graphics-

Light areas in the curvature map indicate positive curvature (concave-up) associated with synformal structures, whereas dark areas indicate negative (concave-down) curvature associated with antiformal structures.

7.5 References and Recommended Reading

Briggs, I.C., 1974, Machine contouring using minimum curvature: *Geophysics*, v. 39, p. 39-48.

Burrough, P.A. and R.A. McDonnell, 1998, *Principles of Geographic Information Systems*: Oxford University Press.

Carr, J.R., 2002, *Data Visualization in the Geosciences*: Prentice Hall.

Davis, J.C., 2002, *Statistics and Data Analysis in Geology (3d ed.)*: John Wiley & Sons.

Fischer, M.P., and M.S. Wilkerson, 2000, Predicting the orientation of joints from fold shape: Results of pseudo-three-dimensional modeling and curvature analysis: *Geology*, v. 28, p. 15–18.

Isaaks, E.H. and R.M. Srivastava, 1989, *An Introduction to Applied Geostatistics*: Oxford University Press.

Middleton, G.V., 2000, *Data Analysis in the Earth Sciences using Matlab*: Prentice Hall.

Scheidegger, A.E., 1991, *Theoretical Geomorphology*: Springer Verlag.

Smith, W.H.F. and P. Wessel, 1990, Gridding with continuous curvature splines in tension: *Geophysics*, v. 55, p. 293–303.

Stewart, S.A. and T.J. Wynn, 2000, Mapping spatial variation in rock properties in relationship to scale-dependent structure using spectral curvature: *Geology*, v. 28, p. 691–694.

8 Digital Signal and Image Processing

8.1 *Mathematica* Packages You Will Need

```
In[1]:= Needs["Statistics`DescriptiveStatistics`"]
        Needs["Statistics`MultiDescriptiveStatistics`"]
        Needs["Statistics`HypothesisTests`"]
        Needs["Statistics`ConfidenceIntervals`"]
        Needs["Statistics`ContinuousDistributions`"]
        Needs["Statistics`MultinormalDistribution`"]
        Needs["Graphics`Graphics`"]
        Needs["CompGeosci`"]
```

Computer Note: The CompGeosci package will load correctly only if it is located in one of the directories in *Mathematica*'s standard file path. Execute the statement **`$Path`** to see a list of the default paths on your computer and place the file CompGeosci.m in one of those directories. The specific file paths may differ from one operating system to another. See Chapter 1 for more information about installing the CompGeosci package.

8.2 The Nature of Periodic Waveforms

Geoscientific data recorded as a function of time, also known as **time series**, are in many cases composed waveforms that repeat themselves periodically. Examples of periodic waveforms include stream discharge that peaks each year during spring runoff, diurnal or annual temperature flucuations in the subsurface (see Chapter 3), seismograms, tidal measurements in bays or estuaries, and deformation of Earth's surface as a result of solid earth tides. As illustrated using Fourier series in Chapter 6, data distributed in space can also be represented by periodic waveforms. Although spatially periodic data are not truly time series, they can be analyzed using the same methods if distance is substituted for time. They should, strictly speaking, probably be referred to as **space series** but throughout this chapter we will use the term time series to apply to variables in either space or time. If a time series consists solely of a periodic waveform with no long term drift or trend, it is said to be **stationary**. If there is a drift or trend in addition to the periodic component, the time series is **non-stationary**. If the trend of a non-stationary time series is not of interest, it can be removed by either fitting a straight line to obtain an equation for the trend and then calculating residuals (see Chapter 7) or, as described further on in this chapter, by calculating first differences. A time series is said to be **homoscedastic** if its variance is constant with time and **heteroscedastic** if its variance changes as a function of time. It is important to realize that time (or space) series do not have to be periodic, although in many geoscientific problems there is an important component of periodicity.

Periodic waveforms are described in terms of **amplitudes**, **frequencies**, and **wavelengths**. In the plot below, for example, the amplitude of the waveform is 0.2.

```
In[2]:= Plot[0.2 Sin[6 π x/18.], {x, 0, 18},
          AxesLabel → {"t", "f(t)"}]
```

```
From In[2]:=
```

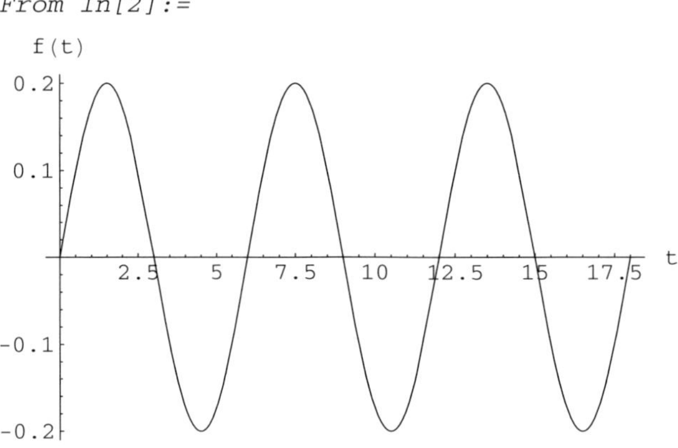


```
Out[2]= -Graphics-
```

The frequency can be written as 1 cycle per 6 units of time (*i.e.*, 1/6) or 3 cycles per 18 units of time (*i.e.*, 3/18). Both equal 1/6 and are therefore algebraically equivalent. Because the frequency is a ratio, it is even possible to specify it in terms of non-integer wavelengths such as 3/4 cycles per 9/2 units of time because that, too, will reduce to a frequency of 1/6.

In[3]:= $\frac{3/4}{9/2}$

Out[3]= $\frac{1}{6}$

Frequencies waves are often expressed as cycles per second using units of Hertz. Because the frequency per unit of time in our example reduces to 1/6 regardless of how it is written, you might be asking what is to be gained by expressing the frequency to wavelength ratio as anything but 1/6. The reason is that in digital signal processing data are commonly presented as a list of dependent variables without any corresponding time coordinate. For example, the sine curve above might be represented as a list of discrete measurements obtained at 0.25 unit intervals.

```
In[4]:= sint = Table[0.2 Sin[6 πx/18.], {x, 0, 18, 0.25}]

Out[4]= {0, 0.0517638, 0.1, 0.141421, 0.173205, 0.193185, 0.2,
    0.193185, 0.173205, 0.141421, 0.1, 0.0517638,
    2.44929 × 10^-17, -0.0517638, -0.1, -0.141421,
    -0.173205, -0.193185, -0.2, -0.193185, -0.173205,
    -0.141421, -0.1, -0.0517638, -4.89859 × 10^-17,
    0.0517638, 0.1, 0.141421, 0.173205, 0.193185,
    0.2, 0.193185, 0.173205, 0.141421, 0.1, 0.0517638,
    7.34788 × 10^-17, -0.0517638, -0.1, -0.141421, -0.173205,
    -0.193185, -0.2, -0.193185, -0.173205, -0.141421, -0.1,
    -0.0517638, -9.79717 × 10^-17, 0.0517638, 0.1,
    0.141421, 0.173205, 0.193185, 0.2, 0.193185, 0.173205,
    0.141421, 0.1, 0.0517638, 1.22465 × 10^-16, -0.0517638,
    -0.1, -0.141421, -0.173205, -0.193185, -0.2, -0.193185,
    -0.173205, -0.141421, -0.1, -0.0517638, -1.46958 × 10^-16}

In[5]:= ListStemPlot[sint, 0.015, AxesLabel → {"t", "f(t)"}]

From In[5]:=
```

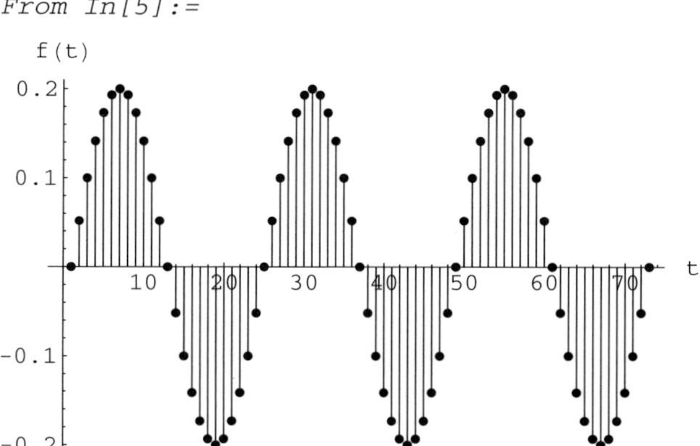


```
Out[5]= -Graphics-
```

The frequency of **sint** at first glance appears to be either 3 (because the waveform repeats itself three times during the length of the time series) or 3/72, which is assumed to have a wavelength of 1. Only by knowing the sampling rate (4 samples per unit of time) and the length of the data set ($n = 72$) will we be able to determine that the frequency is really

```
In[6]:=      3 * 4
        ------------------
        Length[sint] - 1

Out[6]= 1/6
```

The length of **sint** is reduced by 1 because the 73rd element is actually the first sample from the beginning of a fourth repetition of the waveform.

8.3 Discrete Fourier Tranforms

In Chapter 6 we obtained Fourier series coefficients using linear regression and showed that the power spectrum could be calculated from the coefficients of the sine curves. The coefficients can also be obtained using a Fourier transform that is, for a discretely sampled data set of length n, given by

$$f(\omega) = \frac{1}{\sqrt{n}} \sum_{t=1}^{n} F(t) \exp[2\,\pi\,i\,(t-1)(\omega-1)/n]$$

where $f(\omega)$ is a list of results by frequency, $F(t)$ is a list of regularly sampled data, ω is the frequency, and t is time. Different variations of the Fourier transform are used in different fields, and the example above uses *Mathematica*'s default sign convention. Refer to the written or online documentation for more details. The data are said to lie in the **time domain**, whereas the results are in the **frequency domain**. The exponential term on the right-hand side of the Fourier transform equation is equivalent to

```
In[7]:= ExpToTrig[e^(2 π i ω t/n)]

Out[7]= Cos[2 π t ω / n] + i Sin[2 π t ω / n]
```

According to this definition, the wavelength n is the length of the data set.

Why use a Fourier transform when linear regression seems to work well enough? Although it can be a very useful method, particularly when data are not sampled regularly or are otherwise missing, linear regression can also be computationally slow. This was particularly so in the early days of computing. Today, software such as *Mathematica* can perform the least squares calculations very rapidly and the speed difference may not be significant for any but the largest data sets. Still, it is good to have a fast numerical alternative for cases in which speed does matter. Another reason is that many filtering operations are easier when a time series is expressed in terms of its frequencies, or spectral components, than in terms of time. The fast Fourier transforms, or FFT, is an especially efficient method that works when the data are sampled at regular time intervals and the length of the data set is a power of 2. *Mathematica* implements an extremely efficient fast Fourier transform that can accept data sets of any length, but they must be sampled regularly in time or space. Missing values can be approximated by interpolation or by setting them to an arbitrary value such as 0, but must be specified in one way or another. Some Fourier transform routines require that the input data length be a power of 2, and require users to pad the end of the series with zeroes to attain a length that is a power of 2. *Mathematica* automatically takes care of this problem, however, so there is no need for users to pad the input to `Fourier`.

The discrete Fourier transform of `sint` is lengthy because it consists of 73 terms, each with a real and an imaginary component, so the output will be surpressed.

```
In[8]:= fft = Chop[Fourier[sint]];
```

Chop is used to eliminate any very small numerical errors ($< 10^{-10}$) in the result. To illustrate the real and imaginary components, we can look at just one term of the results.

```
In[9]:= fft[[2]]

Out[9]= 0.000369209 + 0.00857387 i
```

This result can be shown to be identical to that obtained by explicitly typing out the definition of the Fourier transform and taking the second element of the result.

```
In[10]:= len = Length[sint];
         Table[Chop[1/Sqrt[len] Sum[(sint[[t]]
           * Exp[2 π i (t - 1) (ω - 1)/len]), {t, 1, len}]], {ω, 1, len}];

In[11]:= %[[2]]

Out[11]= 0.000369209 + 0.00857387 i
```

The real components are multiples of the amplitudes of the cosine terms and the imaginary components are the multiples of the amplitudes of the sine terms in a Fourier series of the form

$$F(t) = \frac{a_0}{2} + \sum_{i=1}^{\infty} a_n \cos(2\,n\,\pi\,t/L) + b_n \sin(2\,n\,\pi\,t/L)$$

The first term in the list returned by **Fourier** contains the value for a_0, so the second term contains values for a_1 and b_1, and so forth. Because of the particular definition of a Fourier transform used by default in *Mathematica*, the amplitudes are found by multiplying each term in the Fourier transform by $2/\sqrt{n}$. In this example, the amplitudes as a function of frequency are:

```
In[12]:= ListStemPlot[2Re[fft]/Sqrt[len], 0.015, PlotRange → All,
           AxesLabel → {"ω", "a(ω)"}, AspectRatio → 1/2]

From In[12]:=
```

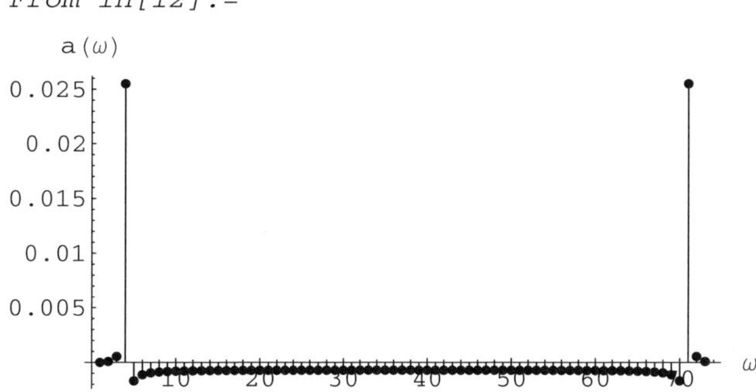


```
Out[12]= -Graphics-
```

and

```
In[13]:= ListStemPlot[2Im[fft]/√len, 0.015, PlotRange → All,
            AxesLabel → {"ω", "b(ω)"}, AspectRatio → 1/2]
```

From In[13]:=

Out[13]= -Graphics-

As expected from the function that created **sint**, the maximum amplitude is for the sine component is 0.2 and occurs at a frequency of 4 – 1 = 3 cycles per data length. The subtraction is necessary because the $a_0$term the first element of the results (**fft[[1]]**); therefore, the *i*th value in the Fourier transform results represents a frequency of *i* – 1. The results are symmetric or antisymmetric about a frequency of 36, which is known as the **Nyquist frequency**. Frequencies above the Nyquist frequency are said to be **aliased** because they contain no new information. An important ramification of the Nyquist frequency is that the highest frequency that can be represented in a discretely sampled signal is *n*/2 cycles per data set length. Thus, sampling should always be planned to that the Nyquist frequency is greater than the frequencies of the phenomena being studied. For example, if temperature varies on a daily basis then a sampling frequency of at least twice a day is necessary to correctly detect the fluctuations without aliasing.

The **power spectrum** (known as the variance spectrum or spectral density function in some fields) is given by the square of the absolute value of the real and imaginary parts of each term, which is the sum of the squares of the real and imaginary parts, divided by the square root of the number of data. We can use the logical == operator to see if the definition of the absolute value is indeed true

```
In[14]:= Abs[fft]^2/√len == (Re[fft]^2 + Im[fft]^2)/√len
```

Out[14]= True

and then plot the power spectrum

```
In[15]:= ListStemPlot[Abs[fft]^2/Sqrt[len], 0.015,
           PlotRange → All, AxesLabel → {"ω", "Power"}]
```

From In[15]:=

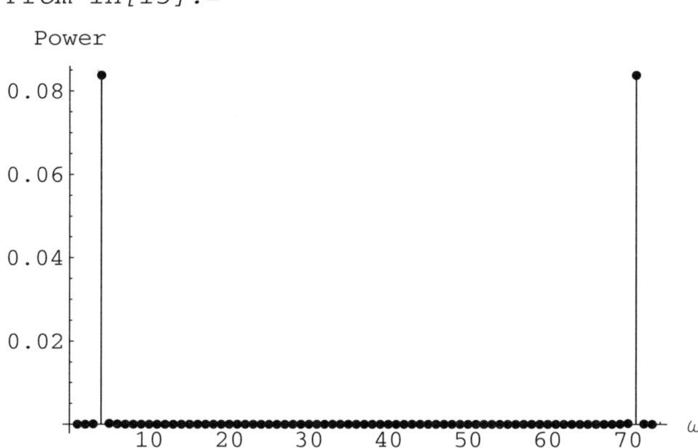

Out[15]= -Graphics-

The sum of squares of the absolute value for each frequency is closely related to the variance of the original data set. In this example, the summation yields

In[16]:= $\frac{1}{\mathbf{len - 1}}\sum_{\mathbf{i=1}}^{\mathbf{len}} \mathbf{Abs[fft[[i]]]}^2$

Out[16]= 0.02

which can be compared to

```
In[17]:= Variance[sint]
```

Out[17]= 0.02

Because of this relationship, the power of each frequency can be interpreted as its contribution to the total variance of the data set, and the significance of any particular frequency can be tested using an F ratio test (see Chapter 4). If the first term of the Fourier transform is not zero, then the first term should be subtracted before attempting to calculate the variance from the sum of powers. A related result is the **amplitude spectrum**, which is (using *Mathematica*'s Fourier transform convention) twice the square root of the power spectrum. Taking the absolute values, the amplitude spectrum is

```
In[18]:= ListStemPlot[2 Sqrt[Abs[fft]^2]/Sqrt[len], 0.015, PlotRange → All,
           AxesLabel → {"ω", "Amplitude"}]
```

From In[18]:=

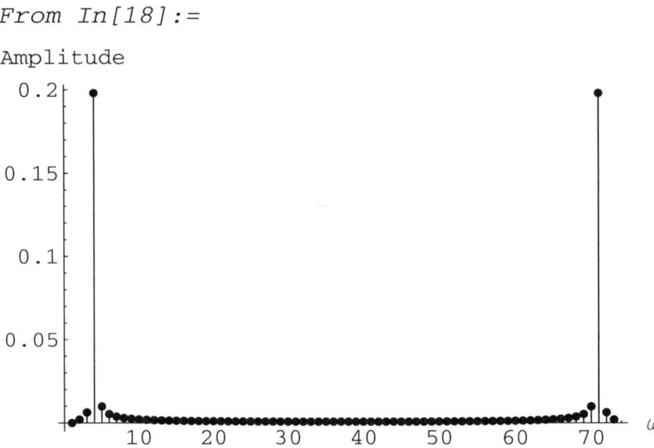

Out[18]= -Graphics-

As discussed by Press *et al.*, (1992), there are several other commonly used definitions of the power spectrum. Ignoring results above the Nyquist frequency, the frequency with the highest power and amplitude is 4 – 1 = 3 cycles per data set length. The results were obvious in this simple example but, as will be shown below, it is not as easy to select the dominant frequency or frequencies in real data. Finally, taking the inverse Fourier transform of **fft** returns the original data.

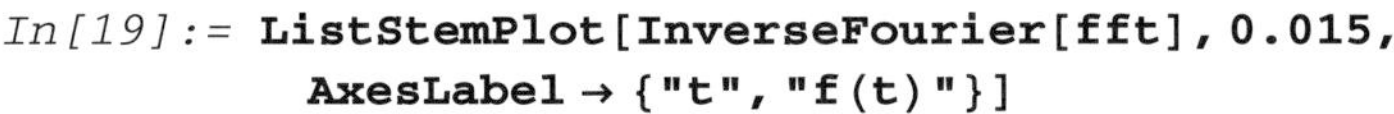

```
In[19]:= ListStemPlot[InverseFourier[fft], 0.015,
           AxesLabel → {"t", "f(t)"}]
```

From In[19]:=

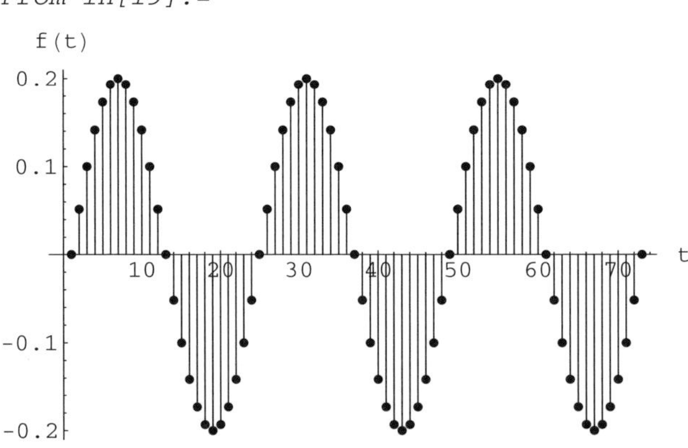

Out[19]= -Graphics-

As implemented in *Mathematica*, **Fourier** and **InverseFourier** can also accept multi-dimensional tables of data, for example digital elevation models.

Real data are not usually as well behaved as our simple sine curve. The data imported below consist of monthly streamflow measurements of the Palouse River near Colfax, Washington collected between January 1956 and December 1963 by the U.S. Geological Survey.

```
In[20]:= data = Import[
           "/Users/bill/Mathematica_Book/palouse.dat", "List"];
```

The original data are in cubic feet per second, and can easily be converted to cubic meters per second (1 cfs = 0.02832 cms).

```
In[21]:= data = 0.02832 data;
```

The length of the data set is

```
In[22]:= len = Length[%]

Out[22]= 96
```

and the time series looks like this:

```
In[23]:= ListStemPlot[data, 0.015, PlotRange → All,
           AxesLabel → {"t", Q (m^3/s)}]

From In[23]:=
```

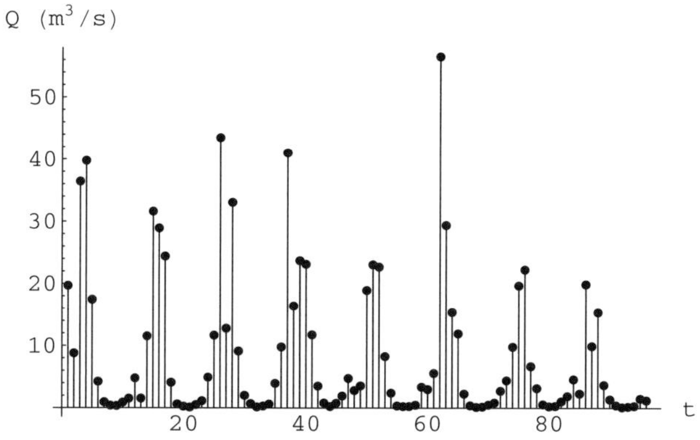

```
Out[23]= -Graphics-
```

The Fourier transform is obtained just as above except that we will subtract the mean value of **data**.

```
In[24]:= fft = Fourier[data - Mean[data]];
```

We are not interested in reproducing the streamflow measurements, but would like to identify the predominant frequency. A reasonable guess might be that it is one cycle per year, To find out if this is correct, plot the amplitude spectrum

$$\texttt{In[25]:= ListStemPlot}\left[\frac{2\sqrt{\texttt{Abs[fft]}^2}}{\sqrt{\texttt{len}}}, \texttt{0.015, PlotRange} \rightarrow \texttt{All,}\right.$$

$$\left.\texttt{AxesLabel} \rightarrow \{\texttt{"}\omega\texttt{", "Amplitude"}\}\right]$$

From In[25]:=

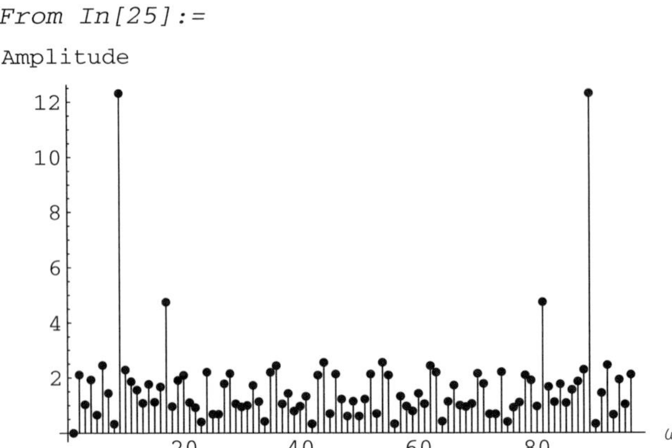

Out[25]= -Graphics-

or the power spectrum for the streamflow data.

```
In[26]:= ListStemPlot[Abs[fft]^2/√len, 0.015,
           PlotRange → All, AxesLabel → {"ω", "Power"}]
```

From In[26]:=

Power

350
300
250
200
150
100
50

20 40 60 80 ω

Out[26]= -Graphics-

The largest amplitude and power are associated with the a frequency of 9 – 1 = 8 cycles per data length, corresponding to the eight years of record. A second prominent but weaker peak in the amplitude spectrum occurs for a frequency of 17 – 1 = 16 cycles per data length, or 2 cycles per year. In many applications the power spectrum is extremely noisy, in part because the effect of high-power frequencies can leak into adjacent lower power frequencies. Although we will not discuss the details, improved power spectra can be generated by using various smoothing processes.

This data set is an example of one with a non-zero mean, but we have already subtracted the mean value and can proceed to calculate the variance.

In[27]:= $\frac{1}{\text{len} - 1}\sum_{i=1}^{\text{len}} \text{Abs}[\text{fft}[[i]]]^2$

Out[27]= 141.125

The result is the same as that calculated using the **Variance** function.

In[28]:= **Variance[data]**

Out[28]= 141.125

If the objective is to calculate the variance of a data set in the simplest way possible, then using **Variance** is much easier than subtracting a mean value, taking a Fourier transform, and then summing squares. But, it is very useful to understand that the range of values plotted in a power spectrum is closely related to the variance of the data set. The significance of the power of any particular frequency, for example, can be expressed as the ratio of the power to the total variance of the data set. The variance associated with a frequency of 8 cycles per data length is in this case

In[29]:= $\frac{\text{Abs}[\text{fft}[[9]]]^2}{\sqrt{\text{len}}}$

Out[29]= 371.367

This is substantially higher than the variance of the entire data set, which is approximately 141. Because we are working with two variances, the null hypothesis that their ratio is not significantly different than 1 can be tested using an F ratio test (see Chapter 4). First, calculate the ratio of variances.

In[30]:= **fratio = %/Variance[data]**

Out[30]= 2.63148

Then, use the ratio in **FRatioPValue**. Notice that the numerator, which is the spectral power for a frequency of 8, has two degrees of freedom because it contains to parts: one real and one imaginary.

In[31]:= **FRatioPValue[fratio, 2, len - 1]**

Out[31]= OneSidedPValue → 0.0772141

There is about an 8% chance of committing a Type I error if we reject the null hypothesis, which is greater than the standard value of 5% level of significance that is considered acceptable in many scientific problems. The power of the 3 cycles per data length frequency seems, however, to be substantially greater than any of the other amplitudes in the power spectrum. Why, then, is its p value low enough that the null hypothesis cannot be rejected at the standard 0.05 level? The reason is that the small number of degrees of freedom associated with the power of each frequency makes the estimate of the variance very uncertain. In this case, the confidence interval for the variance ratio has a very wide range

```
In[32]:= FRatioCI[fratio, 2, len - 1, ConfidenceLevel → 0.95]

Out[32]= {0.686013, 103.91}
```

Thus, if we want to be 95% certain about the F ratio we can only say that it lies somewhere between 0.69 and 104. That is indeed a very uncertain estimate! We can also test for the significance of the smaller peak at a frequency of 16 cycles per data length. Its p value is

```
In[33]:= FRatioPValue[(Abs[fft[[17]]]^2/Sqrt[len])/Variance[data],
           2, len - 1]

Out[33]= OneSidedPValue → 0.322769
```

Therefore, it is unlikely that this frequency is significantly different than the background noise. The confidence interval of this F ratio is:

```
In[34]:= FRatioCI[(Abs[fft[[17]]]^2/Sqrt[len])/Variance[data],
           2, len - 1, ConfidenceLevel → 0.95]

Out[34]= {0.102022, 15.4532}
```

The F ratio confidence intervals can be decreased considerably by smoothing the data.

> **Computer Note:** Using the methods developed in Chapter 6, fit a straight line trend to **`data`** using **`Fit`** (with a subsequent ANOVA) or **`Regress`** to determine if there is a significant linear trend in the streamflow data.

8.4 Autocovariance and Autocorrelation

Autocovariance and **autocorrelation** are two measures of self-association that can be applied to time series. The reflect the degree of similarity or correlation between values of a time series at some value t and another value $t + \Delta t$, where the increment Δt is known as the **lag**. Unless a time series consists of truly random values, it is reasonable to expect that values close to each other in time (small lag) will be more similar to each other than those separated by large lags. The covariance between n pairs of two variables x and y is defined as

$$\mathrm{Cov}(x, y) = \frac{1}{n-1}\sum_i (x_i - \bar{x})(y_i - \bar{y})$$

but can be modified to compare instances of the same variable separated by a lag Δt, in which case it becomes the autocovariance. As is done for the variance, the covariance function uses a denominator of $n - 1$ to produce an unbiased estimate.

$$\text{ACov}(x, y) = \frac{1}{n-1}\sum_i (x_i - \bar{x})(x_{i+\Delta t} - \bar{x})$$

The correlation between two variables is the covariance divided by the product of their standard deviations, or

$$\text{Corr}(x, y) = \frac{1}{n-1}\frac{\sum_i (x_i - \bar{x})(y_i - \bar{y})}{s_x s_y}$$

Thus, the autocorrelation is by analogy

$$\text{ACorr}(x, y) = \frac{1}{n-1}\frac{\sum_i (x_i - \bar{x})(x_{i+\Delta t} - \bar{x})}{s^2{}_x}$$

The *Mathematica* package Statistics`MultiDescriptiveStatistics` includes the functions **Covariance** and **CovarianceMLE**, which calculate unbiased and maximum likelihood (population) estimates of the covariance from a data set, as well as **Correlation** (which we have already used).

Be aware that the use of the terms autocovariance and autocorrelation is not standardized and can be very confusing. Some authors, for example, use autocorrelation to describe a population statistic and serial autocorrelation to describe the corresponding sample statistic. Other authors have used serial correlation to mean the correlation between two different time series, whereas others use the term cross-correlation for the same purpose. The convention here will be to use autocorrelation to refer to both population and sample statistics. For time series with more than a handful of data, the difference between the two will generally be inconsequential.

To illustrate the calculation of autocorrelation values, we will use the Palouse River monthly peak discharge measurements that were previously assigned to the name **data**. The function below calculates the autocorrelation as a function of lag. As in the examples above, we will not explicitly consider the independent variable (time in this case) but will instead use the length of the data set and assign the discharge measurements time values of 1, 2, 3, and so forth. The function below calculates the autocorrelation of data set y at lag Δt

```
In[35]:= AutoCorrelation[y_, Δt_] :=
           Return[ Covariance[y, RotateLeft[y, Δt]] / Variance[y] ]
```

The built-in function **RotateLeft** shifts the values in y one place to the left. The autocorrelation of **data** for a lag of, say, 8 is thus

```
In[36]:= AutoCorrelation[data, 8]

Out[36]= -0.316315
```

Thus, peak flow measurements separated by six months have a negative correlation with each other. What about measurements separated by different lags? Our strategy will be to calculate the covariance between y and itself for $\Delta t = 0$, shift the values one place so that $\Delta t = 1$, calculate another covariance, shift the values one place, and so on to create a list of autocorrelation values. Here is a table of autocorrelation values for **data**:

```
In[37]:= temp = Table[AutoCorrelation[data, k], {k, 0, len - 1}]

Out[37]= {1., 0.515331, 0.25053, -0.0726287, -0.316138, -0.430316,
         -0.464672, -0.431148, -0.316315, -0.0516652, 0.196309,
         0.575729, 0.563856, 0.479987, 0.230943, -0.0972641,
         -0.300231, -0.430843, -0.461767, -0.414054, -0.269887,
         -0.0332377, 0.276985, 0.453023, 0.561094, 0.561085,
         0.179031, -0.122242, -0.33475, -0.434842, -0.461141,
         -0.422388, -0.299621, -0.0740648, 0.267218, 0.443251,
         0.673412, 0.502206, 0.253383, -0.0887864, -0.327184,
         -0.431817, -0.464714, -0.436074, -0.341298, -0.0879607,
         0.206113, 0.472249, 0.510625, 0.472249, 0.206113,
         -0.0879607, -0.341298, -0.436074, -0.464714, -0.431817,
         -0.327184, -0.0887864, 0.253383, 0.502206, 0.673412,
         0.443251, 0.267218, -0.0740648, -0.299621, -0.422388,
         -0.461141, -0.434842, -0.33475, -0.122242, 0.179031,
         0.561085, 0.561094, 0.453023, 0.276985, -0.0332377,
         -0.269887, -0.414054, -0.461767, -0.430843, -0.300231,
         -0.0972641, 0.230943, 0.479987, 0.563856, 0.575729,
         0.196309, -0.0516652, -0.316315, -0.431148, -0.464672,
         -0.430316, -0.316138, -0.0726287, 0.25053, 0.515331}
```

As with most time series data, it is difficult to glean much useful information from this list of numbers. Using **ListStemPlot** to visualize the values produces an autocorrelogram.

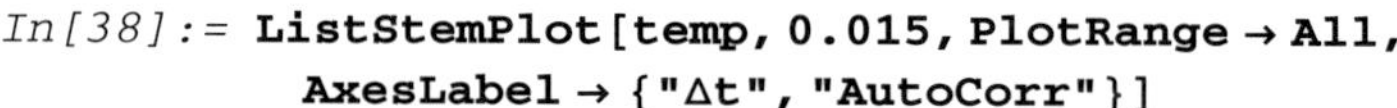

```
In[38]:= ListStemPlot[temp, 0.015, PlotRange → All,
           AxesLabel → {"Δt", "AutoCorr"}]

From In[38]:=
```

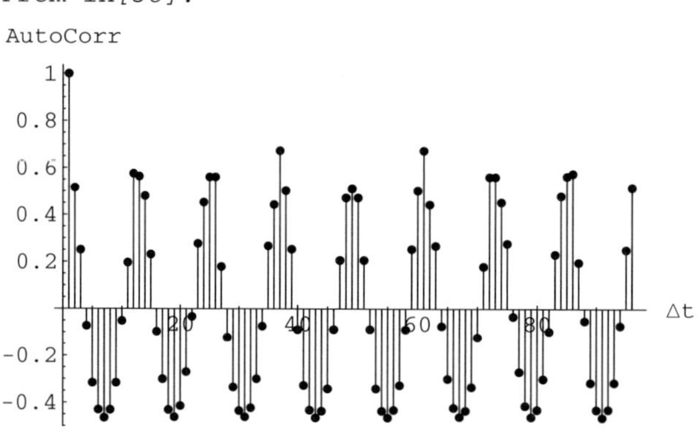


```
Out[38]= -Graphics-
```

The autocorrelation of 1 for Δt means that the peak flow data set is identical to itself at zero lag. The repetitive peaks lie at lags that are multiples of 12 months, reflecting annual cycles of the monthly peak flow measurements. The troughs at values of approximately 6, 18, 30… further show that peak flow measurements taken about six months apart tend to be the most different from each other. The same results can be obtained using *Mathematica*'s **Correlation** function.

```
In[39]:= Clear[AutoCorrelation]

In[40]:= AutoCorrelation[y_, Δt_] :=
           Return[Correlation[y, RotateLeft[y, Δt]]]
```

The approach is the same as in the previous function, except that the variance of *y* is not explicitly calculated in the function. One advantage of the first approach is that, although it is slightly more complicated, the **Covariance** and **Variance** functions can be replaced with **CovarianceMLE** and **AutoCovarianceMLE** if there is a need to distinguish between the population and sample statistics. As shown below, the second implementation produces results identical to the first in this example. In this example the calculation of the results is accomplished within the **ListStemPlot** function.

```
In[41]:= ListStemPlot[Table[AutoCorrelation[data, k],
           {k, 0, len - 1}], 0.015, AxesLabel → {"Δt", "AutoCorr"}]

From In[41]:=
```

AutoCorr

1
0.8
0.6
0.4
0.2
-0.2
-0.4
20 40 60 80 Δt

```
Out[41]= -Graphics-
```

8.5 Filters and Convolution

Digital filters can be used to remove trends, smooth data, accentuate details, and selectively remove groups of frequencies from data such as time series, photographs, or maps. Filters are typically created as lists (for 1-D data) or matrices (for multidimensional data) and then applied using a process known as **convolution**. Convolution is a complicated process that generally involves the use of Fourier transforms to convert data to frequencies, application of the filters to the frequencies, and then the use of inverse Fourier transforms to recover the filtered data. *Mathematica*'s built-in **ListConvolve** and **ListCorrelate** functions simplify the use of digital filters because they eliminate the need to explicitly consider Fourier transforms when performing convolutions.

8.5.1 First Differences

Trends can be removed from time series by calculating **first differences**, which are defined as $x_{t+\Delta t} - x_t$. To illustrate, consider a contrived data set consisting of both a sine wave and a trend.

```
In[42]:= pseudodata = Table[0.1t + Sin[2πt/10.], {t, 0, 100}];
```

```
In[43]:= ListPlot[pseudodata, PlotJoined → True]
```

From In[43]:=

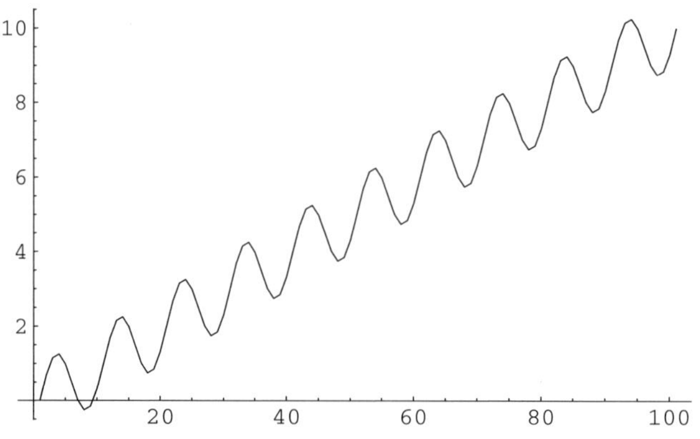

Out[43]= -Graphics-

The most obvious way to calculate first differences in *Mathematica* is probably to create a table of differences, which is plotted below.

```
In[44]:= ListPlot[Table[pseudodata[[i + 1]] - pseudodata[[i]],
           {i, Length[pseudodata] - 1}], PlotJoined → True]
```

From In[44]:=

Out[44]= -Graphics-

Notice that calculating first differences preserves the frequency content, but not the amplitude, of the time series. As such, it will be a useful technique in situations where the goal is to understand the periodicity of a time series without regard to its amplitude(s).

Another way to calculate first differences is to use **ListConvolve** or **ListCorrelate**. The function **ListConvolve**[*k*, *y*] applies the **kernel** *k*, which is a list or matrix, term-by-term to the data set *y* by calculating $\sum_r k_r y_{s-r}$ for each element, *s*, in the data. For example,

```
In[45]:= ListConvolve[{a, b, c}, {x1, x2, x3, x4}]

Out[45]= {c x1 + b x2 + a x3, c x2 + b x3 + a x4}
```

Because the kernel contains three terms, it cannot be applied to the first and last elements of *x* and the resulting list contains only two terms. The related function **ListCorrelate** applies the kernel the the list in a forward direction by calculating $\sum_r k_r y_{s+r}$.

```
In[46]:= ListCorrelate[{a, b, c}, {x1, x2, x3, x4}]

Out[46]= {a x1 + b x2 + c x3, a x2 + b x3 + c x4}
```

The symmetry of **ListConvolve** and **ListCorrelate** can be demonstrated by symbolically comparing the two.

```
In[47]:= ListConvolve[{a, b, c}, {x1, x2, x3, x4}] ==
           ListCorrelate[{c, b, a}, {x1, x2, x3, x4}]

Out[47]= True
```

To calculate first differences using **ListCorrelate**, define the kernel as

```
In[48]:= k = {-1, 1}

Out[48]= {-1, 1}
```

which, after clearing the previous definition of **x**, produces the following result for each element in the data set

```
In[49]:= Clear[x]

In[50]:= ListCorrelate[k, {x_t, x_t+Δt}]

Out[50]= {-x_t + x_t+Δt}
```

Applied to **pseudodata**, the kernel *k* produces an identical plot of first differences

```
In[51]:= ListPlot[ListCorrelate[k, pseudodata],
           PlotJoined → True]
```

From In[51]:=

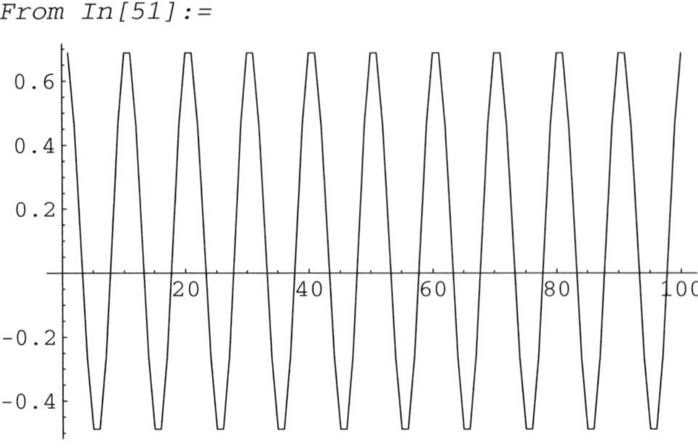


```
Out[51]= -Graphics-
```

> **Computer Note:** Show that reversing the kernel and using **`ListConvolve`** will produce identical results.

The results are the same, so why go to the trouble of using **`ListCorrelate`** or **`ListConvolve`**? The answer is that they are much faster. To illustrate this point, we can repeat the first difference example 1000 times and have *Mathematica* return the elapsed time (performing it only once would return a time of zero)

```
In[52]:= Timing[
            Do[
              ListCorrelate[k, pseudodata], {i, 1000}
            ]
          ]

Out[52]= {0.13 Second, Null}
```

and compare the result to that obtained by calculating a table of differences

```
In[53]:= Timing[
            Do[
              Table[pseudodata[[i + 1]] - pseudodata[[i]],
              {i, Length[pseudodata] - 1}], {i, 1000}
            ]
          ]

Out[53]= {1.28 Second, Null}
```

Thus, in this example (and on my computer) it appears that **`ListCorrelate`** is 11.4 times as fast as the table-filling method. Although it may not mean much for small data sets consisting of tens or even hundreds of data, the time savings can be significant in problems involving digital images or digital elevation models with millions of data.

8.5.2 Moving Averages and Smoothing

Filters can also be used to smooth noisy or rough data sets by calculating *n*-term moving averages, and are therefore a type of **smoothing** or **low-pass** filter. The term low-pass reflects the fact that high frequency noise is eliminated while allowing the low frequency signal to pass through unaffected. The kernel for a simple 3-term moving average kernel is a list of length *n* in which each element has a value of 1/*n*:

```
In[54]:= MovingAvg3 = {1/3., 1/3., 1/3.}

Out[54]= {0.333333, 0.333333, 0.333333}
```

Its use can be illustrated by applying it to a noisy periodic pseudodata set generated by adding random noise to a sine curve.

```
In[55]:= pseudodata2 = Table[Sin[2 π t/10.] +
           Random[Real, {-0.8, 0.8}], {t, 1, 100}];
```

Although there is some indication of periodicity in the plot below, it is difficult to identify the underlying sine curve (signal) because of the noise.

```
In[56]:= ListStemPlot[pseudodata2, 0.015]

From In[56]:=
```

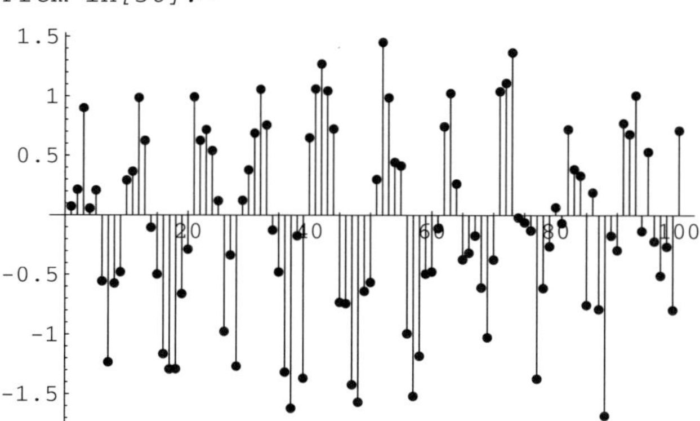


```
Out[56]= -Graphics-
```

The 3-term moving average of **pseudodata2**, which helps to reduce the noise, is

```
In[57]:= ListStemPlot[ListConvolve[MovingAvg3, pseudodata2],
           0.015]
```

From In[57]:=

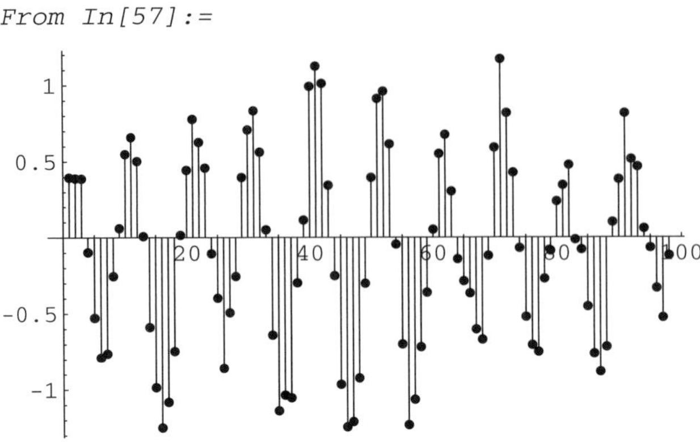

Out[57]= -Graphics-

A five-term moving average can be used to smooth **pseudodata** even more.

In[58]:= **MovingAvg5 = {0.2, 0.2, 0.2, 0.2, 0.2}**

Out[58]= {0.2, 0.2, 0.2, 0.2, 0.2}

In[59]:= **ListStemPlot[ListConvolve[MovingAvg5, pseudodata2], 0.015]**

From In[59]:=

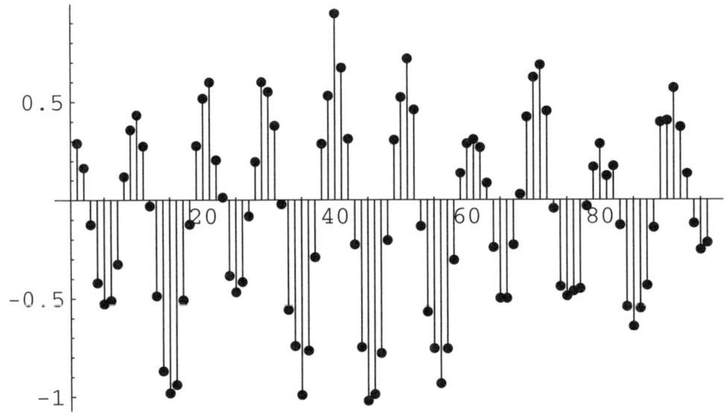

Out[59]= -Graphics-

The 3-term and, especially, the 5-term moving averaged time series give a good indication of the periodicity in the underlying signal. They do not, however, do a very good job of uncovering the amplitude of the signal.

More sophisticated smoothing filters can assign different weights to the adjacent values. One approach is to use a Gaussian bell-shaped curve to assign weights, as illustrated below for a 5-term smoothing filter. Notice that the sum of the elements is unity.

```
In[60]:= Gaussian = Table[PDF[NormalDistribution[0., 1.], x],
           {x, -2, 2}]

Out[60]= {0.053991, 0.241971, 0.398942, 0.241971, 0.053991}
```

Using **GaussianSmooth** on **pseudodata2** yields a result that does a better job of preserving the amplitude of the sine curve used to generate **pseudodata2**.

```
In[61]:= ListStemPlot[ListConvolve[Gaussian, pseudodata2],
           0.015]

From In[61]:=
```

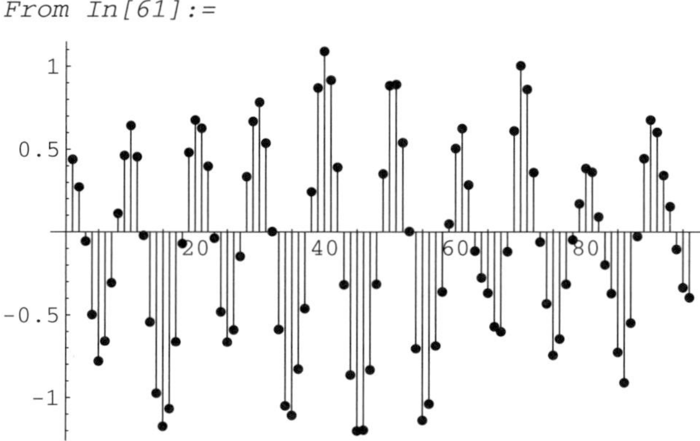


```
Out[61]= -Graphics-
```

Computer Note: The effective width of the Gaussian smoothing filter can be controlled by adjusting the standard deviation in the normal distribution used to generate the filter. Modify the filter above so that operates on 3 and 7 terms, in each case with the outermost terms each having a weight of approximately 0.05.

8.5.3 High-Pass Filtering

The high frequency noise that was removed by the calculation of moving averages can be found by subtracting the averaged results from the original data. Because of the end effects, the 1 or 2 elements at each end of **pseudodata** must be removed before this can be accomplished. To calculate the high frequency component resulting from the 3-term moving average, first remove the first and last elements by using **Take** to remove everything in between.

```
In[62]:= Take[pseudodata2, {2, Length[pseudodata2] - 1}];
```

Next, subtract the averaged results

```
In[63]:= % - ListConvolve[MovingAvg3, pseudodata2];
```

Finally, plot the high frequency component.

```
In[64]:= ListStemPlot[%, 0.015]
```

From In[64]:=

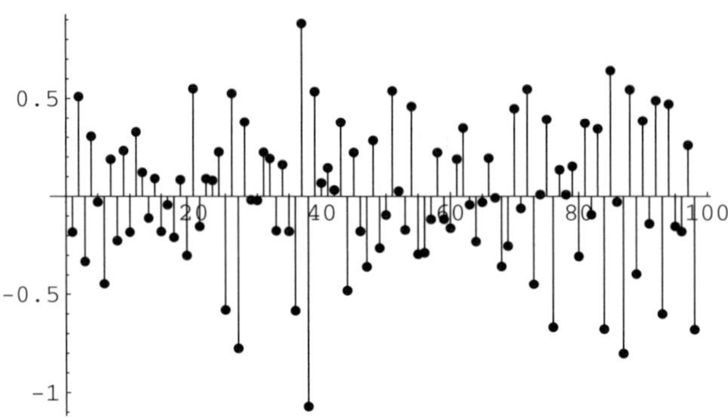

Out[64]= -Graphics-

In this case, the high frequency component is unwanted noise. In other cases, the high frequency component can represent significant features, for example the edges of objects in an image.

8.6 Image Processing

The same techniques used on 1-D time series can also be applied to 2-D arrays of data such as digital images or digital elevation models. *Mathematica*'s **Import** function will automatically recognize and import common raster graphics file formats such as gif, jpeg, tiff, bmp, and png. Although specialized commercial or public domain image processing programs may provide better solutions for some image processing tasks, *Mathematica* provides tools that can be used to explore basic image processing concepts and prototype new techniques. A more complete add-on image processing package for *Mathematica* is also available through Wolfram Research.

8.6.1 Importing Digital Images

The example below uses a false color infrared orthophoto of the southeastern quarter of the Wheeling, West Virginia topographic quadrangle. The original orthophoto had a resolution of 1 m per pixel, but the resolution of the example image has been greatly reduced both to save memory and to ensure that it will fit within the width of one page. Those readers who have only the printed copy of this book will see a grayscale rendering.

```
In[65]:= picture =
    Import["/Users/bill/Mathematica_Book/wheeling.tif"]

Out[65]= -Graphics-
```

Computer Note: Readers using the digital version of this text should change the file path above to reflect the location of the image file on their hard drive or CD.

The size of the multidimensional array containing the image data is found using **Dimensions**. The results show 474 rows and 374 columns in each of three color layers (red, green, and blue). A gray scale image would have only one layer.

```
In[66]:= Dimensions[picture[[1, 1]]]

Out[66]= {474, 374, 3}
```

Because **picture** is a graphics object, it can be shown using **Show**. Readers of the paper copy of this book will see only a gray level image, whereas those reading the digital version will see a reddish false color photograph.

```
In[67]:= Show[picture, ImageSize → {474/2., 374/2.}]

From In[67]:=
```

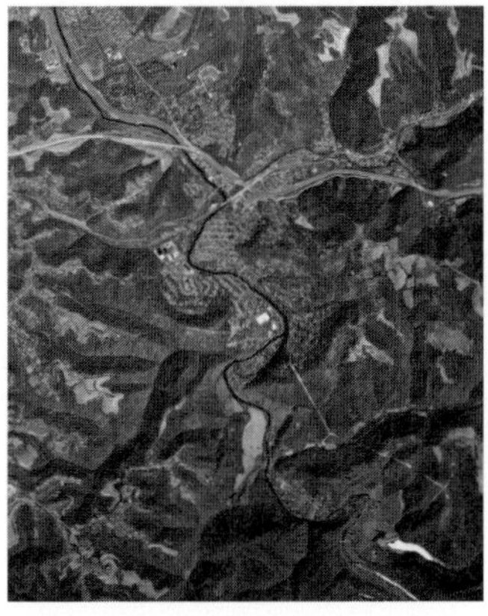

```
Out[67]= -Graphics-
```

The **ImageSize** option is used to control the size of the displayed image, in this case reducing its original size by 1/2, or by clicking on the image and dragging the handles with a mouse. The default **ImageSize** values are the numbers of rows and columns in the image, with each row or column representing one printer's point on the computer monitor or printed page. The option **ImageResolution** can also be used to specify the resolution of the image in pixels per inch. The color image can be displayed as a gray scale image using the option **ColorOutput → GrayLevel**.

In order to perform any image processing, the *Mathematica* graphics object will have to be converted into an array, or list of lists, using the command below:

```
In[68]:= picture = picture /. Graphics → List;
```

The red, green, and blue values for each pixel are contained in the first element of **picture**, and can be extracted and assigned to their own variable name.

```
In[69]:= picturevalues = picture[[1, 1]];
```

As did the original image, **picturevalues** contains 474 rows and 374 columns of red, green, and blue (RGB) values. See Appendix C or the *Mathematica* documentation for more information about working with color in *Mathematica*.

```
In[70]:= Dimensions[picturevalues]

Out[70]= {474, 374, 3}
```

Instead of being graphics objects, though, the elements of **picturevalues** are integers ranging from 0 to 255. The RGB components at row 200 and column 100, for example, are.

```
In[71]:= picturevalues[[200, 100]]

Out[71]= {155, 96, 74}
```

Sections of an image can be isolated using **Take**. For example, the following line extracts a 100 by 100 pixel section of the red channel.

```
In[72]:= Table[picturevalues[[r, c, 1]], {r, 200, 300},
           {c, 200, 300}];

In[73]:= ListDensityPlot[%, Mesh → False, AspectRatio → 1]

From In[73]:=
```

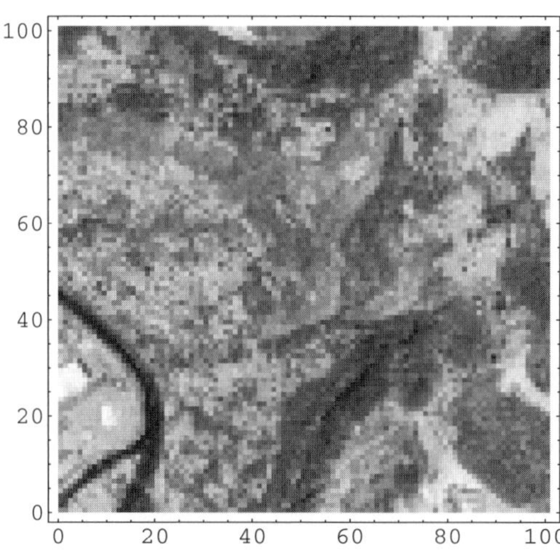


```
Out[73]= -DensityGraphics-
```

To extract any of the three layers, use **picturevalues[[All, All,** ***n*****]]**, where *n* is the value of the channel to be extracted. The intensity of the red channel, for example, can be displayed using **ListDensityPlot**. Dark values represent low red values and light values represent large red values. We will use the red channel in the examples the follow, and readers with the digital version of the book and a copy of *Mathematica* can experiment with the green and blue channels.

```
In[74]:= ListDensityPlot[picturevalues[[All, All, 1]],
           Mesh → False, AspectRatio → 474/374.]
```

From In[74]:=

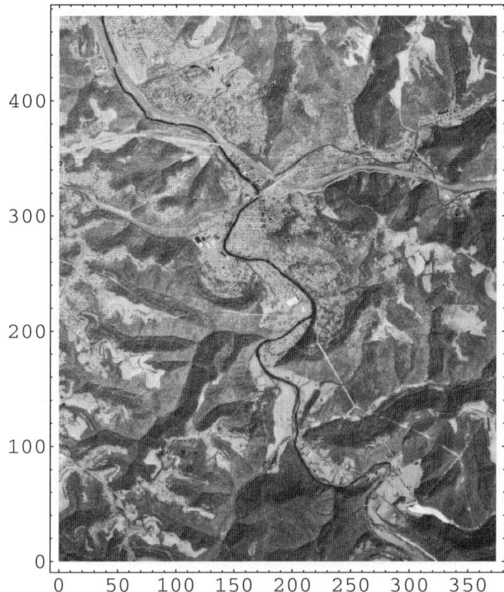


```
Out[74]= -DensityGraphics-
```

As with any other data set, the range of red values can also be visualized with a histogram of the flattened **picturevalues** array.

```
In[75]:= Histogram[Flatten[picturevalues[[All, All, 1]]],
           ColorOutput → GrayLevel]
```

From In[75]:=

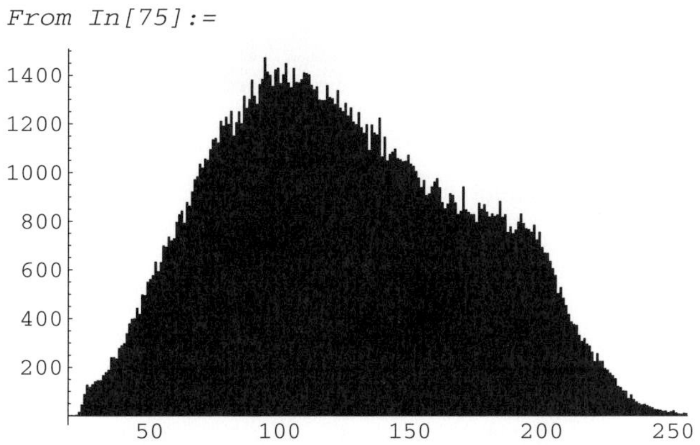


```
Out[75]= -Graphics-
```

The relative contribution of each of the three channels can be compared by placing density plots side-by-side using **GraphicsArray**. Notice that **DisplayFunction → Identity** is used to suppress output until the entire array is assembled, at which point **DisplayFunction → $DisplayFunction** is used within the **Show** function to make the three plots visisble.

```
In[76]:= Show[
           GraphicsArray[{
               ListDensityPlot[picturevalues[[All, All, 1]],
                 Mesh → False, AspectRatio → 474/374.,
                 Frame → False, PlotLabel- > "red",
                 DisplayFunction → Identity],
               ListDensityPlot[picturevalues[[All, All, 2]],
                 Mesh → False, AspectRatio → 474/374.,
                 Frame → False, PlotLabel- > "green",
                 DisplayFunction → Identity],
               ListDensityPlot[picturevalues[[All, All, 3]],
                 Mesh → False, AspectRatio → 474/374.,
                 Frame → False, PlotLabel- > "blue",
                 DisplayFunction → Identity]}],
                 DisplayFunction → $DisplayFunction
         ]
```

From In[76]:=

red

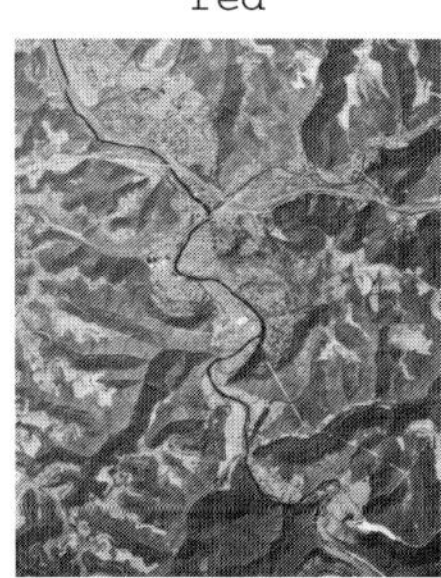

green

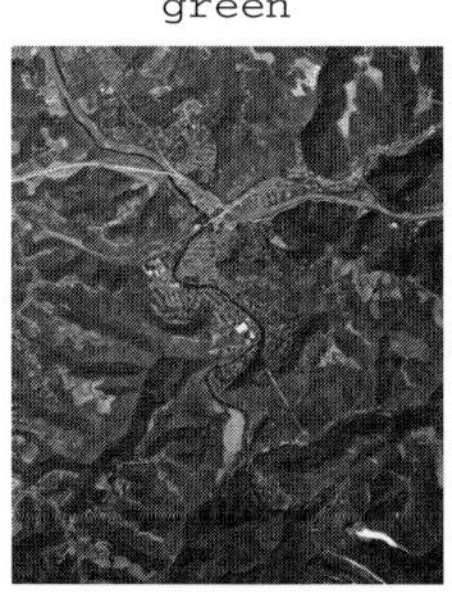

blue

Out[76]= -GraphicsArray-

The lighter (larger) values in the red channel image suggest that the composite RGB image has a reddish color. As anyone who is viewing the original image on a color computer monitor can attest, it is indeed primarily red.

Now that we have converted the image to an array of numbers and disassembled it layer-by-layer, it will be helpful to know how to return it to a graphics object. This can be done by applying the function **RGBColor** to each element in **picturevalues** (which must be divided by 255 because **RGBColor** accepts arguments only in the range of 0 to 1), and then putting the result into a *Mathematica* variable known as a raster array.

```
In[77]:= reconvertedpicture = Graphics[
             RasterArray[Apply[(RGBColor[#1, #2, #3]&),
             picturevalues/255., {2}]],
             AspectRatio → 474/374.
           ]
Out[77]= -Graphics-
```

The # and & characters in the statement above are shorthand notation for a *Mathematica* pure function. In essence, it selectively applies the **RGBColor** function to the values contained in **picturevalues**, assigning the first elements to the red channel, the second to the green channel, and the third to the blue channel. Consult the printed or online *Mathematica* documentation for more information on pure functions. Such a statement might be used after an image is imported, manipulated, and ready to be exported as a graphics file using **Export** or to recombine the three separate channels into one color image. The following statement exports the re-converted image as a jpeg file:

```
In[78]:= Export["/Users/bill/Mathematica_Book/wheeling.jpg",
           reconvertedpicture, "JPEG"]
Out[78]= /Users/bill/Mathematica_Book/wheeling.jpg
```

You will of course want to use a file name and path, as well as file format, of your own choosing. The printed and on-line documentation includes details about the graphics formats supported by *Mathematica* and their options. The jpeg format, for example, includes options to set the color space (RGB or gray level), image quality (default is 75 out of 100), smoothing, and whether or not to create a progressive jpeg file.

Computer Note: The statement

```
Show[Graphics[RasterArray[Apply[
  (RGBColor[#1, #2, #3]&),
  picturevalues/255., {2}]]], AspectRatio→ 474/374.]
```

can be used to convert **picturevalues** into a graphics object and display the result.

8.6.2 Basic Mathematical Operations

Now that the image has been transformed into a set of numbers, any *Mathematica* numerical function can be applied to them. To create a negative image, multiply any of the channels by –1.

```
In[79]:= ListDensityPlot[-picturevalues[[All, All, 1]],
           Mesh → False, AspectRatio → 474/374.]
```

From In[79]:=

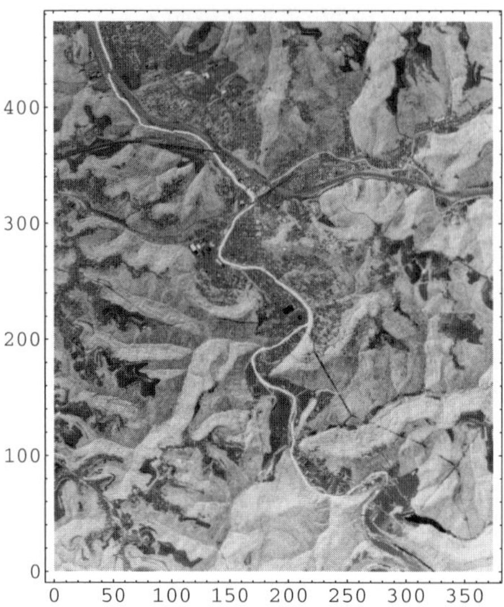

Out[79]= -DensityGraphics-

Similarly, we can square the red channel values to see what effect that operation will have.

```
In[80]:= ListDensityPlot[picturevalues[[All, All, 1]]^2,
           Mesh → False, AspectRatio → 474/374.]
```

From In[80]:=

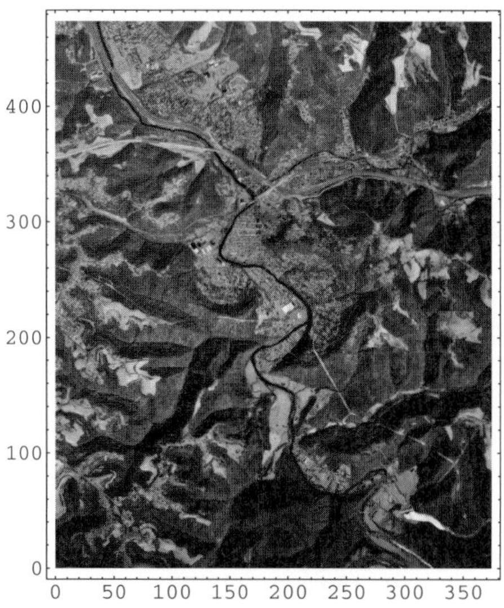

Out[80]= -DensityGraphics-

The result of squaring the red channel is to produce a darker plot, which may be surprising at first glance. **ListDensityPlot** plots large values as light colors, so why does squaring the red channel darken instead of lighten the plot? The reason is that, whereas the numerical values in each channel range from 0 to 255, *Mathematica* scales them to gray scale or RGB values ranging from 0 to 1 before plotting. Thus, squaring a value less than 1 produces a smaller number and the image is darkened. Readers who have their own copies of *Mathematica* may wish to see what effect squaring the red channel has on the color image. This can be done with the following statement:

```
In[81]:= Show[
           Graphics[
             RasterArray[Apply[(RGBColor[#1^2, #2, #3]&),
             picturevalues/255., {2}]], AspectRatio → 474/374.
           ]
         ]
```

From In[81]:=

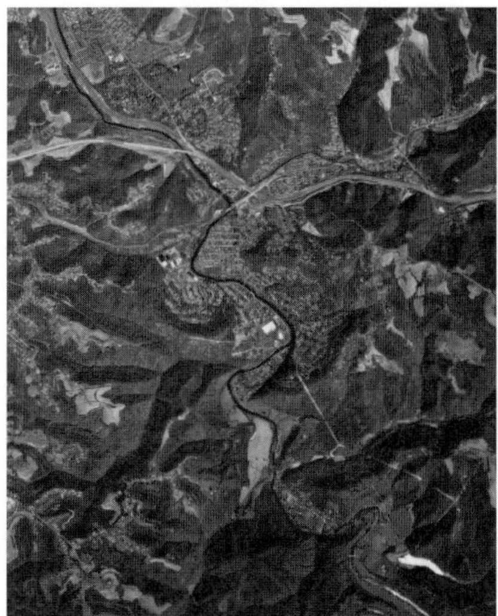

Out[81]= -Graphics-

For those who are reading the paper copy of this book and do not have access to the color images in the digital version, the effect of squaring the red channel was to significantly reduce the red hue of the image. Forested hills that were brownish red are now green, and the red areas appear to be restricted to grassy fields that were bright red in the original image. Areas of bare soil or rock remain light pink to white.

The **PlotRange** option can be used to control the contrast of an image. Decreasing **PlotRange** will increase contrast:

```
In[82]:= ListDensityPlot[picturevalues[[All, All, 1]],
           Mesh → False, AspectRatio → 474/374.,
           PlotRange → {127.5 - 50, 127.5 + 50}]
```

From In[82]:=

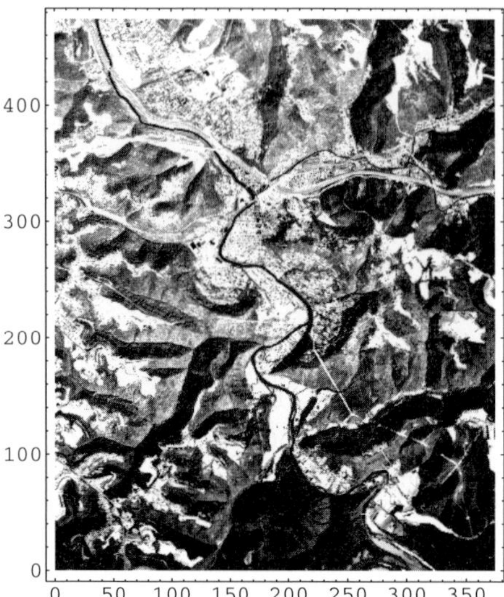


```
Out[82]= -DensityGraphics-
```

It is best to adjust the contrast by changing **PlotRange** symmetrically about the midpoint of the range of values, or 127.5. Doing otherwise will simultaneously lighten or darken the image, and can produce unintended effects. Conversely, increasing **PlotRange** will decrease the contrast.

```
In[83]:= ListDensityPlot[picturevalues[[All, All, 1]],
           Mesh → False, AspectRatio → 474/374.,
           PlotRange → {127.5 - 300, 127.5 + 300}]
```

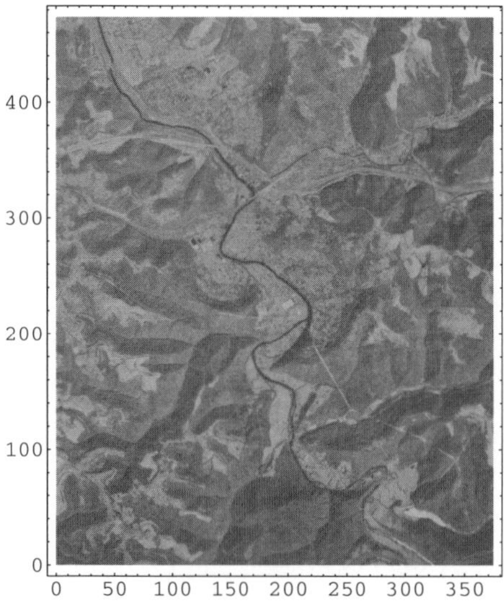


```
Out[83]= -DensityGraphics-
```

8.6.3 Thresholding

One common image processing technique is **thresholding**, in which values below a threshold are all changed to a constant value and those above the threshold are changed to another constant value. For example, the red channel of **picture** can be thresholded in a way such that values below 128 are changed to black (0) and values above 128 are changed to white (1). This will have the effect of changing the continuous tone gray level image into one that is truly black and white. First, create table with the same number of rows and columns as **picturevalues**.

```
In[84]:= thresholdvalues = Table[0., {r, 474}, {c, 374}];
```

Then, compare each element in **picturevalues** to the threshold. We originally set all of the values in **thresholdvalues** to zero, so if the red channel value in **picturevalues** is less than 128 we will leave it as-is. If the value is greater than 128, however, we will change the 0 to 1.

```
In[85]:= Do[
           If[picturevalues[[r, c, 1]] ≥ 128.,
           thresholdvalues[[r, c]] = 1.], {r, 474}, {c, 374}
         ]
```

This approach is similar to the one that we used to produce the landslide hazard maps in Chapter 7. Here is the result of the thresholding:

```
In[86]:= ListDensityPlot[thresholdvalues, Mesh → False,
           AspectRatio → 474/374.]
```

From In[86]:=

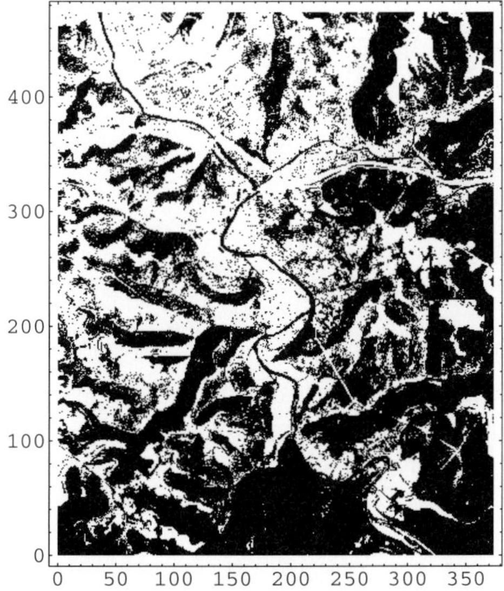

Out[86]= -DensityGraphics-

8.6.4 Smoothing or Blurring

The moving average approach that was introduced for 1-D time series can be extended to 2-D images. A 5 by 5 pixel moving average filter is:

```
In[87]:= k = Table[1/25., {r, 3}, {c, 3}]

Out[87]= {{0.04, 0.04, 0.04}, {0.04, 0.04, 0.04},
          {0.04, 0.04, 0.04}}
```

The filter is applied exactly as it was for the 1-D time series, in this case specifying that only the red channel is to be smoothed. The green and blue channels will remain unchanged. As shown below, the result is a blurred image.

```
In[88]:= ListDensityPlot[ListConvolve[k,
           picturevalues[[All, All, 1]]],
           AspectRatio → 474/374., Mesh → False]

From In[88]:=
```

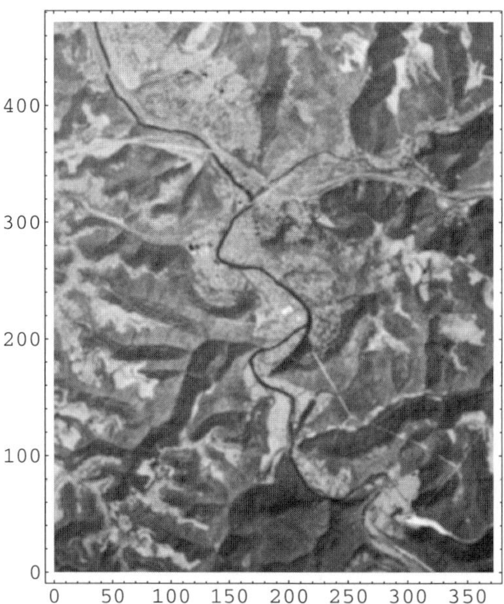


```
Out[88]= -DensityGraphics-
```

The Gaussian smoothing filter can also be extended to 2-D images, and variations are included as Gaussian blur filters in many image processing programs. The *Mathematica* package Statistics`MultinormalDistribution` contains functions that can be used to develop a 2-D Gaussian PDF.

```
In[89]:= k = Table[PDF[MultinormalDistribution[{0., 0.},
           {{1., 1./√3.}, {1./√3., 1.}}],
           {x, y}], {x, -2, 2}, {y, -2, 2}]
```

```
From In[89]:=

Out[89]=
  {{0.0154362, 0.0259108, 0.0097047, 0.000811038, 0.0000151237},
   {0.0259108, 0.103403, 0.0920757, 0.0182942, 0.000811038},
   {0.0097047, 0.0920757, 0.194924, 0.0920757, 0.0097047},
   {0.000811038, 0.0182942, 0.0920757, 0.103403, 0.0259108},
   {0.0000151237, 0.000811038, 0.0097047, 0.0259108, 0.0154362}}
```

The multinormal distribution has two mean values and a covariance matrix instead of a standard deviation. The PDF above is the 2-D equivalent of a standard normal distribution with zero mean and unit variance. Summing the elements in **k**will yield a value greater than 0.98, which is close to the value of 1 that must be obtained for any legitimate PDF. As illustrated in the plot below, the use of a Gaussian smoothing filter produces a result that is visually similar to that produced by a simple 3 by 3 term moving average.

```
In[90]:= smoothplot = ListDensityPlot[ListConvolve[k,
           picturevalues[[All, All, 1]]],
           AspectRatio → 474/374., Mesh → False]

From In[90]:=
```

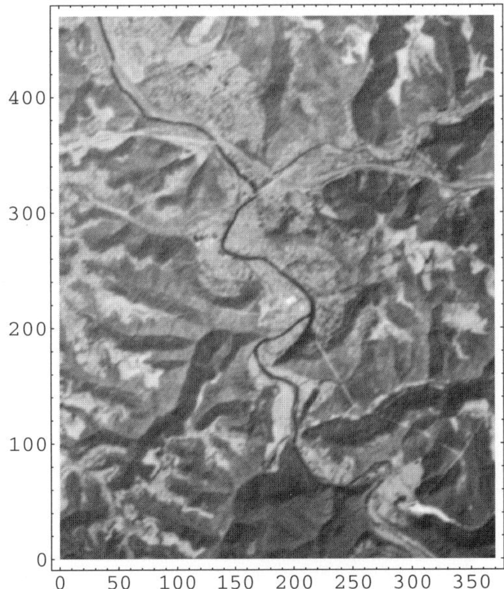


```
Out[90]= -DensityGraphics-
```

Although it may not be obvious why anyone would want to blur a perfectly good image, it turns out that the ability will become very useful. Blurring can be an important part of image sharpening and also used to pre-process noisy images before applying edge detection filters.

8.6.5 Unsharp Masking

Despite its name, unsharp masking is used to increase the sharpness of images. It is based on a technique originally developed for use with photographic film in which a slightly blurred copy (known as an **unsharp mask**) is combined with the original image to increase detail in the shadows. While unsharp masking will sharpen images that are already in focus, it will not improve a completely out-of-focus photograph. Many commercial image processing programs include sophisticated unsharp masks or filters, and the basic mathematics of unsharp masking can be demonstrated using *Mathematica*.

The first step is to obtain a slightly cropped version of the original image. This is necessary because we will be combining it with the result of a convolution of the image with a 2-D Gaussian smoothing kernel, which reduces the number of rows and columns by 4. As above, we will work on the red channel of the Wheeling orthophoto as an example.

```
In[91]:= original = Table[picturevalues[[r, c, 1]],
           {r, 3, 472}, {c, 3, 372}];
```

Next, start to create the unsharp mask using the Gaussian kernel developed in the previous section.

```
In[92]:= smooth = ListConvolve[k, picturevalues[[All, All, 1]]];
```

The results of the Gaussian kernel are the same as those shown in the previous section. The unsharp mask will be some fraction of the difference between **`original`** and **`smooth`**, which will tend to emphasize boundaries and edges. In this case, the unsharp mask with a constant of 0.5 looks like this:

```
In[93]:= ListDensityPlot[0.5 * (original - smooth),
           Frame → False, Mesh → False, AspectRatio → 470/370.]
```

From In[93]:=

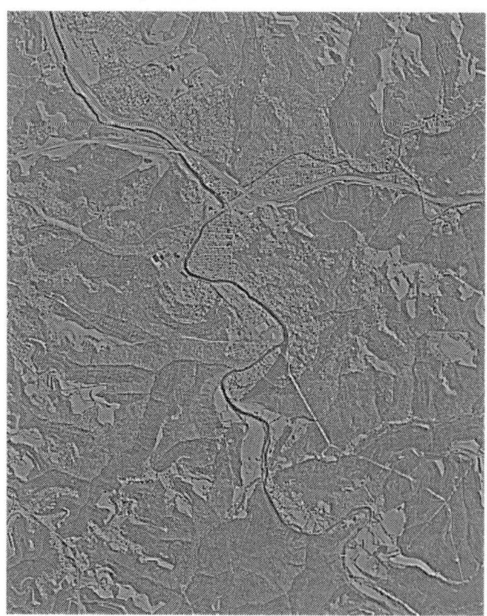

```
Out[93]= -DensityGraphics-
```

Finally, the unsharp mask is subtracted from the original. In this example, it is shown side-by-side with the original image for comparison.

```
In[94]:= Show[
           GraphicsArray[
             {ListDensityPlot[original, Frame → False,
                 Mesh → False, AspectRatio → 474/374.,
                 PlotLabel- > "original",
                 DisplayFunction → Identity],
               ListDensityPlot[original+0.5*(original-smooth),
                 Frame → False,
                 Mesh → False, AspectRatio → 474/374.,
                 PlotLabel- > "sharpened",
                 DisplayFunction → Identity]}],
                 DisplayFunction → $DisplayFunction
         ]
```

From In[94]:=

original

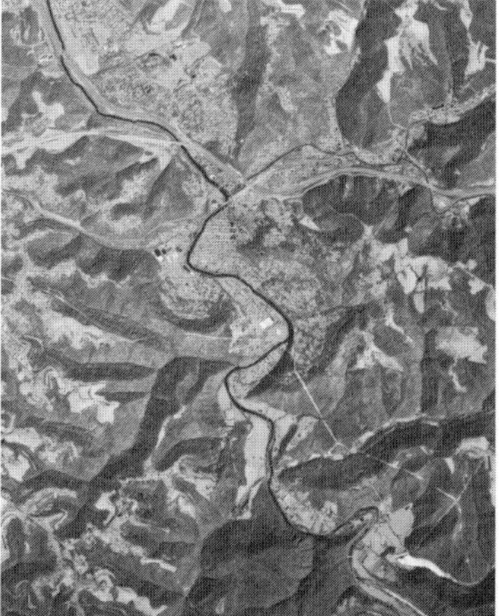

sharpened

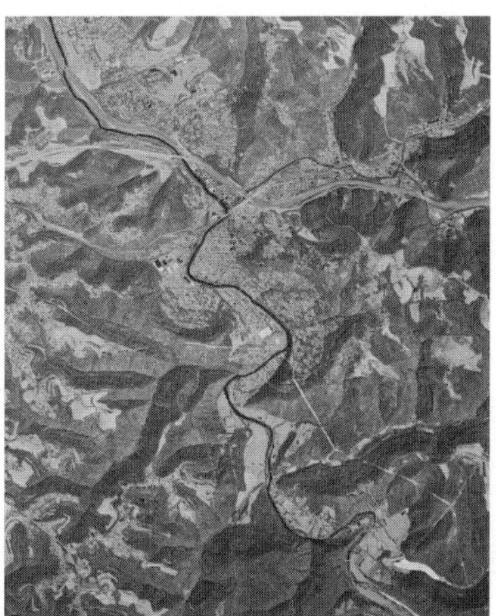

Out[94]= -GraphicsArray-

Look closely at the two images and you will see that the sharpened image has greater detail in the mid-tones and shadows, and well as crisper boundaries between light and dark areas. While unsharp masking will not perform miracles on poorly focussed images, it can add a significant degree of sharpness to images that are already in focus. If you are following these examples on your own computer, click on the image above and drag one of the handles to enlarge the image and examine it in more detail.

> **Computer Note:** Experiment with values other than 0.5 to determine which is best for this image.

8.6.6 Edge Detection

Filters can also be used to detect edges between elements in images by identifying areas where the values in the image or its derivatives change significantly over short distances.

One common edge detection filter is the **Laplacian filter**, which takes its name from the fact that it is the finite difference operator used in numerical solutions of Laplace's equation (see Chapter 3). Recall that Laplace's equation is $\partial^2 f/\partial x^2 + \partial^2 f/\partial y^2 = 0$. Therefore, a Laplacian filter detects edges by delineating narrow zones of zero curvature (sometimes referred to as **zero crossings** in image processing literature) that separate concave from convex portions of the image. When referring to the curvature of the image values, remember that we are referring to the curvature of the array of gray scale values comprising the image and not the topography depicted in the image. Thus, an area of positive curvature would correspond to a dark area surrounded by light areas regardless of its topographic expression (for example, the rivers in the image). Areas of negative curvature correspond to light areas surrounded by dark (for example, the highways in the image).

To apply a Laplacian smoothing filter, redefine the kernel k as

```
In[95]:= k = {{0, 1, 0}, {1, -4, 1}, {0, 1, 0}}
Out[95]= {{0, 1, 0}, {1, -4, 1}, {0, 1, 0}}
```

and then use **ListConvolve** to convolve the kernel with the image.

```
In[96]:= ListDensityPlot[ListConvolve[k,
           picturevalues[[All, All, 1]]],
           AspectRatio → 474/374., Mesh → False]
```

```
From In[96]:=
```

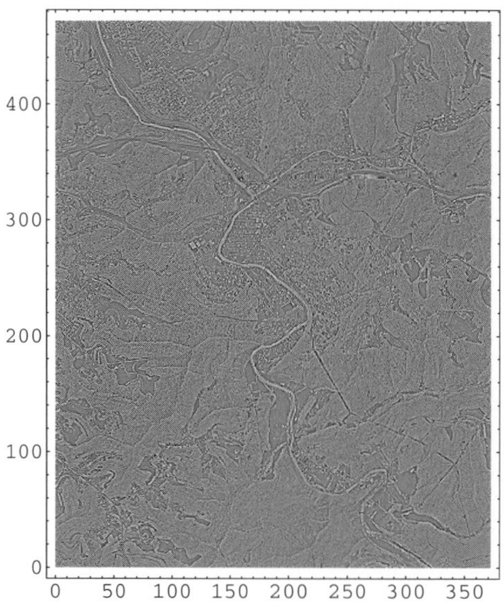


```
Out[96]= -DensityGraphics-
```

Although edges stand out weakly in the filtered image, they are not strong because the original image contains many high frequency details. If the high frequency components were unwanted, they would be called noise. In this case, however, they are desirable and we will call them details. Regardless of the name we choose, smoothing will tend to make the detected edges stronger. This can be illustrated using the array **smooth** that we created for unsharp masking. We will want to make use of the results several times, so the first step will be to apply the Laplace filter to **smooth** and then assign the result to the variable name **smoothlaplace**.

```
In[97]:= smoothlaplace = ListConvolve[k, smooth];
```

The result shows much stronger edges:

```
In[98]:= ListDensityPlot[smoothlaplace, AspectRatio → 474/374.,
           Mesh → False]
```

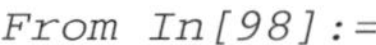
From In[98]:=

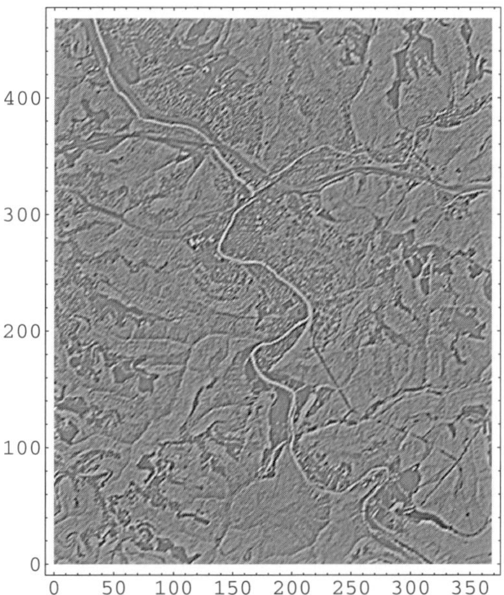


```
Out[98]= -DensityGraphics-
```

Although linear features such as the river and roads stand out on this image, it is because they are either white or black and not because their edges have been clearly delineated. As shown in the histogram below, there are relatively few pixels with strong positive or negative curvature and many with near-zero curvature. In other words, the image contains many edges even though it was smoothed in an attempt to remove details.

```
In[99]:= Histogram[Flatten[smoothlaplace]]
```

From In[99]:=

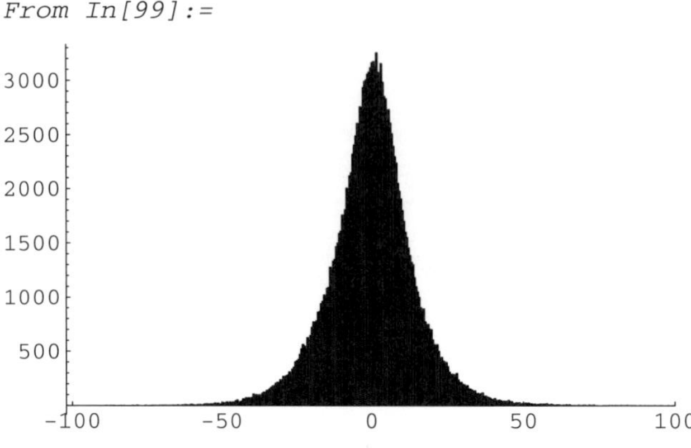


```
Out[99]= -Graphics-
```

In some cases, thresholding an edge-detected image can help to delineate edges. In this example, however, thresholding will not help much.

A more sophisticated edge-detection process is the Sobel filter, which actually consists of several filters applied in sequence. The two kernels below calculate the gradients of the image in the vertical and horizontal directions, respectively.

```
In[100]:= ygrad = {{-1., -2., -1.}, {0., 0., 0.}, {1., 2., 1.}}

Out[100]= {{-1., -2., -1.}, {0., 0., 0.}, {1., 2., 1.}}

In[101]:= xgrad = {{-1., 0., 1.}, {-2., 0., 2.}, {1., 0., 1.}}

Out[101]= {{-1., 0., 1.}, {-2., 0., 2.}, {1., 0., 1.}}
```

Sobel edge detection filtering is accomplished by applying the two gradient kernels in succession.

```
In[102]:= smoothsobel = ListConvolve[ygrad,
            ListConvolve[xgrad, smooth]];
```

The result is an image in which the edges are more pronounced than in the Laplace filter example, although there are some conspicuous diagonal artifacts in the image. The primary reason that the edges stand out so clearly is that the are the highest (lightest pixels) and lowest (darkest pixels) values instead of mid-range values.

```
In[103]:= ListDensityPlot[smoothsobel, AspectRatio → 466/366.,
            Mesh → False]
```

From In[103]:=

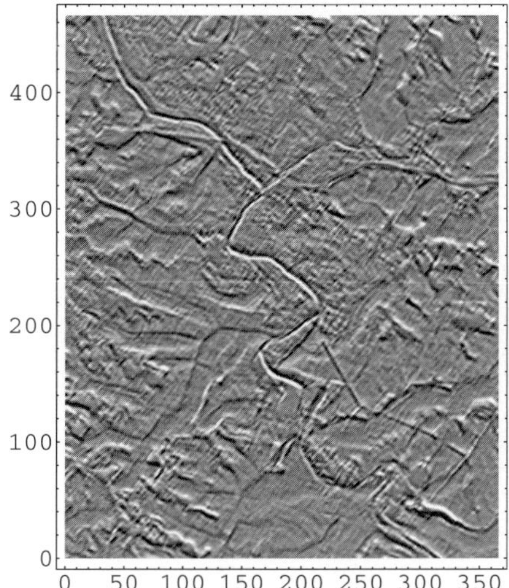

Out[103]= -DensityGraphics-

8.6.7 Using ListInterpolation

The *Mathematica* function **ListInterpolation** can be used to interpolate a smooth surface that passes through each of the pixel values in an image, which can then be operated on as a function rather than an array of discrete values. For example, the statement below interpolates a surface through **smooth** and returns an interpolating object. Notice that the order of rows and columns used for arrays is different than the *x*and *y* order typically used for functions.

```
In[104]:= smoothinterp = ListInterpolation[smooth]

Out[104]= InterpolatingFunction[{{1., 470.}, {1., 370.}}, <>]
```

Here is a density plot of the interpolated surface, which does a good job of reproducing the smoothed image. The column iteration was listed before the row iteration to ensure that the image would appear in the correct orientation.

```
In[105]:= DensityPlot[smoothinterp[r, c], {c, 1, 370}, {r, 1, 470},
            PlotPoints → 400, Mesh → False,
            AspectRatio → 470/370.]
```

From In[105]:=

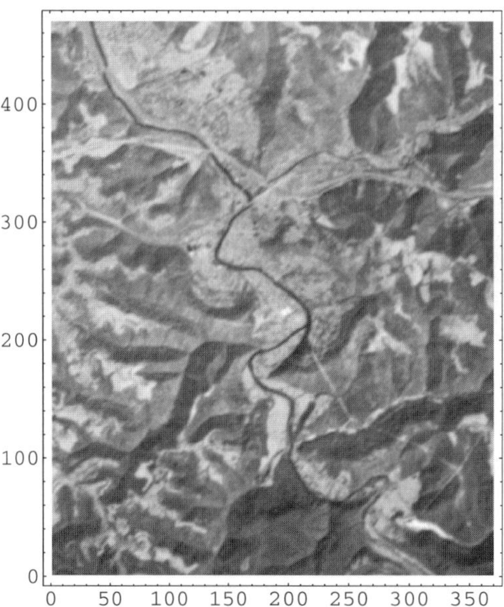

Out[105]= -DensityGraphics-

The equivalent of a Sobel edge detection filter can be applied by differentiating the interpolated surface with respect to the row and column coordinates. Negative signs are included to be consistent with the Sobel filters used above. The resulting image is sharper than that created by the discrete Sobel filters, particularly with regard to the strong diagonal artifacts that were so apparent in the Sobel filtered image. Interpolation also allows the resolution of the image to be increased (although sharpness does not increase because a smooth curve is interpolated between known values). A drawback to the use of interpolated surfaces, however, is that the calculations can be much slower than numerical convolution.

```
In[106]:= DensityPlot[Evaluate[-D[smoothinterp[r, c], c]-
              D[smoothinterp[r, c], r]],
            {c, 1, 370}, {r, 1, 470}, PlotPoints → 500,
            Mesh → False, AspectRatio → 470/370.]
```

```
From In[106]:=
```

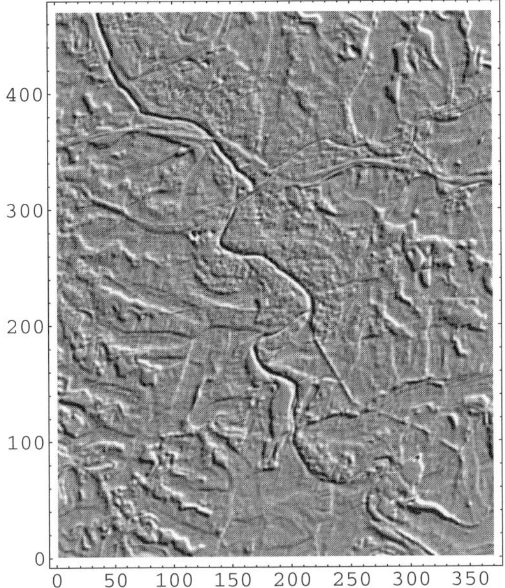


```
Out[106]= -DensityGraphics-
```

8.7 Recommended Reading

Bras, R.L. and Rodgriguez-Iturbe, I., 1993, *Random Functions and Hydrology*: Dover Publications.

Burrough, P.A. and McDonnell, R.A. 1998, *Principles of Geographical Information Systems*: Oxford University Press.

Carr, J.R., 2002, *Data Visualization in the Geosciences*: Prentice Hall.

Davis, J.C., 2002, *Statistics and Data Analysis in Geology (3d ed.)*: John Wiley & Sons.

Gonzales, R.C. and Woods, R.E., 2002, *Digital Image Processing (2d ed.)*: Addison-Wesley.

Gubbins, D., 2003, *Time Series Analysis and Inverse Theory for Geophysicists*: Cambridge University Press.

Hamming, R.W., *Numerical Methods for Scientists and Engineers (2d ed.)*: Dover Publications.

Isaaks, E.H. and Srivastava, R.M., 1989, *An Introduction to Applied Geostatistics*: Oxford University Press.

Middleton, G.V., 2000, *Data Analysis in the Earth Sciences Using Matlab*: Prentice Hall.

Appendix A
Mathematica Functions in the Computational Geoscience Package

A.1 Introduction

The *Computational Geosciences with Mathematica* package, located in the file CompGeosci.m on the CD accompanying this book, contains a number of functions that are too long to be conveniently listed in the text. This Appendix contains a list of all of those functions, a brief description of each, and the chapter in which the function is first introduced. Consult those chapters for details on the the use of the functions.

Mathematica packages can be generated using either a text editor or creating a *Mathematica* notebook and using the Save As Special menu item under File to export them in package format. The CompGeosci package was created using the Save As Special method, and the notebook from which the package was created (BookFunctions.nb) is provided on the accompanying CD. If you are interested in creating your own package, consult *The Mathematica Book* or the online documentation for instructions for authors of *Mathematica* packages.

A.2 Plotting and Calculations

`CircleOverlapArea`: Calculates the overlap of counting circles used in the Kamb contouring method. Used by **`ListKambPlot`**. **Chapter 2**.

`CumFreqPlot`: Creates a cumulative frequency plot from a list of data. **Chapter 4**.

`CumFreqs`: Calculates a list of cumulative frequencies from a data set. **Chapter 4**.

`EqualAreaLinePoint`: Plots a line of specified orientation on an equal area net (projected as a point on a sphere). Used by **`ListEqualAreaPointPlot`**. **Chapter 2**.

`EqualAreaXYCoords`: Calculates the Cartesian coordinates of points on a sphere using an equal area projection. Used by **`EqualAreaLinePoint`**. **Chapter 2**.

`EucDist`: Calculates the Euclidean Distance between two points. Used in **`ListKambPlot`**. **Chapter 2**.

`KSOneList`: Calculates the Kolmogorov-Smirnov statistic between a data list and a normal distribution having the sample mean and standard deviation of the data list. **Chapter 4**.

`KSOneListPlot`: Plots the empirical cumulative distribution of a data list and superimposes it on a cumulative distribution function for a normal distribution having the sample mean and standard deviation of the data list. **Chapter 4**.

`KSProb`: Calculates the approximate probability of committing a Type I error by rejecting the null hypothesis that two distributions with n samples are the same. See Press et al., *Numerical Recipes*: Cambridge Press for details. **Chapter 4**.

`KSTwoList`: Calculates the Kolmogorov-Smirnov statistic between two data lists by comparing their values at a user-specified number of points.

`KSTwoListPlot`: Plots the empirical cumulative distributions of two data lists.

`ListBoxWhiskerPlot`: Creates a box-and-whisker plot from lists of quantile data. **Chapter 2**.

`ListEqualAreaPointPlot`: Creates an equal area plot from a list of line orientation data. **Chapter 2**.

`ListKambPlot`: Creates a Kamb contour plot of lines (projected as points) on an equal area projection. **Chapter 2**.

`ListStemPlot`: Produces a stem plot from a list of data, either as x-$y$pairs or a list of y values, in which case the x values are assumed to be 1, 2, 3... **Chapter 2**.

`ListStereoArcPlot`: Plots the stereographic projection of planes (projected as arcs). **Chapter 2**.

`ListStereoPointPlot`: Plots the stereographic projection of planes (projected as arcs). **Chapter 2**.

`ListRosePlot`: Plots a rose diagram of a uni-directional or bi-directional data set. **Chapter 2**.

`ListTernaryPlot`: Produces a ternary plot of three-phase compositional data. The sum of percentages of all three phases must sum to 1 (100%). **Chapter 2**.

`RGBViewer`: Draws a colored square corresponding the the specified RGB color. **Appendix B**.

`SlopeAngle`: Calculates the slope angle, in degrees, at each point in a square grid of numbers (*e.g.*, elevation values). Uses a finite difference approximation of the maximum gradient. **Chapter 7.**

`SlopeCurvature`: Calculates the slope curvature, in degrees, at each point in a square grid of numbers (*e.g.*, elevation values). Uses a 9 point finite difference approximation of the curvature. **Chapter 7.**

`StereoLinePoint`: Plots a line projected as a point on a stereographic projection. Used by **`ListStereoPointPlot`**.

`StereoPlaneArc`: Plots a plane projected as an arc on a stereographic projection. Used by **`ListStereoArcPlot`**. **Chapter 2**.

`ThinPlateGrid`: Performs thin plate spline gridding of irregularly spaced data. **Chapter 7**.

A.3 Color Functions

The following color functions are also contained in the CompGeosci package. They can be used in contour, density, and 3-D surface plots. See Appendix B (CD only) for details about using color in *Mathematica* and creating your own color functions.

The general syntax of the color functions below is **`colorfunction`**`[`z`]`, where $0 \le z \le 1$ is the dependent variable being plotted.

`Rainbow`
`RainbowReverse`
`BrownGreenCream`
`BrownGreenWhite`
`GreenYellowRed`
`GreenWhiteRed`
`RedWhiteGreen`
`RedYellowGreen`
`RedWhiteBlue`
`BlueWhiteRed`

Appendix B
Working with Color

B.4 *Mathematica* Packages You Will Need

```
In[107]:= Needs["Graphics`Colors`"]
          Needs["CompGeosci`"]
```

B.5 Specifying Colors in *Mathematica*

B.5.1 Hue, Saturation, and Brightness

Mathematica provides three built-in ways to specify colors: **Hue**, **RGBColor**, and **CMYKColor**. **Hue**, which in its simplest form takes a single value between 0 and 1, produces a rainbow-like range of colors. The statement below produces a graphics array consisting of a table of rectangles ranging in hue from 0 (left) to 1 (right).

```
In[108]:= Show[
            GraphicsArray[
              Table[Graphics[{Hue[i],Rectangle[{0, 0}, {1, 1}]}],
                {i, 0, 1, 0.1}]
            ]
          ]
```

```
Out[108]= -GraphicsArray-
```

Although **Hue** is simple to use, one major drawback is that its range of colors begins and ends with red. If a contour or density plot were to be colored using the unmodified **Hue** function, both the lowest and highest contour intervals would be identical in color. This is rarely desirable. One way to avoid this problem is to scale the values that are used in **Hue**. For example, the following statement scales **Hue** by 0.8 and produces a range of colors from red (**Hue[0]**) to violet (**Hue[0.8]**).

```
In[109]:= Show[
            GraphicsArray[
              Table[Graphics[{Hue[0.8i],
                Rectangle[{0, 0}, {1, 1}]}], {i, 0, 1, 0.1}]
            ]
          ]
```

```
Out[109]= -GraphicsArray-
```

To show only the upper portion of the hue spectrum, rescale and shift the argument in **Hue**, as illustrated below, to produce values that range from orange (**Hue[0.1]**) to red (**Hue[1]**).

```
In[110]:= Show[
            GraphicsArray[
              Table[Graphics[{Hue[0.1 + 0.9i],
                Rectangle[{0, 0}, {1, 1}]}], {i, 0, 1, 0.1}]
            ]
          ]
```

```
Out[110]= -GraphicsArray-
```

Any combination of rescaling and shifting of the value passed to **Hue** is allowed as long as the result falls between 0 and 1.

Hue can also be used with three arguments, the latter two specifying the *saturation* and *brightness* of the color. If only one argument is used in **Hue**, *Mathematica* assumes that the saturation and brightness values are both 1. Reducing the saturation in the color bar above to 0.5, for example, produces the following range of colors:

```
In[111]:= Show[
            GraphicsArray[
              Table[Graphics[{Hue[0.1 + 0.9i, 0.5, 1.],
                    Rectangle[{0, 0}, {1, 1}]}], {i, 0, 1, 0.1}]
            ]
          ]
```

```
Out[111]= -GraphicsArray-
```

The grid below shows how changing the saturation (ranging from 0 in the top row to 1 in the bottom row) and brightness (ranging from 0 in the left-most column to 1 in the right-most column) changes the appearance of a rectangle with a hue of 0.7.

```
In[112]:= Show[
            GraphicsArray[
              Table[Graphics[{Hue[0.7, i, j],
                Rectangle[{0, 0}, {1, 1}]}], {i, 0, 1, 0.1},
                {j, 0, 1, 0.1}]
            ]
          ]
```

```
Out[112]= -GraphicsArray-
```

Notice that about half the rectangles, which correspond to small saturation or brightness values, appear to be black or gray. Therefore, it is important to use relatively high saturation and brightness values (say, greater than 0.5) if the resulting colors are to be distinguishable.

B.5.2 Red, Green, and Blue (RGB)

The second way to specify colors is to use the **RGBColor** function, which takes as its arguments the intensity of the red, green, and blue primary colors of transmitted light. Computer monitors and televisions typically create images using RGB colors. As with **Hue**, the value for each component can range between 0 and 1. Pure red, for example, would be **RGBColor[1, 0, 0]** whereas pure blue would be **RGBColor[0, 0, 1]**. Unlike **Hue**, **RGBColor** can also be used to specify black (**RGBColor[0, 0, 0]**) and white (**RGBColor[1, 1, 1]**). The *Mathematica* statement below, which is a two-dimensional version of the statements used above to illustrate the **Hue** function, shows an array of colors for the red and green components ranging from 0 (top row and left-most column) to 1 (bottom row and right-most column) in increments of 0.1 while holding the blue component fixed at 0.5. Thus, the dark blue rectangle in the upper left-hand corner was drawn using **RGBColor[0, 0, 0.5]** and the yellow rectangle in the lower right-hand corner was drawn using **RGBColor[1, 1, 0.5]**.

```
In[113]:= Show[
            GraphicsArray[
              Table[
                Graphics[{RGBColor[i, j, 0.5],
                Rectangle[{0, 0}, {1, 1}]}],
                {i, 0, 1, 0.1}, {j, 0, 1, 0.1}]
            ]
          ]
```

```
Out[113]= -GraphicsArray-
```

> **Computer Note:** Modify the previous statement to draw a series of color grids, similar to that above, in order to illustrate the complete range of RGB colors. For example, you might let all three of the color components vary from 0 to 1 in increments of 0.2.

Guessing the RGB components of a color that you might want to use in a plot can be a tricky process. One way to obtain the color you want is to make your own RGB color chart (use the instructions in the preceeding Computer Note), or to find a color chart in a computer graphics book or web site. Note that many color charts will show colors with RGB components ranging in value from 0 to 255. To recreate these colors in *Mathematica*, just divide each value by 255 before using it as an argument in **RGBColor**. Another way to explore RGB colors is to use the simple user-defined function that draws a rectangle having the specified RGB components.

```
In[114]:= RGBViewer[r_, g_, b_] :=
            Show[Graphics[{RGBColor[r, g, b],
            Rectangle[{0, 0}, {1, 1}]}]
            ]
```

The RGB combination 0.7, 0.2,0.8, for example, produces the bright purple color shown below.

```
In[115]:= RGBViewer[0.7, 0.2, 0.8]
```

```
Out[115]= -Graphics-
```

The add-on package Graphics`Colors` contains a list of 193 RGB color specifications with names like **CinnabarGreen**, **DarkOrchid**, **GeraniumLake**, **VenetianRed** and **PapayaWhip**. Once the Colors package is loaded, typing in a color name will return its RGB specification. For example,

```
In[116]:= VenetianRed

Out[116]= RGBColor[0.829997, 0.099994, 0.119999]
```

To see the complete list of the predefined colors, enter the variable name **AllColors**. You can preview these predefined colors with a one-line statement that is very similar to the **RGBViewer** function defined above.

```
In[117]:= Show[Graphics[{VenetianRed,
            Rectangle[{0,0}, {1,1}]}]]
```

```
Out[117]= -Graphics-
```

Computer Note: *Mathematica* reads lists of graphics specifications from left to right, so the color must be specified *before* the rectangle is drawn. To illustrate this, try switching the order of **VenetianRed** and **Rectangle** in the statement above.

B.5.3 Cyan, Magenta, Yellow, and Black (CMYK)

The third method of specifying a color is to use its cyan, magenta, yellow, and black components that comprise reflected light. These correspond to the four inks used to print full color images on paper and, as such, CMYK colors are most commonly used when making color separations for printed materials. Modern ink jet printers and papers can be calibrated to produce very high quality output by translating RGB colors on a computer screen to CMYK colors on paper, so it is unlikely that CMYK color specifications will be of much concern to most geoscientists using *Mathematica*. The built-in function **CMYKColor** works similarly to **Hue** and **RGBColor** except that four color components must be given. The statement below shows a rectangle with the CMYK components 0.3, 0.4, 0.6, and 0.2.

```
In[118]:= Show[
            Graphics[{CMYKColor[0.3, 0.4, 0.6, 0.2],
               Rectangle[{0, 0}, {1, 1}]}]
          ]
```

```
Out[118]= -Graphics-
```

> **Computer Note:** Write a user-defined function to preview CMYK colors, using as an example the **RGBViewer** shown in the previous section.

B.5.4 Other Color Systems

The Graphics`Colors` package also supports the use of three other kinds of color specifications: CMY (cyan, magenta, yellow), YIQ (NTSC video format), and HLS (hue, lightness, and saturation). None of these are likely to arise in most geoscientific work, although they are available for use if necessary.

B.6 Using Color in Plots and Graphics

B.6.1 Plot and ListPlot

Two-dimensional plots such as those produced by **Plot** and **ListPlot** can be colored using a **PlotStyle** option, as shown below.

```
In[119]:= Plot[Sin[x], {x, 0, 2 π}, PlotStyle → Hue[0.6]]
```

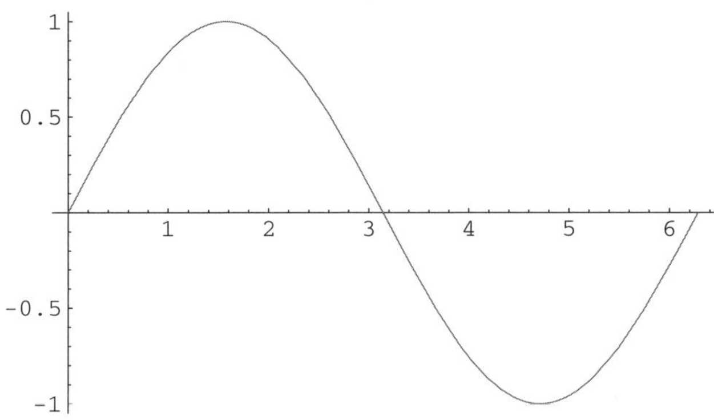

```
Out[119]= -Graphics-
```

To draw the entire plot in a specific color, change the **DefaultColor** option from black (the default value) to the color of your choice.

```
In[120]:= Plot[Sin[x], {x, 0, 2 π}, DefaultColor → Hue[0.6]]
```

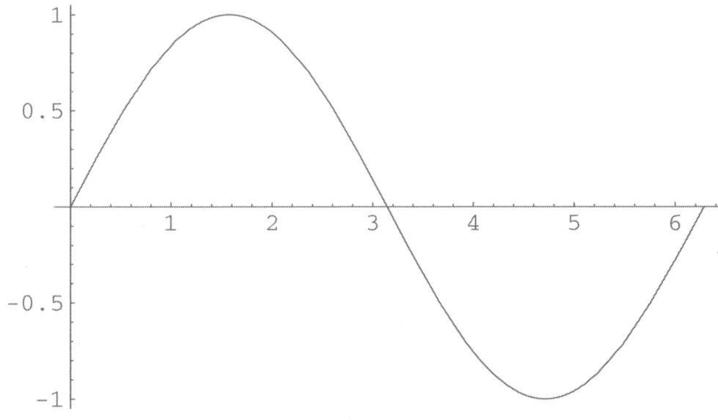

```
Out[120]= -Graphics-
```

Any valid HSB, RGB, or CMYK color (or gray level) specification could have been used in these examples. Color specifications can be combined with plot options such as **Dashing**, **Thickness**, or, in the case of list plots, **PointSize** to produce a variety of effects. Similarly, axis, frame, and background colors can be specified using **AxesStyle**, **FrameStyle**, and **Background**.

```
In[121]:= Plot[Sin[x], {x, 0, 2 π}, Frame → True,
            FrameStyle → VenetianRed, Background → PapayaWhip,
            PlotStyle → {CinnabarGreen, Thickness[0.008],
                Dashing[{0.02}]}]
```

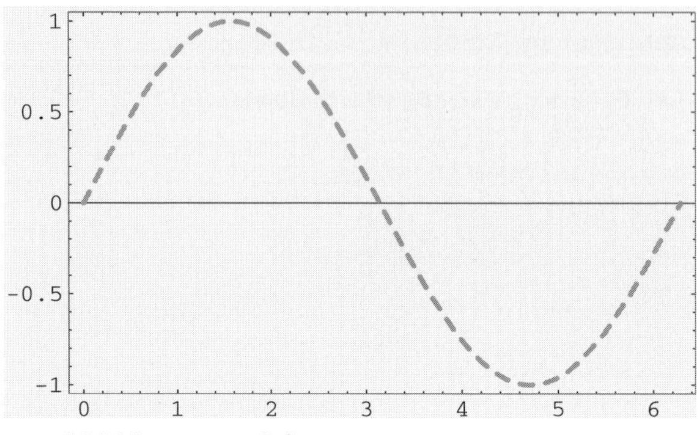

```
Out[121]= -Graphics-
```

B.6.2 Contour and Density Plots

Options such as **AxesStyle** and **FrameStyle** also apply to contour and density plots, and it is possible to specify contour line styles. For example, here is a contour plot with red contour lines and a blue frame:

```
In[122]:= ContourPlot[Sin[x] Sin[y], {x, 0, 2 π}, {y, 0, 2 π},
            ContourStyle → RGBColor[1, 0, 0],
            FrameStyle → RGBColor[0, 0, 1], PlotPoints → 40]
```

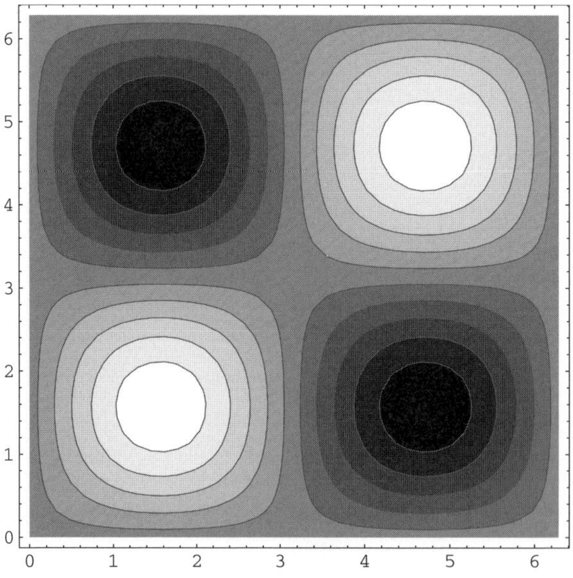

```
Out[122]= -ContourGraphics-
```

A default contour or density plot is filled with a range of gray levels, but color can be used by specifying a *color function*. The simplest way to color a contour or density plot is to use **Hue** without an argument.

```
In[123]:= ContourPlot[Sin[x] Sin[y], {x, 0, 2 π}, {y, 0, 2 π},
            ColorFunction → Hue, PlotPoints → 40]
```

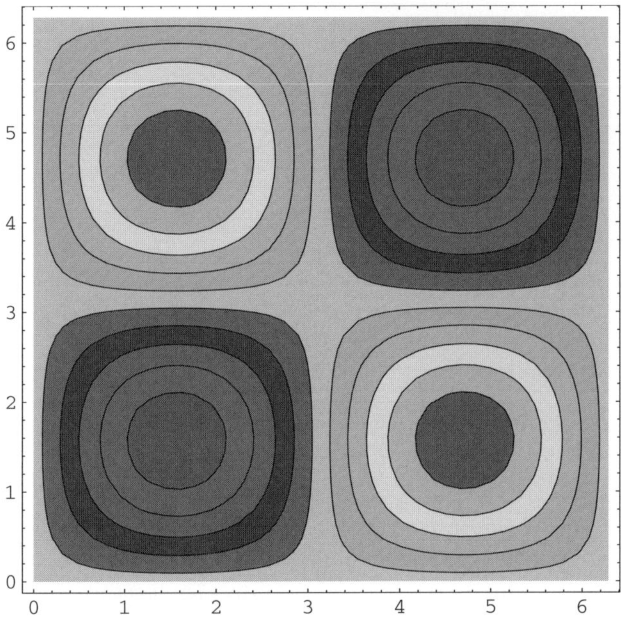


```
Out[123]= -ContourGraphics-
```

As illustrated above, though, using **Hue** produces a plot in which both the lowest and highest contour intervals are colored red. **Hue** can be scaled or shifted, as described in the first section of this appendix, to alleviate the problem. To do so requires that **Hue** be incorporated into a color function that we will call **ScaledHue**.

```
In[124]:= ScaledHue[z_] := Hue[0.8 z]
```

As above, any argument used in a color function must be within the range of 0 to 1 and values outside of this range will produce an error message. The newly defined color function can now be used as an option in **ContourPlot** or **DensityPlot**.

```
In[125]:= ContourPlot[Sin[x] Sin[y], {x, 0, 2 π}, {y, 0, 2 π},
            ColorFunction → ScaledHue , PlotPoints → 40]
```

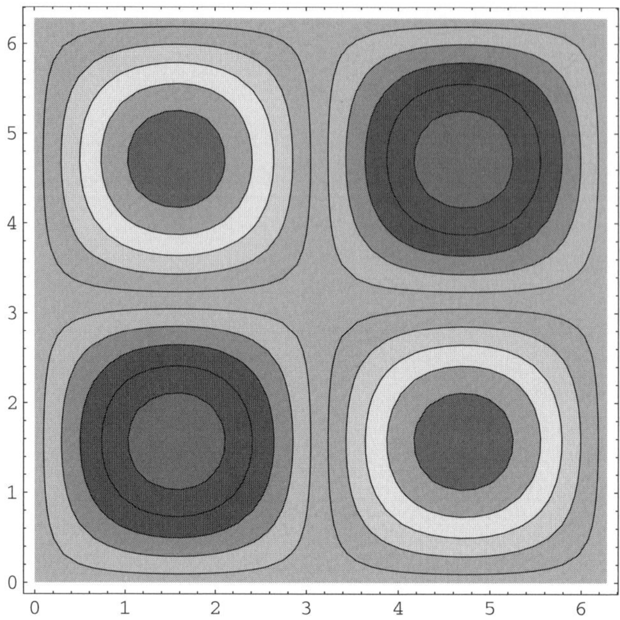

Out[125]= -ContourGraphics-

> **Computer Note:** There are alternative ways in which to define a color function (or any function, for that matter) in *Mathematica*. **ScaledHue = Function[z, Hue[0.8z]]** and **ScaledHue = (Hue[0.8#]&)** will produce the same results as the **ScaledHue** function defined above. Refer to the *Mathematica* documentation for a detailed discussion of the definition and use of functions.

Virtually any color sequence can be defined using HSB, RGB, or CMYK colors. For example, consider a simple example that ranges from green for low values to red for high values.

```
In[126]:= GreenRed[z_] := RGBColor[z, 1 - z, 0];

In[127]:= ContourPlot[Sin[x] Sin[y], {x, 0, 2 π}, {y, 0, 2 π},
            ColorFunction → GreenRed, PlotPoints → 40]
```

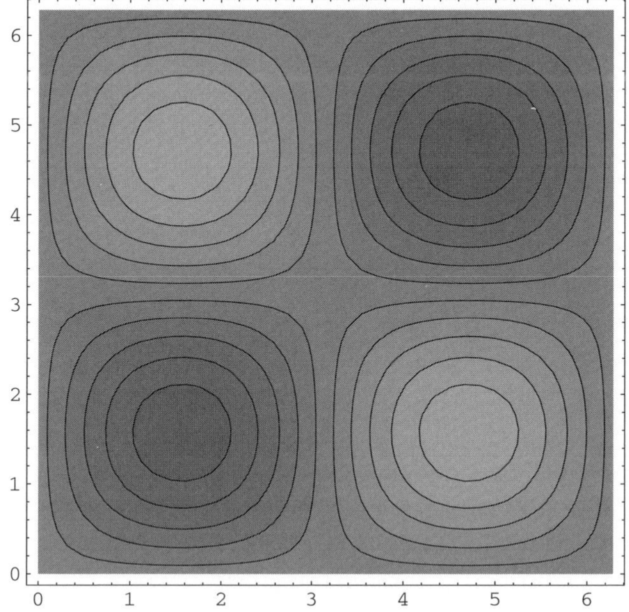


```
Out[127]= -ContourGraphics-
```

More complicated color functions can be derived by specifying the RGB (or HSB or CMYK) colors for several values of the color scale and then using **Interpolate**. Say that you would like to use a red to green color function, similar to that above, but replace the middle brown with white. Therefore, the smallest values shown would have a color of **RGBColor[0, 1, 0]**, the middle values would have a color of **RGBColor[1, 1, 1]**, and the largest values would have a color of **RGBColor[1, 0, 0]**. The first step in defining the color function is to interpolate a series of three curves, which we will call **RedCurve**, **GreenCurve**, and **BlueCurve**, by specifying each of the three components for values of $z = 0$, $z = 0.5$, and $z = 1$.

```
In[128]:= RedCurve = Interpolation[{{0., 0.}, {0.5, 1.}, {1., 1.}},
             InterpolationOrder → 1]

Out[128]= InterpolatingFunction[{{0., 1.}}, <>]
```

The option **InterpolationOrder → 1** is used for two reasons. First, the default value of 3 cannot be used if there are only three data points, because n points are required to interpolate a polynomial of order $n - 1$. Second, although it would have been possible use an interpolation order of 2, in this case the result would be a parabolic curve with values that exceed 1 between the three data points. Therefore, the best strategy is generally to use linear interpolation. The form of **RedCurve** can be illustrated by plotting it.

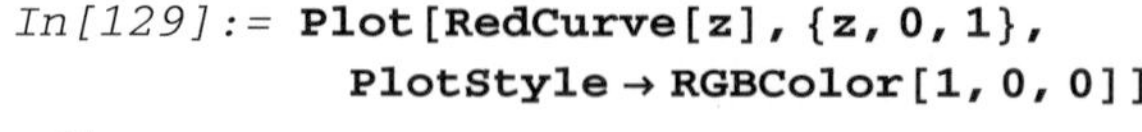

```
In[129]:= Plot[RedCurve[z], {z, 0, 1},
            PlotStyle → RGBColor[1, 0, 0]]
```

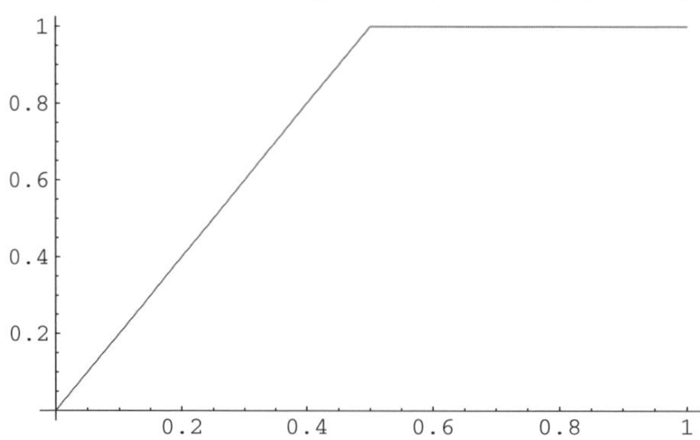


```
Out[129]= -Graphics-
```

A similar process can be used to obtain the green curve.

```
In[130]:= GreenCurve =
            Interpolation[{{0., 1.}, {0.5, 1.}, {1., 0.}},
            InterpolationOrder → 1]

Out[130]= InterpolatingFunction[{{0., 1.}}, <>]

In[131]:= Plot[GreenCurve[z], {z, 0, 1},
            PlotStyle → RGBColor[0, 1, 0]]
```

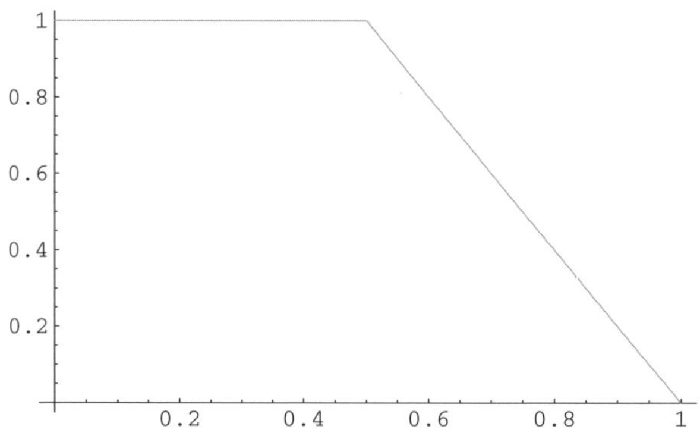


```
Out[131]= -Graphics-
```

Neither red green contain any blue, so the blue curve will have a value of zero for input values of 0 and 1. The middle input value of 0.5 is white, so the red, green, and blue curves must all have a value of 1.

```
In[132]:= BlueCurve =
            Interpolation[{{0., 0.}, {0.5, 1.}, {1., 0.}},
            InterpolationOrder → 1]

Out[132]= InterpolatingFunction[{{0., 1.}}, <>]
```

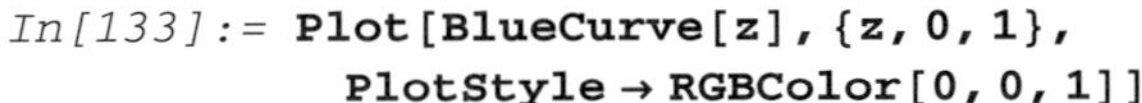

```
In[133]:= Plot[BlueCurve[z], {z, 0, 1},
            PlotStyle → RGBColor[0, 0, 1]]
```

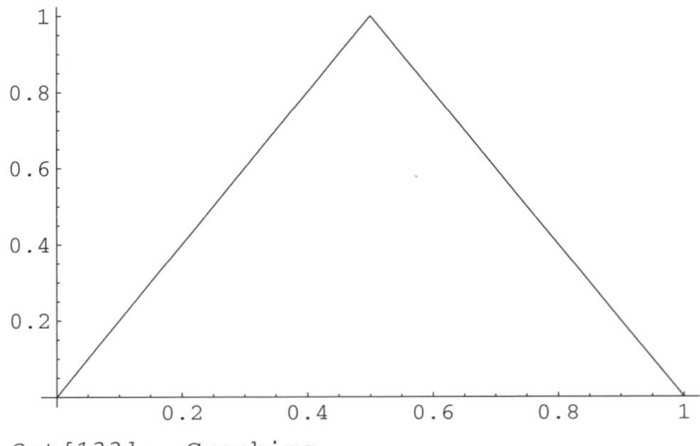

Out[133]= -Graphics-

The resulting color function is:

```
In[134]:= GWR[z_] := RGBColor[RedCurve[z], GreenCurve[z],
            BlueCurve[z]]
```

and a contour plot made using **GWR** as a color function looks like this:

```
In[135]:= ContourPlot[Sin[x] Sin[y], {x, 0, 2 π}, {y, 0, 2 π},
            ColorFunction → GWR, PlotPoints → 40]
```

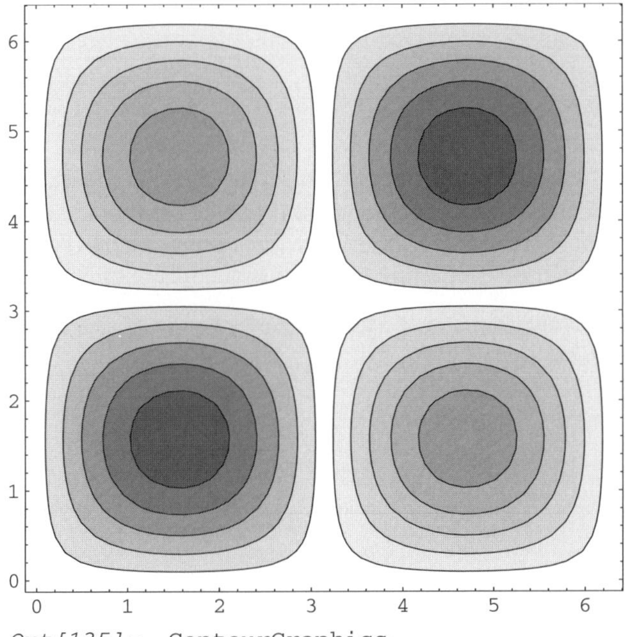

Out[135]= -ContourGraphics-

The *Mathematica* package accompanying this book contains eight pre-defined color functions that may be useful to geoscientists. They are listed below, and can be used anytime after the ComputationalGeoscience package is loaded. **Rainbow** plots the visible spectrum starting with red for the lowest values and ranging to violet for the highest values, and **ReverseRainbow** does the opposite. **BrownGreenCream** and **BrownGreenWhite** are particularly useful for giving a cartographic effect to plots of digital elevation model data. The others are self-explanatory.

Computer Note: The following are the color functions contained in the ComputationalGeoscience package. If you have not yet done so, load the package and use the color functions to produce some contour or density graphics similary to the sin$x$ sin$y$ plots above.

```
Rainbow              RainbowReverse
BrownGreenCream      BrownGreenWhite
GreenYellowRed       RedYellowGreen
GreenWhiteRed        RedWhiteGreen
```

B.6.3 Surface Plots and Graphics

There are two ways to color 3-D plots or graphics objects: by using simulated color light sources and by using a color function. The color function might convey the height of a 3-D surface or some other variable. Each of these two methods is discussed below.

Colored Light Sources

The default coloration for surface plots such as those created by **Plot3D**, **ListPlot3D**, and **SurfaceGraphics** is a range of pastel colors that are controlled by three simulated light sources. For example, the three dimensional version of the sin x sin y contour plot we have been using to demonstrate the use of color is:

```
In[136]:= Plot3D[Sin[x] Sin[y], {x, 0, 2 π}, {y, 0, 2 π},
            PlotPoints → 40]
```

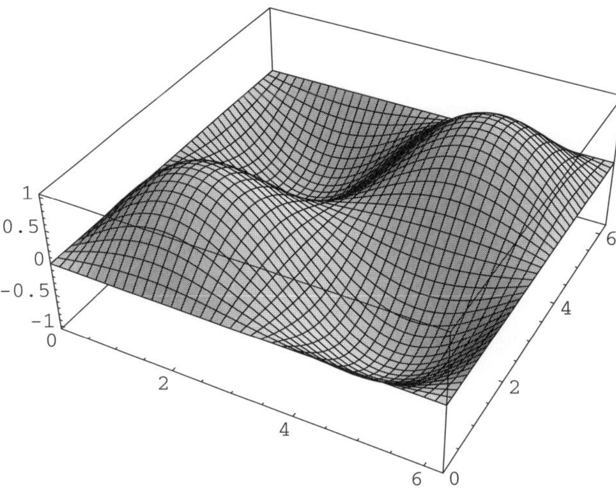


```
Out[136]= -SurfaceGraphics-
```

The coloration of the surface has nothing to do with the z value (height of the surface). It is controlled solely by the relationship of different portions of the surface to the three light sources. The default lighting sources for all 3-D graphics are:

```
In[137]:= Options[Graphics3D, LightSources]

Out[137]= {LightSources → {{{1., 0., 1.}, RGBColor[1, 0, 0]},
            {{1., 1., 1.}, RGBColor[0, 1, 0]},
            {{0., 1., 1.}, RGBColor[0, 0, 1]}}}
```

The first part of each light source is a directional variable given in coordinates *relative to the final display area, not the coordinate axes of the plot*, with x and y in the plane of the image and z perpendicular to it (with $+z$ to the front of the image). The second part of each light source is a color, which can be specified using **Hue**, **RGBColor**, or (for grayscale images) **GrayLevel**. Make a quick sketch illustrating the placement of the default light sources. A surface plot lighted with only a single red source located at {1., 1., 1.} looks like this:

```
In[138]:= Plot3D[Sin[x] Sin[y], {x, 0, 2π}, {y, 0, 2π},
            PlotPoints → 40, LightSources → {{{1., 1., 1.},
            RGBColor[1, 0, 0]}}]
```

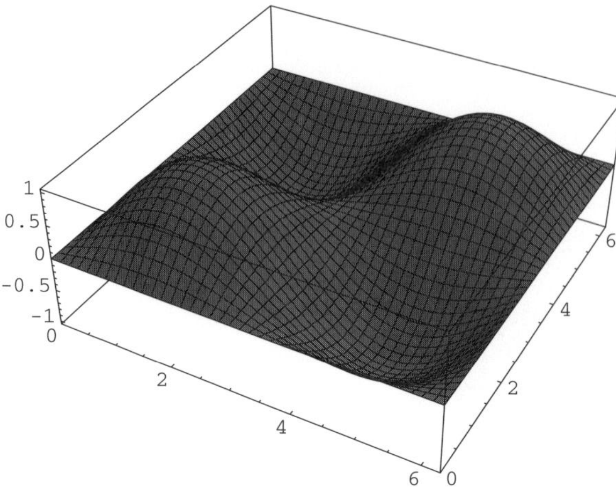

Out[138]= -SurfaceGraphics-

Notice that, because there is a single discrete light source, the surface contains shadows. We can add a blue light source from the other direction to see what effect it will have on the plot.

```
In[139]:= Plot3D[Sin[x] Sin[y], {x, 0, 2 π}, {y, 0, 2 π},
            PlotPoints → 40, LightSources → {{{1., 1., 1.},
              RGBColor[1, 0, 0]}, {{0., 0., 1.},
              RGBColor[0, 0, 1]}}]
```

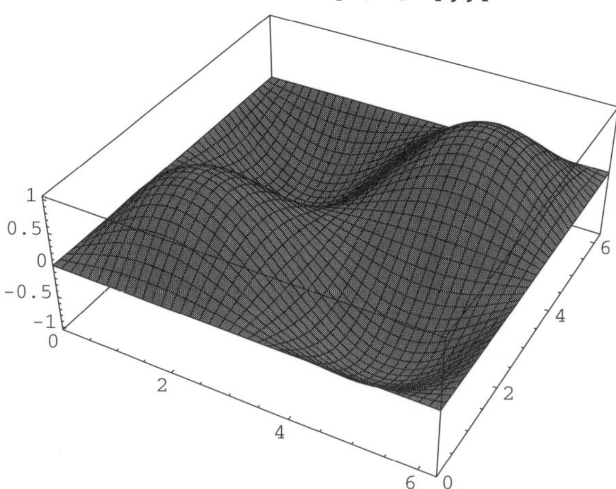

Out[139]= -SurfaceGraphics-

Three-dimensional graphics objects can also be lighted with a uniform, or ambient, light source that is specified as a single hue, RGB color, or (for grayscale graphics) gray level. The plot below shows the $\sin x \sin y$ plot with ambient red light, which produces a slightly different effect than the use of a single discrete red light source. Here is a surface plot with ambient red light (the default light sources are be turned off by giving the option **LightSources** as an empty list).

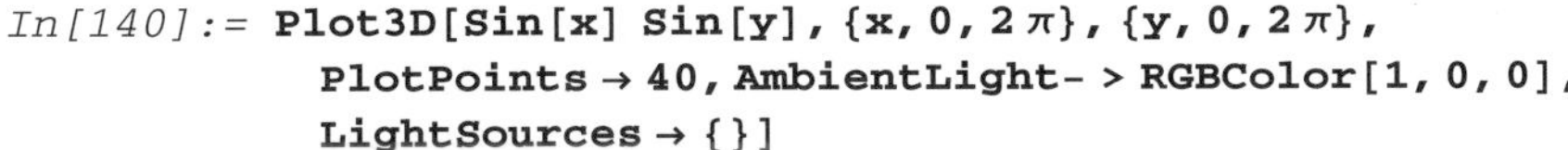

```
In[140]:= Plot3D[Sin[x] Sin[y], {x, 0, 2 π}, {y, 0, 2 π},
            PlotPoints → 40, AmbientLight- > RGBColor[1, 0, 0],
            LightSources → {}]
```

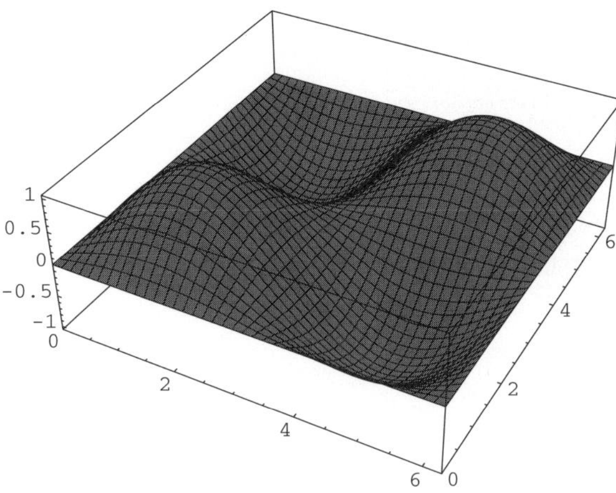

Out[140]= -SurfaceGraphics-

Adding a single white light source from the upper right front {1., 1., 1.} counteracts the ambient light and adds some shadows.

```
In[141]:= Plot3D[Sin[x] Sin[y], {x, 0, 2 π}, {y, 0, 2 π},
            PlotPoints → 40, AmbientLight- > RGBColor[1, 0, 0],
            LightSources → {{{1., 1., 1.}, RGBColor[1., 1., 1.]}}]
```

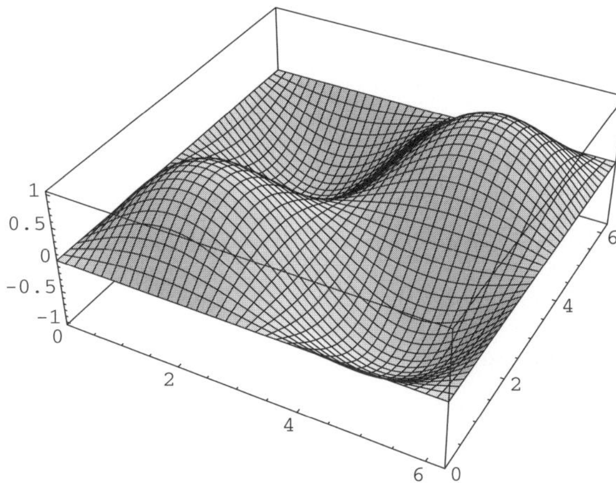

Out[141]= -SurfaceGraphics-

Finally, restoring the three default light sources produces this lighted by three different color discrete sources as well as ambient red light.

```
In[142]:= Plot3D[Sin[x] Sin[y], {x, 0, 2 π}, {y, 0, 2 π},
            PlotPoints → 40, AmbientLight- > RGBColor[1, 0, 0]]
```

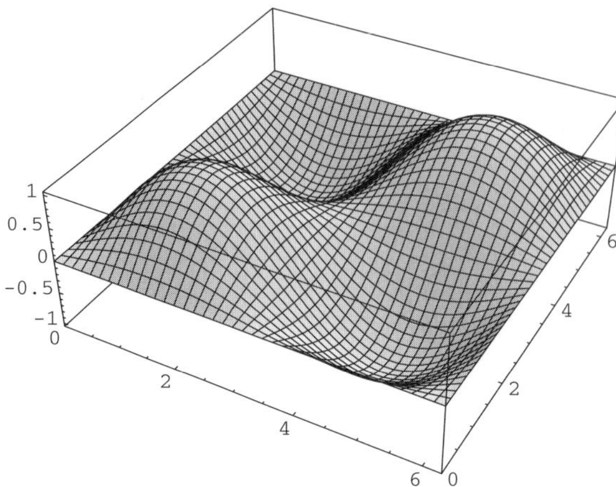

Out[142]= -SurfaceGraphics-

Surface Color

Plot3D and **ListPlot3D** belong to a simplified class of 3-D graphics, known as **SurfaceGraphics**, in which each *x*-*y* coordinate can be represented as a single *z* value. This excludes sufaces that are folded, which would have more than one possible *z* value for each *x*-*y* coordinate. The surface color of a **SurfaceGraphics** plot can be specified by a color function that is related to the height of the surface, for example the color function **RainbowReverse** from the ComputationalGeoscience package.

```
In[143]:= Plot3D[Sin[x] Sin[y], {x, 0, 2 π}, {y, 0, 2 π},
            PlotPoints → 40, ColorFunction → RainbowReverse]
```

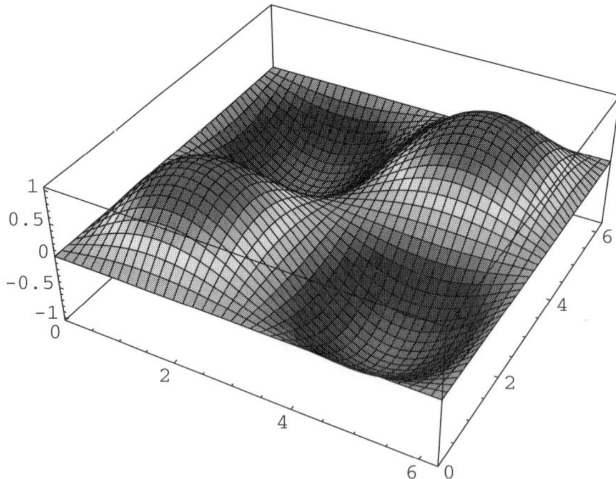

Out[143]= -SurfaceGraphics-

When implemented as in the example above, the color function is applied to the height of the 3-D surface. This is not the only option. It is also possible to specify that the value of a completely different function be used to color the surface. This is done by supplying **Plot3D** with a list of two functions, the second of which is used to color the surface according to the specified color function. In the example below, the height of the surface is given by the usual sin x sin y but the coloration reflects the slope or gradient of the surface. First, define a variable called **SurfaceGradient**.

```
In[144]:= SurfaceGradient =
```

$$\sqrt{(\partial_x (\mathrm{Sin}[x]\ \mathrm{Sin}[y]))^2 + (\partial_y (\mathrm{Sin}[x]\ \mathrm{Sin}[y]))^2}$$

Out[144]= $\sqrt{\mathrm{Cos}[y]^2\ \mathrm{Sin}[x]^2 + \mathrm{Cos}[x]^2\ \mathrm{Sin}[y]^2}$

Next, use **Plot3D** with two functions, the second of which is the color function with **SurfaceGradient** as an argument.

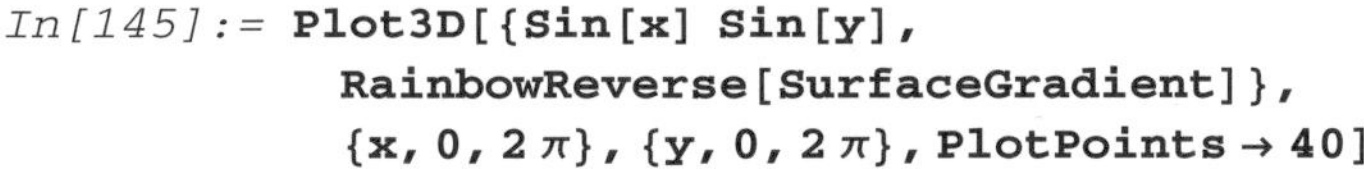

```
In[145]:= Plot3D[{Sin[x] Sin[y],
             RainbowReverse[SurfaceGradient]},
             {x, 0, 2 π}, {y, 0, 2 π}, PlotPoints → 40]
```

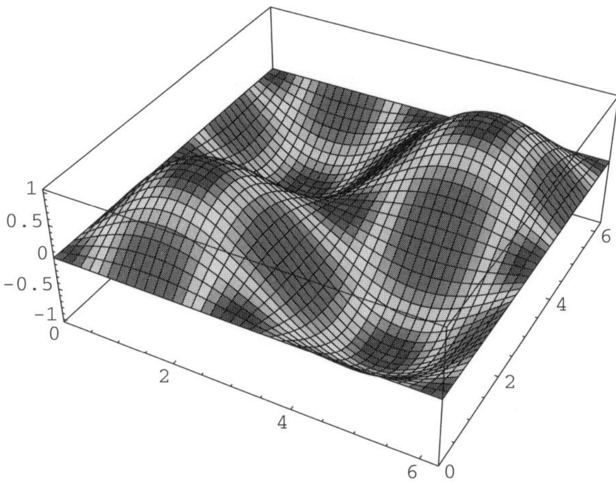


```
Out[145]= -SurfaceGraphics-
```

Similar combinations can be plotted using **ListPlot3D**, in which case the two functions are replaced by two arrays of numerical values, although some extra steps are required. First,, the array containing the color values must be normalized so that its range is 0 to 1. Then, the color function must be applied to each element in the array. **Array1**, defined below, contains a set of values following the sin x sin y function used in previous examples. **Array2** contains the color values. The *Mathematica* documentation warns that **Array2** must have one less row and column than **Array1**, because the height values are treated as points and the color values as polygons between the points. As shown below, however, current versions of *Mathematica* will interpolate color values between points if **Array1** and **Array2** are of the same dimensions.

```
In[146]:= Array1 = Table[Sin[x] Sin[y], {x, 0, 2π, π/20.},
              {y, 0, 2π, π/20.}];
```

Notice that the color function must be applied to each element within **Array2**, not simply to the array as a whole.

```
In[147]:= Array2 = Table[RainbowReverse[SurfaceGradient],
              {x, 0., 2π, π/20.}, {y, 0., 2π, π/20.}];
```

The resulting plot is:

```
In[148]:= ListPlot3D[Array1, Array2]
```

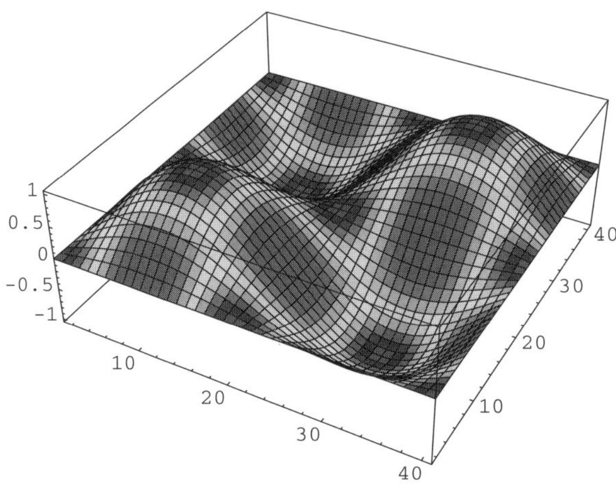


```
Out[148]= -SurfaceGraphics-
```

B.6.4 Graphics3D

Surface Color

The specification of surface colors for general 3-D graphics objects cannot be done using a simple color function because, in general, a 3-D surface does not follow a simple functional relationship. General 3-D surface color can, however, be specified using the function **SurfaceColor**. Using **SurfaceColor[GrayLevel[*n*]]** will specify the albedo (reflectance) of a 3-D surface. **SurfaceColor[RGBColor[*r,g,b*]]** will produce a surface over which the color is given by the specified RGB color multiplied by the cosine of the angle between the surface and the light source.

Graphics3Dobjects are, in general, created as combinations of 3-D lines and polygons. It is also possible to convert a **SurfaceGraphics** object, for example as created by **Plot3D** or **ListPlot3D**, into a **Graphics3D** object. The statement below transforms the sin x sin y plot from a **SurfacegGraphics** object into a **Graphics3D** object and gives it the name **ExampleSurface**.

```
In[149]:= ExampleSurface =
            Graphics3D[Plot3D[Sin[x] Sin[y], {x, 0, 2π},{y, 0, 2π},
              PlotPoints → 40]]
```

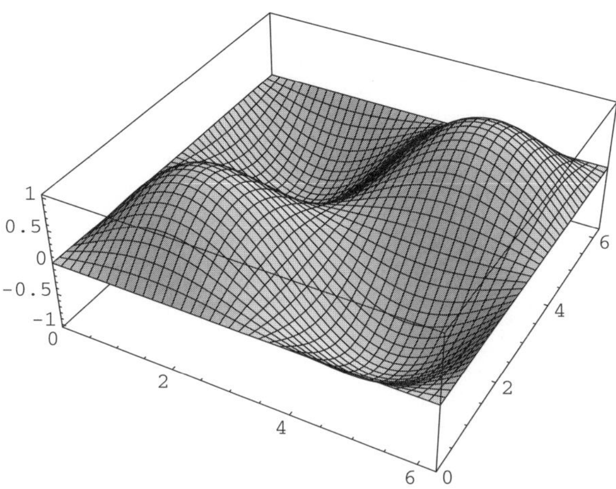


```
Out[149]= -Graphics3D-
```

The default surface color value, which is illustrated above, is **SurfaceColor [GrayLevel [1]]**. This corresponds to the albedo of a sheet of plain white paper. Reducing the albedo produces a much darker plot, as shown below.

```
In[150]:= Show[
            Graphics3D[{SurfaceColor[GrayLevel[0.6]],
              ExampleSurface[[1]]}]
           ]
```

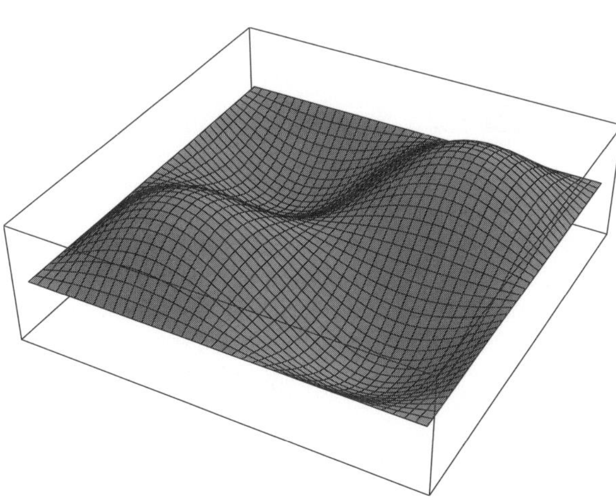

```
Out[150]= -Graphics3D-
```

Notice that **SurfaceColor** must come *before* the object to be drawn, not after it. Also, note that the graphics object is given as **ExampleSurface[[1]]**. Three dimensional graphics objects are actually lists, the first element of which is the graphic described in terms of 3-D primitives such as polygons. The second element contains graphics options.

> **Computer Note:** Type **ExampleSurface[[2]]** to see the list of graphics primitives comprising the sin x sin y plot and **ExampleSurface[[2]]** to see a list of the options.

Specifying **SurfaceColor** as an RGB color has the same effect of lighting a **SurfaceGraphics** plot with a red light. For example, specifying red as the surface color produces this surface:

```
In[151]:= Show[
            Graphics3D[{SurfaceColor[RGBColor[1, 0, 0]],
               ExampleSurface[[1]]}]
          ]
```

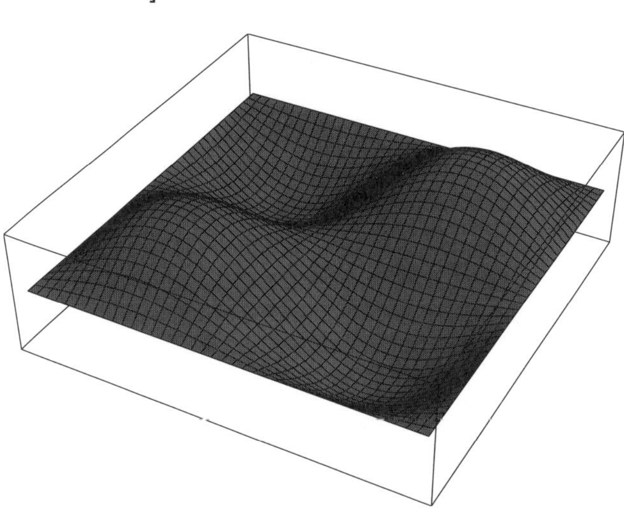

```
Out[151]= -Graphics3D-
```

Another way to investigate the relationship between surface color and lighting in 3-D graphics is to create a white object, for example a pyramid, and then experiment with different lighting options. Here is the *Mathematica* statement to create, but not show, a white pyramid:

```
In[152]:= WhitePyramid = {
              SurfaceColor[RGBColor[1, 1, 1]],
              Polygon[{{0.5, 0.5, 0}, {0.5, -0.5, 0}, {0, 0, 1}}],
              SurfaceColor[RGBColor[1, 1, 1]],
              Polygon[{{0.5, -0.5, 0}, {-0.5, -0.5, 0}, {0, 0, 1}}],
              SurfaceColor[RGBColor[1, 1, 1]],
              Polygon[{{-0.5, -0.5, 0}, {-0.5, 0.5, 0}, {0, 0, 1}}],
              SurfaceColor[RGBColor[1, 1, 1]],
              Polygon[{{-0.5, 0.5, 0}, {0.5, 0.5, 0}, {0, 0, 1}}]};
```

With ambient white light produces a white pyramid even though we have not disabled the default discrete light sources

```
In[153]:= Show[Graphics3D[WhitePyramid],
            ViewPoint- > {2.771, 0.998, 8.505}, Boxed → False,
            AmbientLight → GrayLevel[1.]]
```

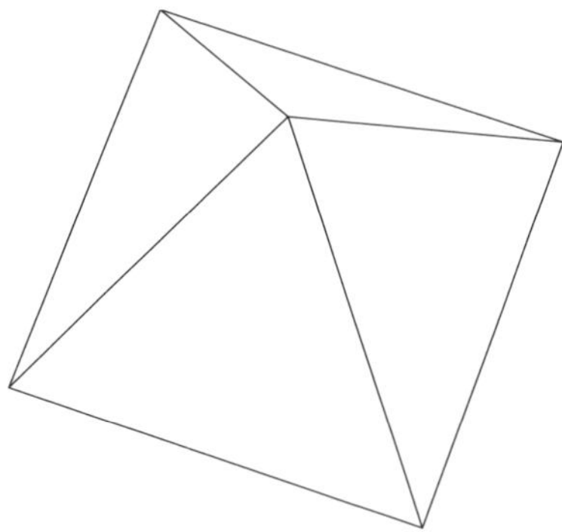

```
Out[153]= -Graphics3D-
```

Using **AmbientLight → RGBColor[1, 1, 1]** would have produced the same result. Turning off the ambient light allows the effects of the default discrete light sources to be seen.

```
In[154]:= Show[Graphics3D[WhitePyramid],
            ViewPoint- > {2.771, 0.998, 8.505}, Boxed → False]
```

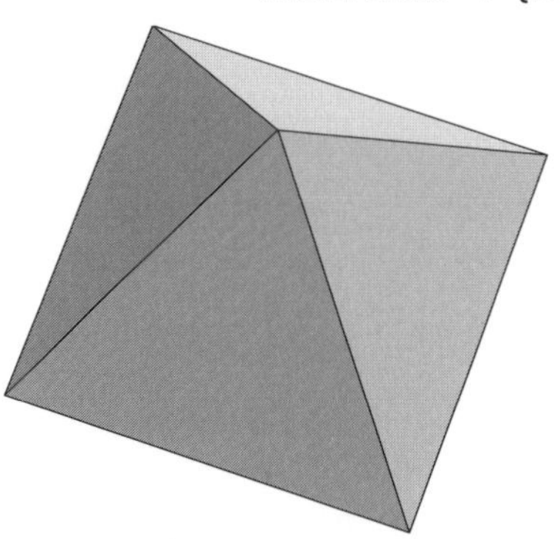

```
Out[154]= -Graphics3D-
```

Using something other than white ambient light results in surface colors controlled by a combination of the default discrete light sources and the ambient light color. In this case, the use of red ambient light gives the pyramid a red to pink tint.

```
In[155]:= Show[Graphics3D[WhitePyramid],
            ViewPoint- > {2.771, 0.998, 8.505}, Boxed → False,
            AmbientLight → RGBColor[1, 0, 0]]
```

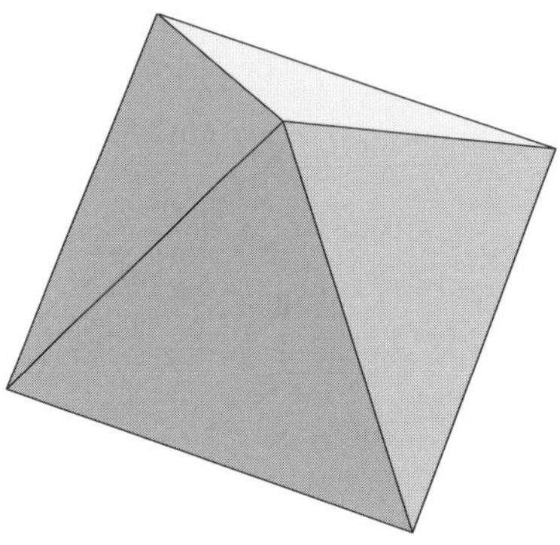

```
Out[155]= -Graphics3D-
```

To create a pure red pyramid, specify **LightSources** as an empty list. Each of the pyramid faces will be a uniform red color.

```
In[156]:= Show[Graphics3D[WhitePyramid],
            ViewPoint- > {2.771, 0.998, 8.505}, Boxed → False,
            LightSources → {}, AmbientLight → RGBColor[1, 0, 0]]
```

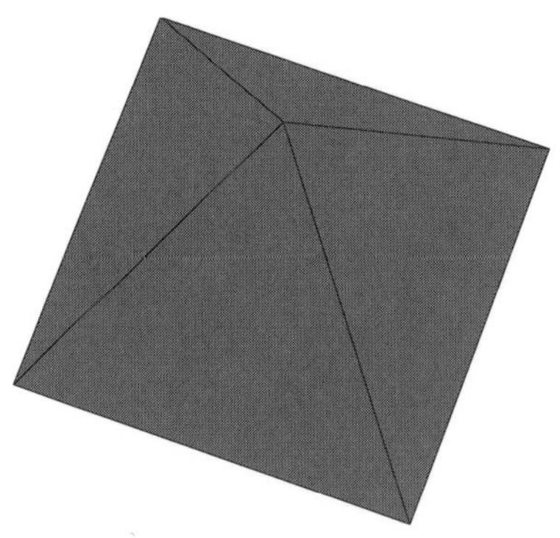

```
Out[156]= -Graphics3D-
```

Illuminating the pyramid with a single red light source located to the right, in front of, and above the pyramid produces a different result.

```
In[157]:= Show[Graphics3D[WhitePyramid],
            ViewPoint- > {2.771, 0.998, 8.505}, Boxed → False,
            LightSources → {{{3.5, -2.4, 5.}, RGBColor[1, 0, 0]}}]
```

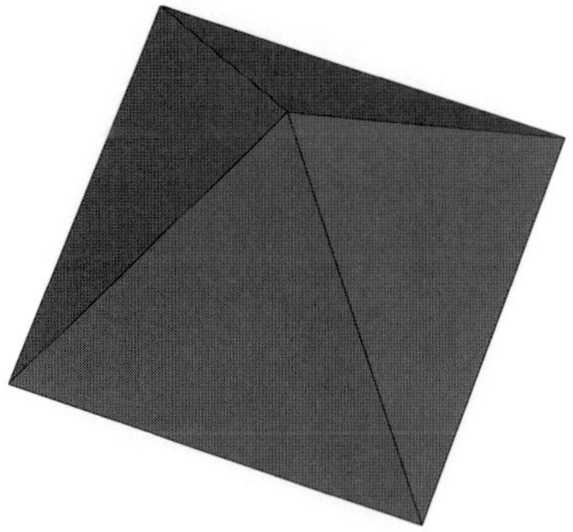

Out[157]= -Graphics3D-

Here is the same pyramid with a discrete white light source added to the left, behind, and above the pyramid.

```
In[158]:= Show[Graphics3D[WhitePyramid],
            ViewPoint- > {2.771, 0.998, 8.505}, Boxed → False,
            LightSources → {{{3.5, -2.4, 5.}, RGBColor[1, 0, 0]},
                {{-3.5, 2.4, 2.}, RGBColor[1, 1, 1]}}]
```

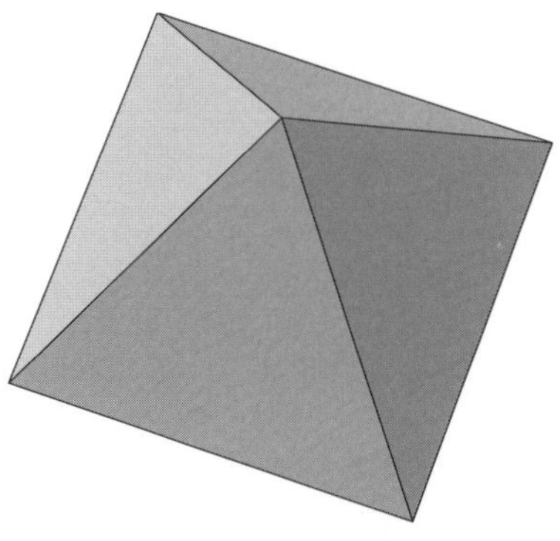

Out[158]= -Graphics3D-

Finally, turning off both the ambient light and the discrete light sources (by specifying them as an empty list in order to override the default values) produces a black pyramid.

```
In[159]:= Show[Graphics3D[WhitePyramid],
            ViewPoint- > {2.771, 0.998, 8.505}, Boxed → False,
            LightSources → {}]
```

```
Out[159]= -Graphics3D-
```

Remember that all of the pyramids shown so far have had white surfaces! The addition of color to the surface, which can be specified using **SurfaceColor**, adds more complexity. As described in the *Mathematica* documentation, surface colors can be specified using one, two, or three terms. If only one term is used, it must be a gray level or RGB color and specifies that the surface is a diffuse reflector of light with the color as specified. If two terms are used, the second term specifies a specular reflectance component of the specified color. It imparts shininess to the surface. An optional third term is a specular reflectance exponent (the default value is 1). The example below uses only a diffusive surface color.

```
In[160]:= ColoredPyramid = {
              SurfaceColor[RGBColor[1, 0, 0]],
              Polygon[{{0.5, 0.5, 0}, {0.5, -0.5, 0}, {0, 0, 1}}],
              SurfaceColor[RGBColor[0, 1, 0]],
              Polygon[{{0.5, -0.5, 0}, {-0.5, -0.5, 0}, {0, 0, 1}}],
              SurfaceColor[RGBColor[0, 0, 1]],
              Polygon[{{-0.5, -0.5, 0}, {-0.5, 0.5, 0}, {0, 0, 1}}],
              SurfaceColor[RGBColor[1, 1, 0]],
              Polygon[{{-0.5, 0.5, 0}, {0.5, 0.5, 0}, {0, 0, 1}}]};
```

Here is the colored pyramid in ambient white light:

```
In[161]:= Show[Graphics3D[ColoredPyramid],
            ViewPoint- > {2.771, 0.998, 8.505}, Boxed → False,
            AmbientLight → RGBColor[1, 1, 1]]
```

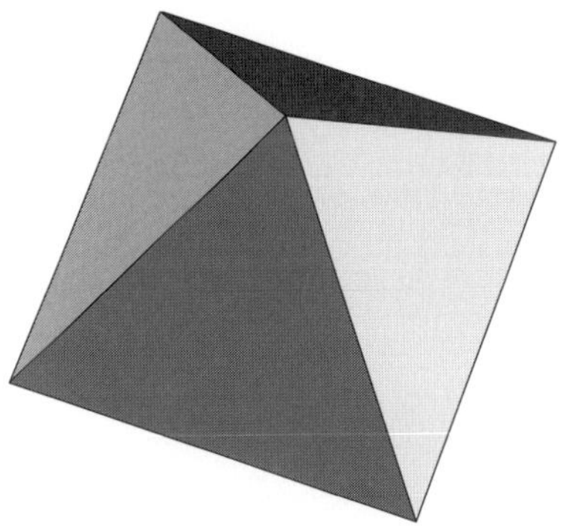

```
Out[161]= -Graphics3D-
```

The same result is obtained regardless of whether discrete light sources are used. If the ambient white light is removed, the default discrete light sources produce this result:

```
In[162]:= Show[Graphics3D[ColoredPyramid],
            ViewPoint- > {2.771, 0.998, 8.505}, Boxed → False]
```

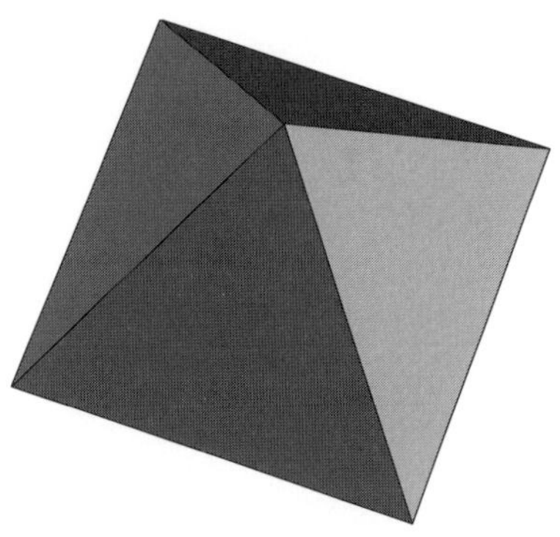

```
Out[162]= -Graphics3D-
```

As with the white pyramid, removing both the discrete light sources and the ambient light will produce a black pyramid.

```
In[163]:= Show[Graphics3D[ColoredPyramid],
            ViewPoint- > {2.771, 0.998, 8.505}, LightSources → {},
            Boxed → False]
```

```
Out[163]= -Graphics3D-
```

If the discrete light sources are removed and something other than white ambient light is used, the results will depend on the surface colors. In the example below, red ambient light colors the red and yellow faces of the pyramid (because both contain red; yellow is **RGBColor[1, 0, 1]**). The blue and green faces, neither of which contain any red and therefore will not reflect red light, appear black.

```
In[164]:= Show[Graphics3D[ColoredPyramid],
            ViewPoint- > {2.771, 0.998, 8.505}, Boxed → False,
            LightSources → {}, AmbientLight → RGBColor[1, 0, 0]]
```

```
Out[164]= -Graphics3D-
```

Likewise, changing the ambient light to pure blue will blacken all but the blue face.

```
In[165]:= Show[Graphics3D[ColoredPyramid],
            ViewPoint- > {2.771, 0.998, 8.505}, Boxed → False,
            LightSources → {}, AmbientLight → RGBColor[0, 0, 1]]
```

Out[165]= -Graphics3D-

Lighting the surface with other primary colors will produce results that depend on the RGB content of both the light and the surface (and specular component if one is used). Below is the colored pyramid illuminated with the *Mathematica* color **DarkTurquoise**.

```
In[166]:= Show[Graphics3D[ColoredPyramid],
           ViewPoint- > {2.771, 0.998, 8.505}, Boxed → False,
           LightSources → {}, AmbientLight → DarkTurquoise]
```

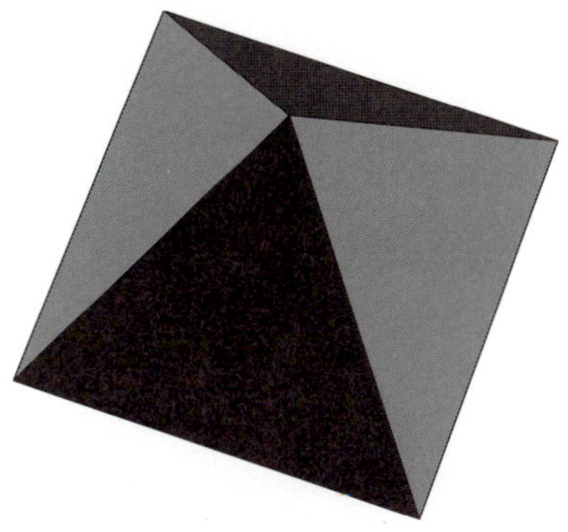

Out[166]= -Graphics3D-

Judging from the results, we can infer that **DarkTurquoise** contains significant amounts of blue and green (because the blue and green faces above closely resemble those rendered using white light), but no red. This can be confirmed by checking the RGB composiiton of **DarkTurquoise**.

```
In[167]:= DarkTurquoise

Out[167]= RGBColor[0., 0.807794, 0.819605]
```

Printing: Mercedes-Druck, Berlin
Binding: Stein + Lehmann, Berlin